Mobile Networking with WAP

Edited by SCN Education B.V.

Mobile Networking with WAP

Edited by SCN Education B.V.

This series of books cover special topics which are useful for a business audience. People who work or intend to work with Internet - at a management, marketing, sales, system integrating, technical or executive level - will benefit from the information provided in the serie

These books impart how new technologies and sales & marketing trends on Internet ma be profitable for business. The practical knowhow presented in this series comes from authors (scientists, research firms and industry experts, a.o.) with countless years of experience in the Internet area.

The HOTT Guides will help you to:

- Enlarge your knowledge of the (im)possibilities of Internet and keep it up-to-date
- Use Internet as an effective Sales & Marketing tool by implementing new technologie as well as future-oriented strategies to improve your business results
- Facilitate decision making on a management level
- Reduce research costs and training time

These books are practical 'expert-to-manager' guides. Readers will see a quick 'return on investment'.

The Publishing department of SCN Education B.V. was founded in 1998 and has built a solid reputation with the production of the HOTT Guide series. Being part of an international IT-training corporation, the editors have easy access to the latest informatio on IT-developments and are kept well-informed by their colleagues. In their research activities for the HOTT Guide series they have established a broad network of IT-specialis (leading companies, researchers, etc) who have contributed to these books.

Books already in print:
Webvertising
ASP – Application Service Providing

Mobile Networking with WAP

The Ultimate Guide to the Efficient Use of Wireless Application Protocol

Edited by SCN Education B.V.

vieweg GABLER

Die Deutsche Bibliothek - CIP-Cataloguing-in-Publication-Data
A catalogue record for this publication is available from Die Deutsche Bibliothek
(http://www.ddb.de).

Mobile Networking with WAP

Trademarks
All products and service marks mentioned herein are trademarks of the respective owners mentioned in the articles and or on the website. The publishers cannot attest to the accuracy of the information provided. Use of a term in this book and/or website should not be regarded as affecting the validity of any trademark or service mark.

1st Edition 2000

Softcover reprint of the hardcover 1st edition 2000

Vieweg is a company in the specialist publishing group BertelsmannSpringer.

Printed on acid-free paper

ISBN-13: 978-3-322-86792-6 e-ISBN-13: 978-3-322-86790-2
DOI: 10.1007/978-3-322-86790-2

Preface

Inspiration for the wireless revolution

The big Geneva Telecom 99 exhibition will probably go down in history as the moment when Wireless Application Protocol (WAP) made its debut in the world. A slew of vendors launched their first WAP phones and carriers announced their first WAP services. And this time it's Europe that's the frontrunner. 'For once I'm not proud to be an American', said Larry Ellison, the CEO of softwaremaker Oracle, in a speech delivered at Telecom 99. Europe is at least 18 months ahead of the United States in deploying WAP.
The Telecom 99 show could mark the beginning of a bright, very bright future for WAP. Analists predict that a significant part of Internet access will be via wireless devices within a few years.

It's important not to get carried away. When the computer industry gets hold of a new concept like WAP it becomes the 'Next Big Thing'. Vendors pre-announce products and the mass media prints hype.

When the revolution fails to materialize during the following months the hype will die soon. But this time it's different. The business case is overwhelming, the numbers are staggering, and it's the 'cash loaded' telecom industry that is delivering the key products. The market for wireless information services and mobile data has always been there. And now there's a standard that has been accepted by over 200 members of the WAP Forum, gateways are going into networks, phones will be in the shops by the time this Hott Guide is printed, and the first WAP services should have appeared.

There is no single 'killer app' but mobile e-mail and financial services are the most promising . E-mail has always been possible via SMS, but the functionality is limited and

Preface

the interface is somewhat clumsy. With a WAP browser interface it is easier to read, reply, forward and even create e-mail messages.

The use of mobiles to pay for small purchases is also possible. A key feature for this service is GSM's built-in encryption. The latest (and one of the funniest) announcement is a mobile commerce (m-commerce) application known as the GSM Chocolate Service brought to you by telecom operator Sonera.

With the Sonera mobile phone subscription you can order, send and pay for Fazer Sininen, which is a famous chocolate bar (in Finland). It can be ordered by phone and is sent to the recipient together with a greeting. The purchase is billed on the mobile phone invoice. An inspiring application.

Hopefully this 'HOTT Guide' will be as inspiring and will help you in your business.

Ing. Adrian Mulder
Content Editor

Adrian Mulder is an Internet journalist who writes for major business computing magazines. He combines a technical background with a vast experience in the computer and business trade magazine industry.

Acknowledgements

Many people and professionals have contributed directly or indirectly to this book. To name them all would be practically impossible, as there are many. Nevertheless the editors would like to mention a few of those who have made the production of this book possible.

Executive Editor for SCN Education B.V.: *Robert Pieter Schotema*
Publishing Manager: *drs. Marieke Kok*
Marketing Coordinator: *Martijn Robert Broersma*
Content Editor: *ing. Adrian Mulder*
Editorial Support: *Dennis Gaasbeek, Rob Guijt, Richard van Winssen*
Interior Design: *Paulien van Hemmen, Bach.*

Also, we would especially like to thank dr. Roland van Stigt for laying a solid foundation for the HOTT Guide series to grow on.

Contents

Contents

Contents

Chapter 1: The Future with WAP

What is Wap?

Title: *What is Wap?*
Author: *WAPsight.com*
Abstract: *WAP stands for Wireless Application Protocol and is the basis for the Mobile Internet. Thanks to WAP, you are able to access the internet and generally keep in touch with your world, anywhere and anytime, via your micro browser equiped wireless phone.*

What is WAP?

Imagine that you are in a meeting with a customer and he asks you a question about his order. You call up his account through a secure connection and you tell him "your order shipped out this morning at 11:23am. Would you like the tracking number?"

Now you are on your way to the airport to catch a flight. You are wondering if your flight is running on time, or is delayed again. You pick up your WAP phone, and bring up your airline's real time flight information. You find out that a mechanical problem has forced a one hour delay. No need to speed, maybe there is even time to stop off for a quick lunch on the way.

Or imagine that your Day Trading record has been going downhill lately because you've missed a few key price breaks. While you wait for your flight, you stay on top of the business news feeds, keep up with real time quotes, and you place your sell order when you judge that the time is right.

These examples might seem blue sky today, but in reality, this is what WAP is ready to bring us. The world of information, not only on our desktops, but literally at our fingertips. The possibilities are truely endless. Some are calling this the "Mobile Internet Revolution".

The slightly more technical answer:

The WAP protocol specification was developed by a consortium of companies involved with the wireless telecommunications industry. This group, the WAP Forum, consisted of four companies at the time of its founding (Phone.com, Nokia, Ericsson, and Motorola), and has since grown to over 250 members.

The purpose of the WAP specification was to provide a standard method for small, limited resource devices such as cellular telephones, to access the internet. Because of the resource and bandwidth constraints, it was deemed necessary by the WAP Forum to set up a special protocol optimized for this device class.

Sitting between a WAP Phone and the Internet, is a WAP proxy. The function of the WAP proxy is to provide a gateway between the wireless environment and the internet environment. The WAP proxy tokenizes the data before transmitting it to the phone, thus achieving a measure of data compression.

WAP defines a markup language WML (wireless markup language). Developers of WAP viewable sites will use WML in much the same way as HTML is used for web sites. WML is based on the XML standard.

Mobile network Evolution to Multimedia Messaging

Title: *Mobile network Evolution to Multimedia Messaging*

Author: *Nokia Networks*

Abstract: *With Multimedia Messaging Service (MMS) it will be possible to combine the conventional short messages with richer content types: photograph, images, voice clips and eventually also video clips. MMS is a key application within the wireless messaging business, and one of the enablers of the Mobile Information Society in which an increasing part of all personal information transmission will take place wirelessly. MMS builds on the user experience of using SMS (Short Message Service). The popularity of SMS and the emergence of an instant culture suggest there is already significant demand for personal communication enchanced by visual content. Success is dependant on investing in the right technology, creating the right applications, and starting with a Multimedia Messaging strategy now.*

Multimedia Message Service (MMS) will be a key application within the wireless messaging business, and one of the enablers of the Mobile Information Society, in which an increasing part of all personal information transmission will take place wirelessly. Nokia aims to lead the way towards the Mobile Information Society by supporting open platforms that enable wide market adoption and stimulate growth. Nokia's approach is based on dynamic development in messaging; MMS, the most versatile messaging service, is a natural continuation of Short Message Service (SMS) and Picture Messaging. In addition to wireless voice and text transmission, visual content can be exchanged. With Multimedia Messaging it is possible to combine the conventional short messages with much richer content types – photograph, images, voice clips, and eventually also video clips. In addition to sending messages mobile-to-mobile, it is possible to send messages mobile-to-email and later also email-to-mobile. This all means new and exciting possibilities especially for person-to-person communication.

Based on Nokia's preliminary market studies and information sharing with users, network Operators as well as service providers, a very strong demand for MMS exists. Innovators, the early adopters of MMS, will mainly be the same people who are the heavy users of SMS. Other mobile phone users are also likely to be interested in MMS.

MMS will most likely be introduced in several phases; the combination of text and photo, for instance, will be adopted first followed by strong needs to combine other Multimedia Messaging elements such as video and voice clips. The price perceptions and storage possibilities for MMS will be of high importance when MMS becomes available. Multimedia Messaging will offer extensive added value especially for person-to-person messaging and also for person-to-group messaging which will be available in the near future.

As the demand for messaging between users and applications grows along with new bearers and the Wireless Application Protocol (WAP), the importance of WAP enabled Multimedia Messaging applications will grow significantly. This will bring network Operators and third party developers wide business opportunities. It is estimated, however, that up to 80–90 % of messaging will include person-to-person and person-to-group messaging which is currently the case for SMS. To meet the high demand for a new era of messaging, Nokia will provide complete solutions for Multimedia Messaging based on mobile device and infrastructure expertise. Nokia's end-to-end solution will include mobile devices supporting Mobile Multimedia, and a comprehensive solution for network Operators to offer Multimedia Messaging. General Packet Radio Service (GPRS) network will be ideal for mobile data networking services. In addition, Nokia will work together with 3rd party developers and other parties to develop and deliver MMS. The Nokia Artus product family will offer network Operators viable possibilities to make an early move in offering attractive MMS.

Moving toward the Mobile Information Society

Multimedia Messaging will mean enhanced personal communication for users, facilitating the new communication styles and needs of the Mobile Information Society. Mobile communication and connectivity are important elements of the Mobile Information Society, especially when enhanced with visual content. Multimedia Messaging will bring richer content to mobile communication and messaging. New forms of communication and a wide range of value-added services will play a dominant role in the Mobile Information Society in which mobile users will be able to access a variety of information and services easily and for their specific, personal needs. Multimedia Messaging will increase wireless data usage which already now is doubling every year in advanced markets. Many mobile Operators already earn over 5 percent of their revenues from data traffic.

Introduction to Multimedia Messaging

Recently, Short Message Service (SMS) has proven to be a tremendous success in many countries. Operators in these countries have often also provided their subscribers with possibillities to personalize their mobile phones with ringing tones and graphical icons which have proven extremely popular. The growth in this area will serve as a valuable path to new and interesting ways for using the mobile phone, in ways yet unseen in the history of wireless communication. As users become accustomed to the easy use of SMS, the opportunity to send multimedia messages will mean new and easy ways for personal communication. Along with this evolution it is important to realize that users will not care about the new technologies they will be using; they will be interested in applications and services. Nokia's Multimeda Messaging applications can be used for various purposes, which will deliver a broad range of user benefits, from emotional sharing and fun to rational utility. Multimedia Message Service (MMS) will be able to utilize picture messages, electronic postcards, audio messages, instant images and video clips. MMS does not require users to learn a new technology. It is a natural consequence of the messaging evolution.

Messaging evolution

Nokia's migration path to Multimedia Messaging builds on the well-established SMS paradigm by adding new functionalities and new content step by step, along with the introduction of new technologies such as Wireless Application Protocol (WAP) and General Packet Radio Service (GPRS). After SMS, the application migration path comprises Picture Messaging and MMS. Wireless communication is rapidly expanding from ears to eyes. In addition to wireless voice and text transmission, visual content can be exchanged between mobile devices. SMS has already proven extremely successful, for instance, in the Nordic countries and in many Asia Pacific countries. Teenagers in these markets often send over 100 SMSs per month and even more. In Finland, more than half of a teenager's mobile phone bill is made up of charges for short messages.
Picture Messaging, already introduced with phones such as Nokia 3210, Nokia 8850 and Nokia 8210, comprises a capability of sending a simple picture message from

Figure 1: Multimedia Messagging migration path

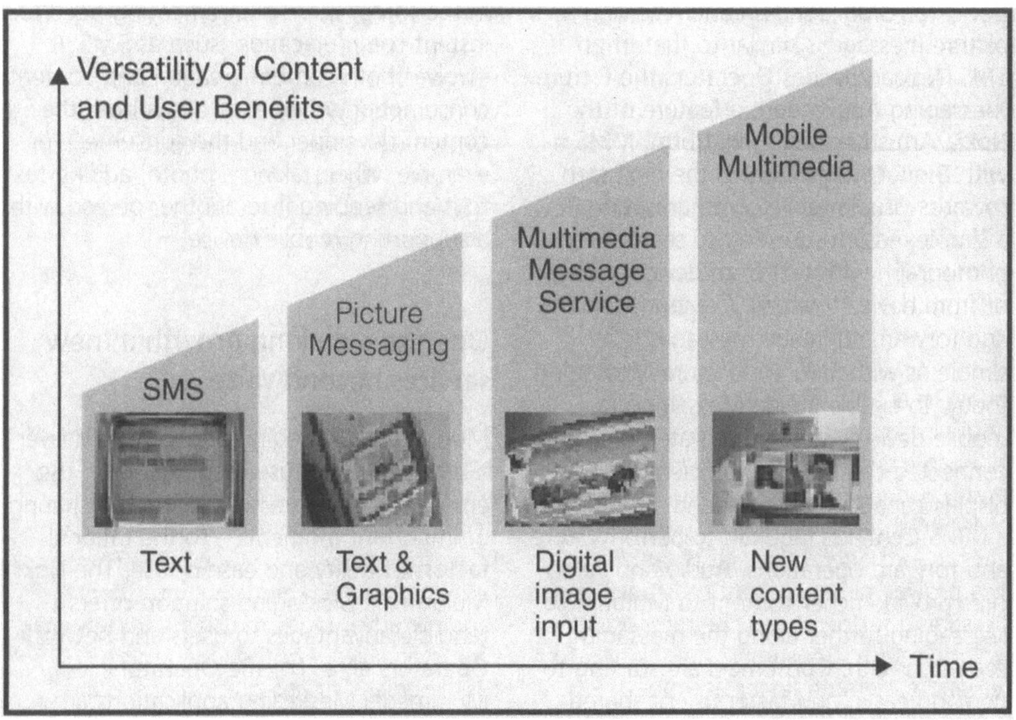

Figure 2: Evolution of messaging content versatility

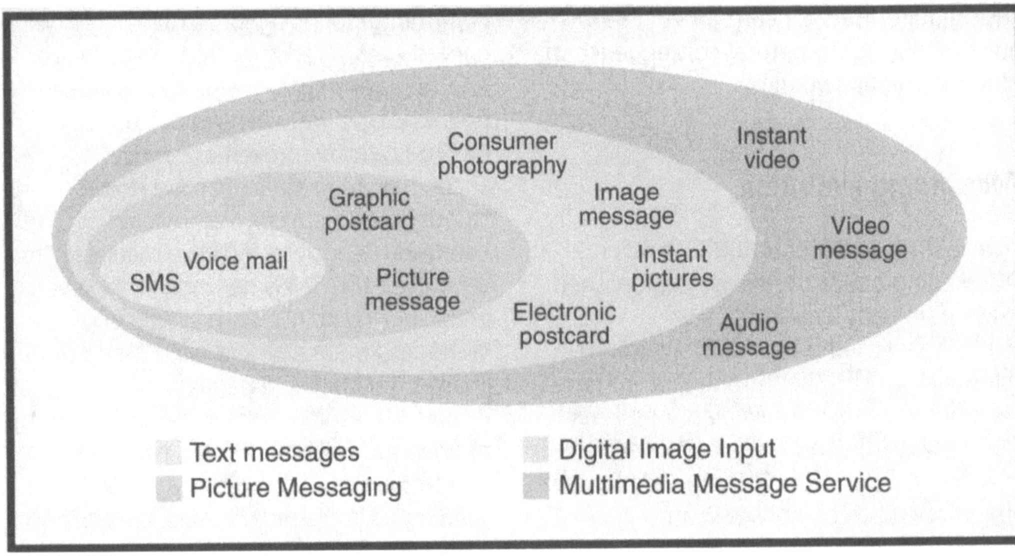

device to device or from a web site to device via SMS. Sending and receiving a picture message is similar to that of an SMS. Nokia provides Operators the Picture Messaging Application, a feature of the Nokia Artus Messaging Platform. MMS with Digital Image Input is the next step towards visual mobile communication. It is a simple, easy-to-use way to send a photograph with text from device to device or from device to email. Creating, sending and forwarding image messages is as simple as with SMS and Picture Messaging today. To enable Image Messaging, a mobile device with an integrated or connected camera and sufficient image display capabilities are needed. In addition, a MMS Center is required to perform store and forward operations. Audio and video clips provide richer content to Multimedia Messaging further along the messaging evolution path. Consumers are starting to demand easier and faster use of shared images, independent of location and time. In the MMS, there is an emerging need for instant communication, such as SMS. It involves both content creation and content consumption where the user is both the content developer and the consumer. For example, when taking a photo, adding text to it and sending it to another person with a supporting mobile device.

Operators seeking growth in new services beyond voice

Multimedia Messaging applications deliver a broad range of user benefits, from the emotional – entertainment and the sharing of important moments – to the rational factors of utility and ease of use. The Nokia Multimedia Messaging solution offers tangible advantages to users and network Operators alike. For the Operators, Multimedia Messaging applications are

Figure 3: Multimedia Message Service (MMS)

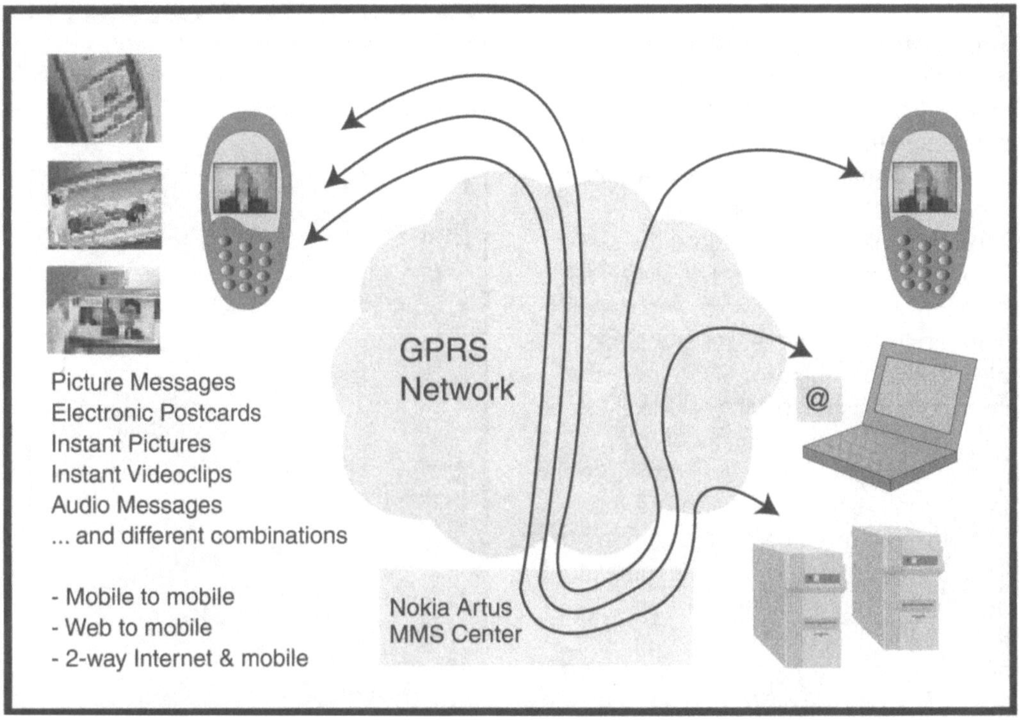

important drivers for continuous growth in new services beyond voice. MMS offers Operators new business opportunities since value-added services and personalized applications for data are estimated to be important revenue creators in the next few years. MMS helps Operators differentiate from competition. Differentiation will be possible by implementing a clear messaging strategy that includes the offering of rich content services. MMS is seen as a messaging service for person-to-person communication where the network Operator plays a major role in the business.

Early demand for MMS

An "instant" culture with new communications styles and needs is emerging. The popularity of SMS, the emergence of the new mobile phone generation, the popularity of sending traditional postcards, and the rising Internet and E-mail penetration and usage suggest a huge market potential for Multimedia Messaging.
For this reason it is important to understand who the early adopters are, the innovators of MMS, since without the innovators the diffusion of MMS to mass markets and thus the standardization of new technology will not take place so easily.

Innovators for MMS

Strong demand exists for Multimedia Message Service (MMS). This is true especially for markets where the mobile phone penetration is high enough, and users are acquainted with SMS. According to preliminary market studies, the innovators mainly appear to be the same people that are heavy users of SMS. That is, people 15 to 35 years old, and who live in an urban area. The target group for MMS is broader than for SMS. Users that are most interested in MMS are likely to be high educated, have a relatively high income, they are mobile, and often use the ringing tone service. They also use the Internet more often compared to users that are less interested in MMS. The innovators tend to own more advanced mobile phones, and they send more SMSs. The innovators are extroverts, pleasure-loving, interested in the latest technology, and are more likely to think that mobile phones are fun. Generally, younger users seem to be more interested in MMS.

Nature and extent of MMS demand

Notable is the amount of potential users for Multimedia Messaging. It is very likely that MMS will eventually be adopted by all segments meaning that it will bring value to all users; business people, youngsters, men and women alike. This is mainly due to the fact that MMS offers not only fun and sharing, but also utility. Numerous usage possibilities make it extremely exciting, interesting, personal and also very useful. Multimedia Messaging will be popular, for instance, when sending an electronic postcard, an invitation, a shopping list, or when showing one's boss how the job is proceeding. A wide range of usage situations exists as the following list indicates.

Possible usage situations for Multimedia Messaging:

- travelling abroad
- a person you are trying to reach is busy
- shopping list or memo
- no time to write
- communicating with blind people
- sending a Happy Birthday song
- sending a love song
- calling up a meeting
- substitute for postcards
- getting to know someone
- getting a quick response for medical extract
- substitute for a fax
- when shopping and needing someone's opinion
- when something extraordinary happens and needs to be perpetuated
- describing the beach on vacation
- keeping project members or customers up-to-date
- inviting friends to a party
- sending "new address" cards
- sending Christmas cards.

Users that have adopted SMS are likely to find added value especially in Photo Message Service, meaning text and photo combined. Photo Message Service would be used the same way people use E-mail, or as a substitute for postcards. A digital camera, a scanner, and a fax in a mobile phone would be very useful. Apparently, SMS is the only service that succeeds by itself, all the other services, meaning Photo, Audio and Video Message Service, will require each other to be successful. Strong demand exists for the combination of SMS and Photo Message Service, whereas Audio Message Service is likely to be useful when combined with Video Message Service or Photo Message service. Older users or business people, are likely to adopt to different combinations of MMS since they find the combinations extremely useful. Younger users, in particular, like the continuance from SMS to the combination of SMS and Photo Message Service. The demand for SMS will remain high also when MMS becomes popular. Additionally, the demand for group messaging will increase fast when MMS is introduced. This means that in certain situations users will want to send a message to several recipients at once. The question of pricing and storing Multimedia Messages should not be ignored. It will be important to know how much users value different services. Viable pricing may take the following shape, for example:

- pricing is transaction based, the user pays according to service and transaction made
- pricing is based on content
- conducted as with SMS, that is, the sender pays.

It will be important to offer users the possibility to store Multimedia Messages. Not only in the mobile device but also in a "Multimedia Message library" offered by an Operator.

Demand for person-to-person Multimedia Messaging

The Nokia Multimedia Messaging solution facilitates new styles of communication. The emerging "instant" culture demands immediate connections, involving both content creation and content consumption. Messaging is becoming the most natural area of personal communication. With advanced digital technologies as enablers, user emphasis is shifting from ears to eyes. Person-to-person messaging will continue its success as a popular means for communication between mobile users. SMS has already proven its strength; it has been widely adopted and is extremely popular especially among young mobile phone users who often send more than 100 short messages per month. Picture Messaging is making person-to-person messaging even more attractive by offering mobile phone users the possibility to send both text and graphics to another user. Now, Multimedia Messaging will bring new and exciting opportunities to messaging by offering even richer content. Although applications are becoming more and more popular and attractive, about 80 to 90 % of personal messaging will most likely continue to include person-to-person or person-to-group messaging. This will mean a huge challenge for Operators, content providers and manufacturers to meet the user demands for messaging. A major driver for the high share of person-to-person messaging is due to the fact that MMS will become a mass-market service. Strong demand for Multimedia Messaging exists, but at the same time, SMS will stay extremely popular, and will most likely remain the most popular single form of messaging compared to other messaging, such as Photo, Audio and Video Message Service. For SMS, up to 80 % or often even 90 % of messaging consists of person-to-person messaging, and approximately 10 to 20 % of messaging is covered by communication between users and applications. There are strong indications that Multimedia Messaging will continue this path with the only difference being that the major share of messaging will most likely consist of both person-to-person and person-to-group messaging. Person-to-person communication will remain a 'killer application', and Multimedia will enhance messaging.

Demand for applications

Although Multimedia Messaging will mainly be person-to-person messaging, communication between users and applications should not be underestimated. Currently, about 10 % of messaging is related to Internet based applications, for example, ringing tones. The demand for Internet based applications is growing fast along with new bearers and WAP. Since Multimedia is based on WAP, the WAP enabled Multimedia Messaging applications will be of great demand in the future bringing a huge opportunity especially to 3rd party developers and Operators. Notable is that MMS will require

Figure 4: Person-to-person and person-to-group messaging

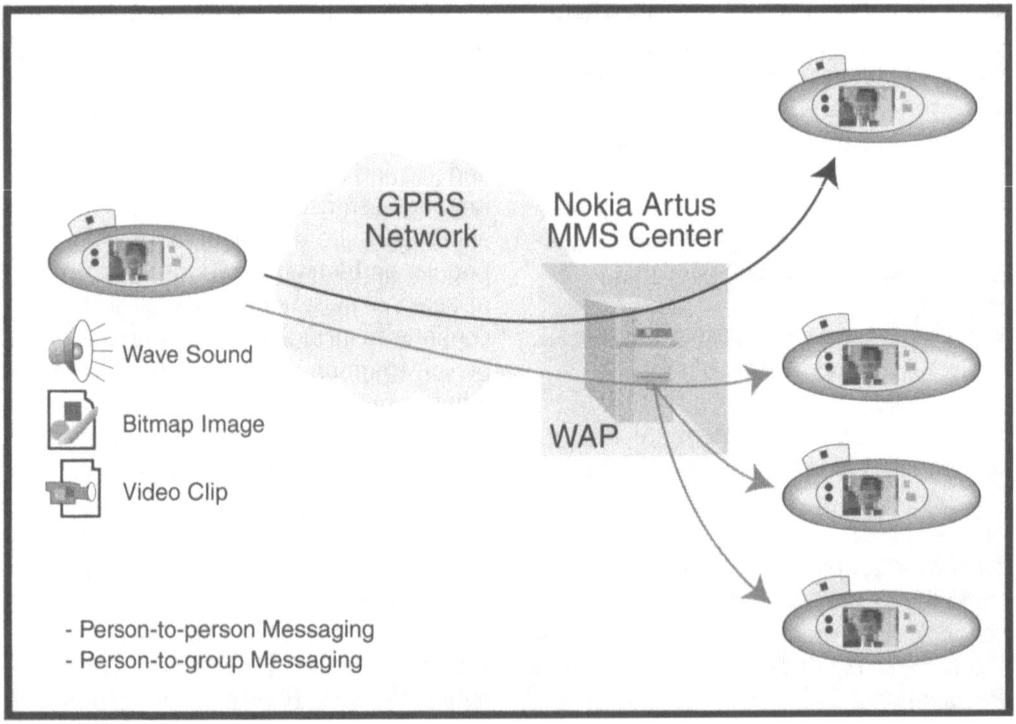

high speed networks that will provide the bandwidth necessary for transmitting messages that include significantly more data compared to current messages. GPRS will be an essential stepping stone to personal multimedia services, and an ideal platform for mobile data networking services. GPRS can be viewed as an enabler of new wireless data services and an optimizer of the radio interface for bursty packet mode traffic. GPRS creates a platform for new applications and provides the ability to stay online" for long periods of time, which is not possible with circuit switched data. What also highlights the importance of GPRS is the fact that MMS will most likely become a mass-market service, and definitively involve transmission of large amounts of data.

Nokia Artus product family as a major driver for messaging evolution

Nokia Artus product family will enable Operators to add extensive value by introducing MMS to the users. The solution will complement the current offering of Nokia Artus products for value-added services (VAS) which already include the Nokia Artus SMS Center, the Nokia Artus USSD Center and the Nokia Artus Messaging Platform, which is Nokia's WAP gateway solution for the Operator markets.

Figure 5: Communication between user and application

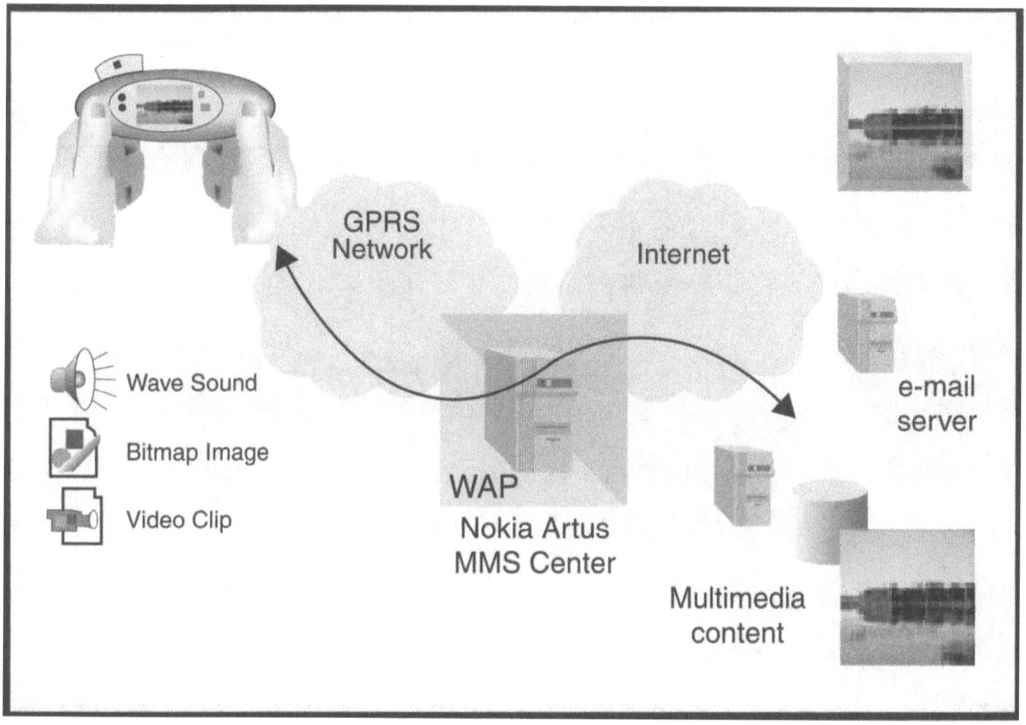

The key element in MMS network architecture will be the Nokia Artus MMS Center, based on WAP technology.
The MMS Center enables multimedia messages to be sent with various content types from device to device, with instant delivery. It will support flexible addressing – to both familiar phone numbers (MSISDN) and E-mail. MSISDN addressing offers ease of use, and helps an Operator to manage the business. In the Nokia solution, Operators can use transaction-based billing. The Nokia Artus solution will include the interface to applications, for device-to-application and application-to-device. An Operator is able to add value on top of MMS.

Conclusions

With Multimedia Messaging it will be possible to combine the conventional short messages with much richer content types – photograph, images, voice clips, and eventually also video clips. In addition to sending messages mobile-to-mobile, it is possible to send messages mobile-to-email, and eventually also email-to-mobile.
This means new and exciting possibilities especially for the person-to-person

Figure 6: Nokia Multimedia Messaging architecture with Nokia Artus MMS Center

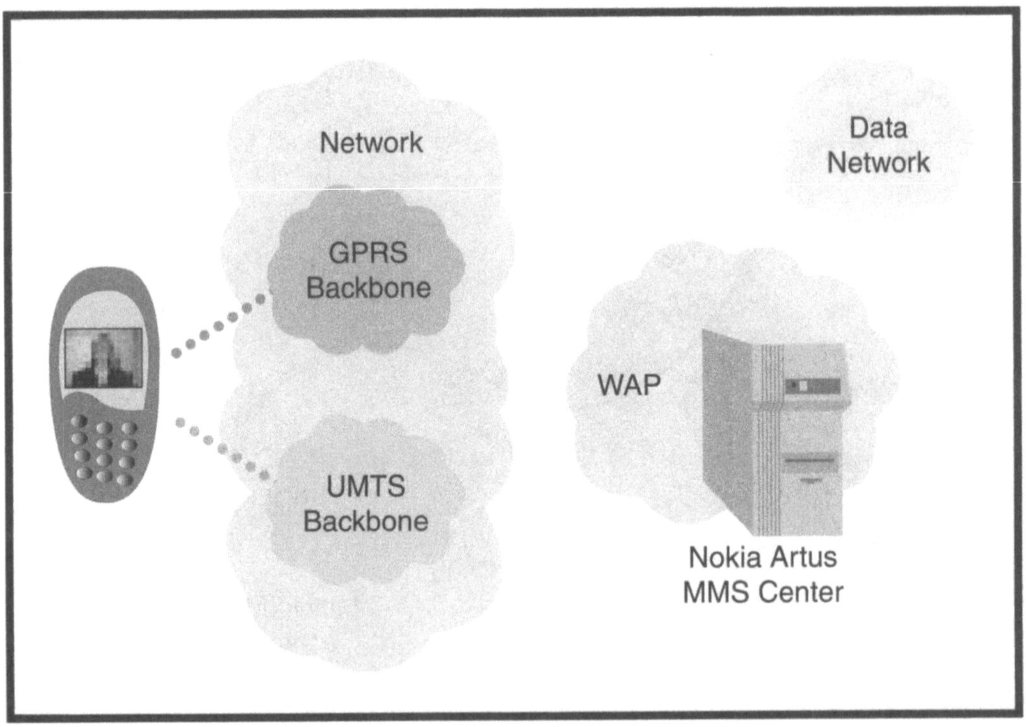

communication. MMS is a key application within the wireless messaging business, and one of the enablers of the Mobile Information Society in which an increasing part of all personal information transmission will take place wirelessly. Nokia aims to lead the way towards the Mobile Information Society by supporting open platforms that enable wide market adoption and stimulate growth. MMS is expected to become a true mass-market service. Nokia's MMS will comprise a complete end-to-end solution for person-to-person mobile messaging, from device to device or to the Internet. Nokia develops WAP solutions to support MMS, and within 3GPP Nokia develops a bearer service including an optimal support for Multimedia Messaging. WAP allows Multimedia Messaging in all product categories and it is a global, bearer independent solution. MMS will require high speed networks that will provide the bandwidth necessary for transmitting messages that include significantly more data compared to current messages. GPRS network is an ideal platform for offering Multimedia Messaging and other mobile data networking services. Nokia's approach in Multimedia Messaging builds on the user experience of using SMS. One key issue is that the user experience of Multimedia Messaging is similar to that of Short Messaging, in terms of mobile device user interface and ease of use. Markets having strong growth in person-to-person messaging are best positioned to take full advantage of the new services building established methods of communication, such as SMS. MMS helps Operators differentiate from the competition by providing richer content to messaging. The popularity of SMS and the emergence of an instant culture suggest there is already significant demand for personal communication enhanced by visual content. Success is dependent on investing in the right technology, creating the right applications, and starting with a Multimedia Messaging strategy now.

Early Demand for Nokia Multimedia Messaging Service: a Case and Market study of MMS

Title: *Early Demand for Nokia Multimedia Messaging Service: a Case and Market study of MMS*
Author: *Nokia Networks*
Abstract: *The purpose of this study was to identify the early adopters for Multimedia Messaging Service (MMS). The potential users of MMS are found among today's heavy users of Short Message Service (SMS), that is, people under 35 years who live in an urban area. The analysis of the results included comparison of the mobile phone experience, technical background, media habits, and psychographic as well as personality characteristics of an early adoptor (innovator) with those of a non-innovator. A strong demand for MMS seems to exist. Some barriers might exist, however. Users probably do not realize how much more capacity is needed for transferring a photo from one terminal to an other than for transferring plain text. Other possible barriers may include the quality of services and the availability of multimedia mobile phones: people do not quite believe that, for instance, photo and video message could be of high quality, and a fact is that users need multimedia mobile phones for MMS.*

Executive summary

The purpose of this study was to identify the early adopters for Multimedia Message Service (MMS). The focus was in person-to-person and person-to-multiperson messaging. Person-to-application messaging was excluded. Also, the nature and extent of the early demand for MMS was examined. More specifically, the opinions of innovators on usage situations, preferences, price perceptions, storage, and most preferred group sizes for Group Message Service were studied. The study was conducted in the capital area of Finland among 15-35-year-old men and women in June 1999. The sample consisted of 800 people, of which 174 returned the questionnaire.

There were two driving assumptions in this study. Firstly, it was assumed that the innovators MMS are found among today's heavy users of Short Message Service (SMS), that is, people under 35 years who live in an urban area. Secondly, since only 22 % of the sample returned the questionnaire and up to 92.5 % of them were interested in MMS, it was assumed that they were all early adopters of MMS. They are called the innovators. Those respondents (7.5 %), who did not show any interest towards MMS, were called non-innovators.

The analysis of the results included comparison of the mobile phone experience, technical background, media habits, and psychographic as well as personality characteristics of an innovator with those of a non-innovators. It was found that the younger the respondent, the more he or she is likely to be interested in MMS. Furthermore, the innovators tended to be more often single and more likely to travel abroad when compared with the non-innovators. The innovators were also more likely to send more short messages, and to be better aware of Smart Messaging or Mobile Media Mode than the non-innovators. Likewise, they were more likely to think that mobile phone is both useful and fun, to be interested in the latest technology, to be trendy and venturesome, and to have opinion leadership compared to the non-innovators.

The innovators were divided into two groups by age and mobile phone experience. The respondents aged between 15 and 24 years were called the youngster segment, whereas those between 25 and 35 years were called the adult segment. There were more women than men in the youngster segment. The majority had a high school education, and up to 73 % of them were still students, which mainly explains the low annual incomes of the youngsters. 96 % of them were single, and 85 % travel abroad at least once a year. The youngster segment was very familiar with the Internet. The mobile phone penetration was 94 %, and 15 % of the phones were Smart phones. The SMS penetration, instead, was 93 %, and the average amount of mobile originated SMSs was 12. TV, newspapers, magazines, as well as the Internet all seemed to be good media to reach the youngster segment by advertising.

Among the adult segment, there were more men than women. The adults were highly educated, and had a high annual income. Almost half of them were single, and 90 % travel abroad at least once a year. Also, the adult segment is very familiar with the Internet and mobile phones. The SMS penetration, however, is lower than among youngsters being 86 %, and the average amount of mobile originated SMSs is only 7.
The adult segment had a high exposure to TV, newspapers, and magazines.

The most preferred services were the combination of SMS and Photo Message Service, and plain SMS. It can be concluded that consumers demand separate, non-packaged services. The plain SMS would be used in the same situations as today, and the new multimedia message would mainly substitute the traditional postcard. If the mobile phone contained also a digital camera, it would substitute cameras, scanners, and faxes. Both segments agreed that the richer the content of the service, the more they are ready to pay for it. A multimedia message was perceived five times as valuable as a short message. When it comes to group messaging, the same relation occurs.

It seems that the youngsters demand longer storage times for mobile terminated short and audio messages than the adults do; youngsters would prefer to store them for about one day and adults for only a couple of hours. Both segments would save photo, video, and multimedia messages for a few days. There were no differences in the demanded storage time for mobile originated messages, though. Both groups would destroy a short message or an audio message after a couple of hours, and photo, video, and multimedia messages after one day.

Introduction

Recently, Short Message Service (SMS) has proven to be a tremendous success in many countries. Operators in these countries have often also provided their subscribers with possibilities to personalize their mobile phones with ringing tones and graphical icons, which have proven extremely popular. The growth in this area will certainly serve as a valuable path to new and interesting ways for using the mobile phone, in ways yet unseen in the history of wireless communication.

As users have gotten accustomed to the easy use of SMS, the opportunity to send multimedia messages will mean new and easy ways of utilizing new handsets. With continuance of the path that Short Messaging and now Picture Messaging have opened, Multimedia Message Service (MMS) will add extensive value to the current offerings of the operator.

The concept of MMS is quite similar to Picture Messaging available with supporting Nokia phones, such as the 3210. With MMS conventional short messages can be combined with other elements – photographs, voice clips, and eventually also video clips – with the same simple way as SMS. In addition to sending messages mobile-to-mobile, it is possible to send them mobile-to-email. This means new and exciting possibilities for the person-to-person and person-to-multiperson communication, on which this study concentrates.

The arguments presented in this paper derive from a market study conducted by Nokia. The market study concerned the extent and nature of the early demand for MMS. Answers were sought to the following questions:

- Do the heavy users of Short Messaging find any added value in Multimedia Message Service?
- What are the usage situations for multimedia services?
- Should different characters (text, audio, photo, and video) be offered separately or together?
- What is the reasonable price for those services?
- Is there a need to store messages?
- Do the innovators of Multimedia Message Service find any added value in Group Message Service?

To find out the early demand for MMS, the early adopters must first be identified. In this study, the early adopters are simply called innovators, whereas the late adopters are called non-innovators.

The market study was conducted in Finland, which can be seen as one of today's leading geographical areas of the adoption and diffusion of mobile technologies.

Conducting the market study

The driving assumption in the market study was that without innovators the diffusion of MMS to mass markets and thus standardization of new technology will not take place. To identify the innovators of MMS, a quantitative questionnaire was delivered to a sample group of 800 people who were between 15 and 35 years old and lived in Helsinki, Espoo, Vantaa, or Kauniainen. In other words, the target group was the heavy users of SMS. The names and addresses were collected from the Center of Population
Register. Inside the previously mentioned frames the sampling method was random. The questionnaires were delivered and

collected in June 1999. 174 people responded.
A Short Message Service (SMS) was described to the consumers as follows: "You are able to send short messages by a multimedia mobile phone. A future short message can contain three times as much characters as the existing one. With a short message you can send, for example, a business card to your customer, or an important note or joke to your friend."
With an Audio Message Service "you can record speech, music and other voice clips to your mobile multimedia terminal and forward them to other terminals as an audio message. By using Audio Messaging you do not have to leave the same message to several voice mail boxes; you record the message only once and send to as many terminals as you wish."
The Photo Message Service was described as follows: "The multimedia mobile phone works as a digital camera as well. You can take a picture, for example, of your family on your vacation, and send the photo to your friends and relatives or to your PC as a photo message. The mobile phone/camera works also as a scanner. You can scan an important document to your phone and send it to your boss, for instance. Or you can scan a picture of your best friend and add it to your phonebook. It is also possible to load pictures to your mobile phone from the Internet and forward them to other multimedia mobile phones."
A Video Message Service was described as follows: "The multimedia mobile phone contains a videocamera, too. You can film your child's first steps or a negotiation at your office, and send a video message to some other multimedia mobile phone. Accordingly, you can load a music video or some animation from the Internet to your multimedia mobile phone and send it forward to some other terminal."
A Multimedia Message Service (MMS) was described to the consumers like this: "By a multimedia mobile phone you are able to send messages that contain some or all of the previously mentioned elements; text, sound, photo and video. By using a multimedia message you are able to, for example, add text to the photo you took of your family on your vacation. You are also able to add text to your 'postcard'. In work-related situations, you are able to take a photo or film the design, for instance, of Japanese cars, add your speech or written comments to it, and send it to your colleagues in another country."
A Group Message Service was described as follows: "If you have to send the same short, audio, photo, video or multimedia message to several people, you can define a group of recipients. By using a group message, sending a message becomes cheaper than by sending it separately to each member of the group. In addition, you save a lot of time and energy. If you, for example, have to inform your football team that the training begins one hour later than it was supposed to begin Group Messaging is a very convenient and quick way to do this."

Next the results of the questionnaire are examined.

Early Demand for Nokia Multimedia Messaging Service: a Case and Market study of MMS

General overview of the sample

The mobile phone penetration in the sample was 94 %, which indicates a successful sample especially when considering that the penetration in the whole country is about 60 %. 32 % owned their first mobile phone, 34 % their second, and 16 % their third one. 87% had used Short Messaging, of which 43 % sent more than 10 and 18 % more than 20 SMSs a week. The average mobile originated SMSs sent per week was 10. 33 % of the sample did not have to pay their own mobile phone bills, but got them paid by their employer, parents, or by some other relative. 27 % of the sample had used the ringing tone service.

The first assumption in this study was that the innovators for the MMS are found among the heavy users SMS, that is, people less than 35 years old, and who live in an urban area. The assumption is based on the fact that MMS is continuous innovation; MMS does not radically change the behavior of users nor does it require them to learn a new technology, but is a natural consequence of messaging evolution.

The second assumption was that since only 22 % of the sample returned the questionnaire and up to 92.5 % of them were at least a little interested in MMS (see Figure 1), they are all most likely to be innovators for MMS. On the whole, the interest in MMS was quite positive with 34.5 % of the sample being very much interested and 29.9 % quite interested in MMS. Only 7.5 % of the sample were not

Figure 1: MMS interest

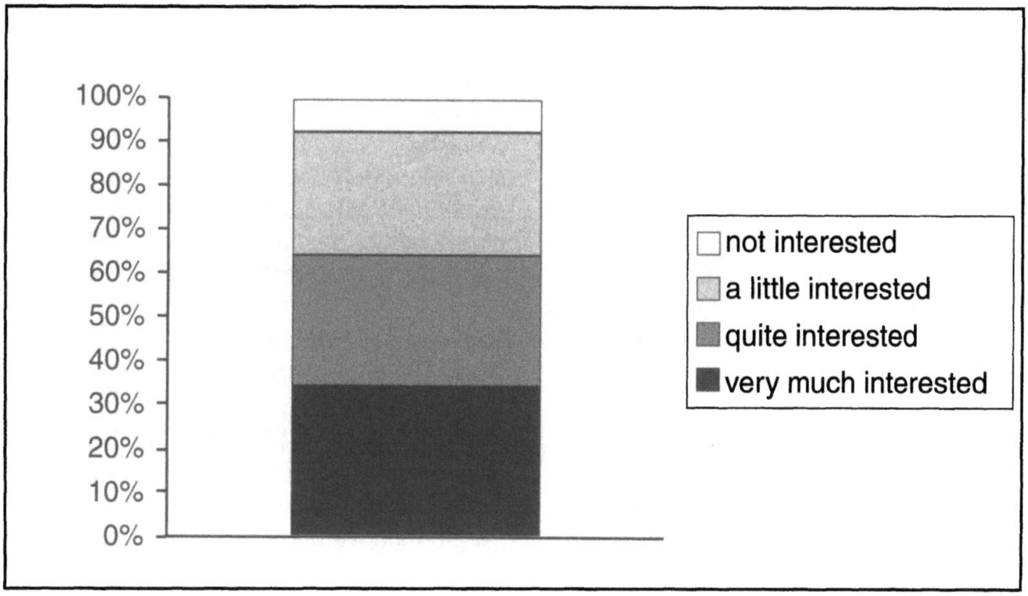

interested at all in MMS. Although this group now feels no need for MMS, their attitude might certainly change as MMS is widely adopted. History proves that this has happened also before, for example, with mobile phones and SMS. Thus, those 7.5 % will be considered as late adopters, that is, the non-innovators. To find out the early demand for MMS, the innovators must first be identified. The next chapter presents some characteristics of an MMS innovator, and compares them with those of an MMS non-innovator.

Differences between innovators and non-innovators

Psychographic characteristics
53.1 % of innovators are between 15 and 24 years old, whereas 46.9 % are between 25 and 34 years old. For non-innovators the figures are 15.4 % and 84.6 %, respectively. Thus, the younger the respondent, the more he or she is likely to be interested in MMS (1). Furthermore, 51.3 % of the innovators are men, whereas 48.7 % are women. Of the non-innovators 30.8 % are men and 69.2 % are women. The influence of gender on MMS interest is not significant (2).

51.3 % of the innovators and 84.6 % of the non-innovators have either a college or a university degree. Hence, the, non-innovators are higher educated than the innovators (3). 28.2 % of the innovators are either managers or professionals and 41.0 % are students, whereas 15.4 % of the non-innovators are either managers or professionals and 15.4 % are students (4). 76.4 % of the innovators and 46.2 % of the non-innovators are single. Thus, innovators are more often single (5) and more likely to travel abroad (6). Annual income (7) does not influence the MMS interest.

Mobile phone experience
The brand or model of a mobile phone (8) or the frequency of how often one has changed his or her mobile phone to a newer model (9) does not affect the respondents interest in MMS. The mobile phone penetration among the innovators is higher (94.9 %) than among non-innovators (80.0 %). The proportion of Smart phones among the innovators is 16.8 % and among non-innovators 12.5 %. One can be interested in MMS even if he or she has no mobile phone; up to 80.0 % of those who own no mobile phone are interested in MMS, 50.0 % of them are very interested in MMS. SMS penetration among the innovators is 89.7 % and 50.0 % among the non-innovators. 44.2 % of the innovators send at least 10 SMSs a week, and 18.6 % more than 20. The respective figures for the non-innovators are 8.3 % and 0.0 %. The innovators send more SMSs than the non-innovators do (10) . No difference is found whether a respondent pays his or her own mobile phone bills or gets them paid by someone else (employer, parents, etc.) (1)1.

Technical background
Firstly, the innovators are more often aware of Smart Messaging (56.2 %) than the non-innovators are (30.8 %) (12). Secondly, the innovators tend to be better aware of Mobile Media Mode (30.8 %) compared to the non-innovators (25.0 %) (13). Furthermore, 51.6 % of the innovators and

38.5 % of the non-innovators have sent an electronic postcard. Nevertheless, there are no differences in the Internet (14), E-mail (15) or ringing tone service (16) usage between the two groups.

Media habits
There are no significant differences in TV (17), newspaper (18) or magazine (19) exposure between the groups.

Personality characteristics
Next, a closer look is taken at the personality characteristics of the innovators and the non-innovators, in order to find some differences between the groups. See table 1 on the next page.

It can be concluded from the table that the innovators are more likely to think that mobile phone is both useful and fun, to be interested in the latest technology, to be trendy and venturesome, and to have opinion leadership. Note that none of the non-innovators considers to be an opinion leader. In other words, everyone who is an opinion leader also seems to be an innovator of MMS.

Innovator segments
Since age is a significant variable that influences MMS interest, groups. In this further analysis, users aged between 15 and 24 years are called the youngster segment. Consumers between 25 and 35 years old, in contrast, are called the adult segment.

Youngster segment
To identify the youngster segment, a clearer look is taken at the youngsters' psychographic and personality characteristics, technical background, mobile phone experience, and media habits.

Psychographic characteristics
In the youngster segment, there are a slightly more women (53 %) than men (47 %). The majority (44 %) of youngsters has a high school education. 12 % have a college and 13 % a university level education. Up to 73 % of youngsters are students, and 96 % are single. Mainly because of the huge amount of students among the youngster segment, 83 % earn annually less than FIM 100.000, whereas only 17 % earn more than FIM 100.000. 85 % of youngsters travel abroad at least once a year (see Figure 2).

Figure 2: Annual trips abroad according to the youngster segment

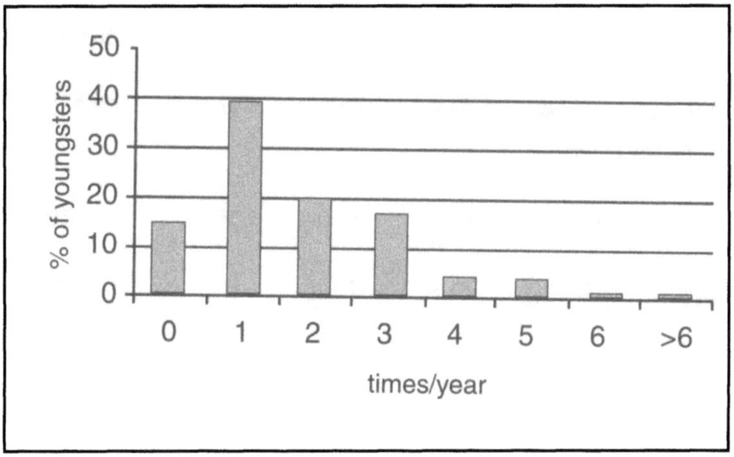

Table 1: Personality characteristics comparison of the innovators and the non-innovators

Characteristic	Innovator	Non-innovator	Difference	Pearson X^2	Pearson's R.	Significance
Extrovert	93.2	92.3	0.9	17.525	0.066	0.131
Brand loyal	74.2	61.5	12.7	7.078	0.043	0.852
Thinks that mobile phone is useful	98.8	92.3	6.5	37.379	0.374	0.000
Sporty	87.0	76.9	10.1	12.080	0.091	0.439
Economical	73.9	76.9	3.0	18.092	-0.008	0.113
Interested in the latest technology	91.9	46.2	45.7	46.707	0.402	0.000
Impulsive	76.7	84.6	7.9	15.275	0.111	0.227
Pleasure-loving	95.6	84.6	11.0	13.396	0.091	0.341
Spends freetime at home	74.8	76.9	2.1	14.383	-0.089	0.277
Trendy	71.7	38.5	33.2	28.348	0.200	0.005
Thinks that mobile phone is fun	78.8	30.8	48.0	33.650	0.381	0.001
Venturesome	80.0	50.0	30.0	25.760	0.240	0.012
Opinion leadership	61.9	0.0	61.9	37.176	0.299	0.000

Technical background
43 % of youngsters use the Internet less than one hour a week (see Figure 3). 36 % use it 1-5 hours a week, 12 % 5-10 hours a week, and 7 % more than 10 hours a week. 81 % send less than 20 E-mails a week, 11 % 20-40, 6 % 40-60, and 2 % more than 80 (see Figure 4). 57 % of the youngster segment have sent an electronic postcard.

Mobile phone experience
94 % of the youngsters own a mobile phone. 15 % of the phones are Smart phones and 85 % are legacy phones. The SMS penetration is 93 %, and the average amount of mobile originated SMSs in a week is 12. 59 % of youngsters send 1-10 SMSs and 24 % 11-20 SMSs a week (see Figure 5). As many as 17 % send weekly more than 20 SMSs. 70 % of youngsters pay their own mobile phone bills, whereas 30 % get them paid either by their parents or employer. Although only 15 % of the youngsters own a Smart phone and 7 % have tried it at least once, 58 % of the youngsters are aware of Smart Messaging. 34 %, instead, were aware of Mobile Media Mode (MMM), and 10 % were interested in it. 30 % use the ringing tone service.

Media habits
46% of youngsters watch TV less than seven hours a week (see Figure 6), which makes less than one hour a day. However, Up to 88 % watch TV more than two hours a week, which makes TV an efficient medium to reach the youngster segment.

Figure 3: Internet usage of the youngster segment

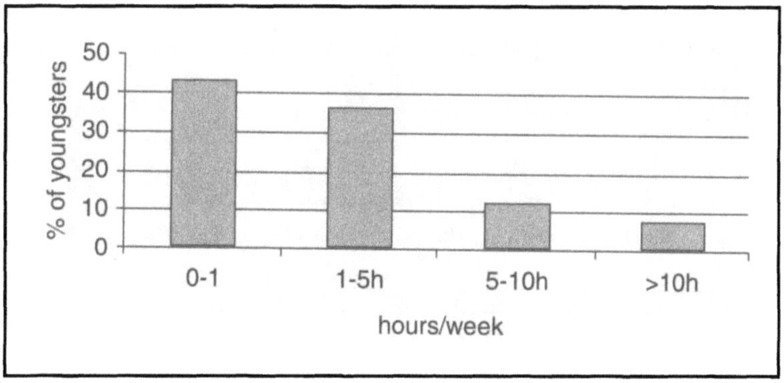

Figure 4: E-mail usage of the youngster segment

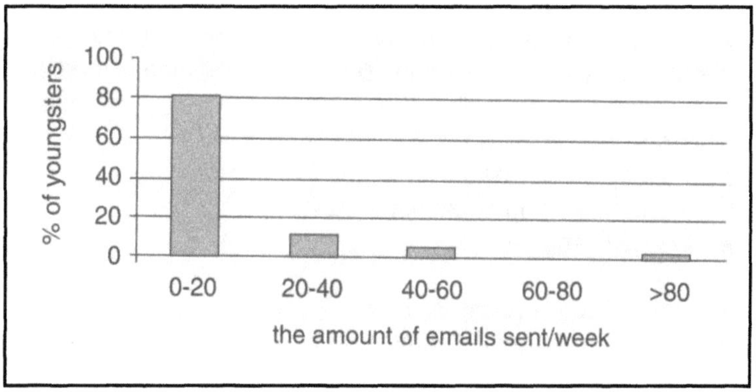

Figure 5: SMS usage of the youngster segment

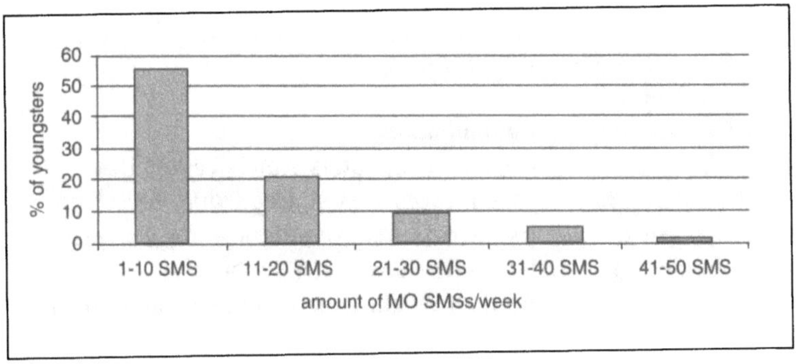

Figure 6: TV exposure in the younger segment

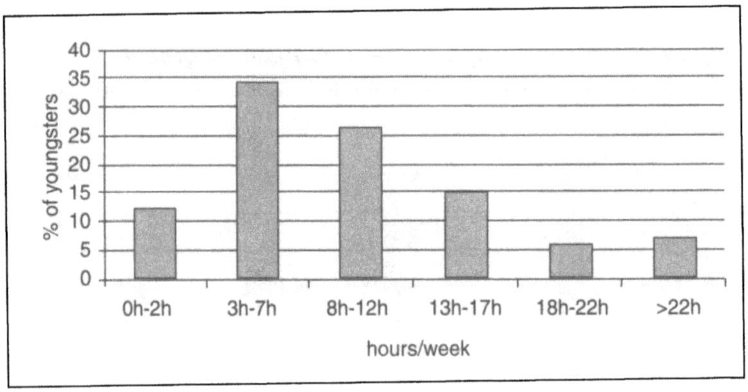

On average, the youngsters read newspapers about 3 hours and 30 minutes a week. The respective figure for all innovators is 3 hours and 40 minutes. However, the most typical time spent on reading newspapers is 1-2 hours (see Figure 7). As many as 77 % of the youngsters read newspapers at least one hour a week, which makes newspapers also a good marketing channel to reach the youngster segment.

Only 54 % of the youngsters read magazines more than one hour a week (see Figure 8). On average, the youngsters read magazines 2 hours and 8 minutes a week compared to 2 hours and 6 minutes for the innovators. That is, youngsters read magazines more than eight hours a month, on average, which means that magazine advertising is also a good marketing communications channel.

Personality characteristics
The youngster segment tends to think that mobile phone is both useful (20) and fun (21). Furthermore, the youngsters seem to be very interested in the latest technology (22), pleasure loving (23), outward (24), sporty (25), venturesome (26), economical (27), brand loyal (28), trendy (29), and impulsive (30). 61% of youngsters state that they have opinion leadership in mobile phone category in their reference group.

Figure 7: Newspaper exposure of the youngster segment

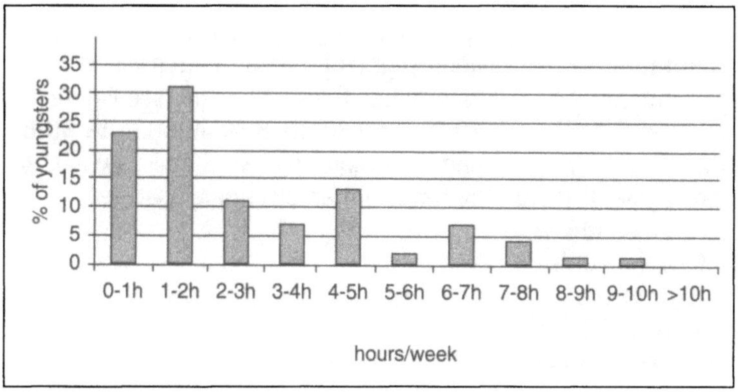

Figure 8: Magazine exposure of the youngster segment

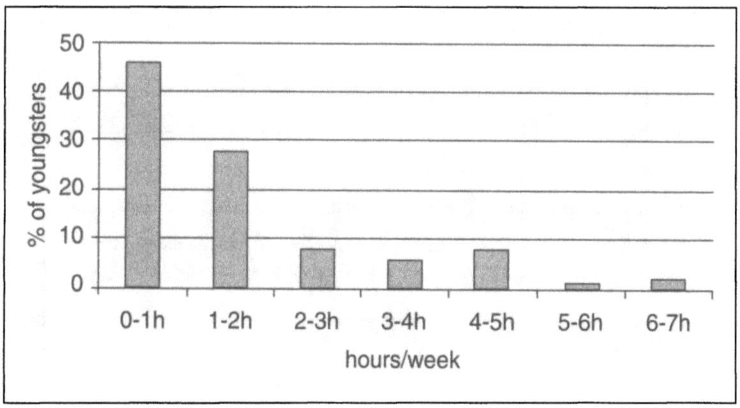

Adult segment

Next a closer look is taken at the characteristics of the adult segment.

Psychographic characteristics
Compared to the youngster segment, there are more men (56 %) than women (44 %) in the adult segment. 46 % of the adult segment have a college and 35 % a university level education. Many (47 %) of the adults are professionals, and 4 % are in a managerial position. 47 % of them are single. Only 24 % of the adult segment earn annually less than FIM 100.000, 59 % FIM 100.000-200.000, and 18 % FIM 200.000-300.000. 90 % of the adult segment travel abroad at least once a year (see Figure 9).

Technical background
34 % of the adult segment use the Internet at least one hour a week. 40 % use it 1-5 hours a week, 17 % 5-10 hours a week and 10 % more than 10 hours a week. Since 1-5 hours is the most typical time spent on the Internet (see Figure 10), the Internet can be seen as an efficient medium to reach the adult segment. 64 % of adults send weekly less than 20 E-mails (see Figure 11). 19 % send 20-40, 8 % 40-60, 4 % 60-80, and 4 % more than 80 E-mails a week. 47 % have sent an electronic postcard.

Figure 9: Annual trips abroad according to the adult segment

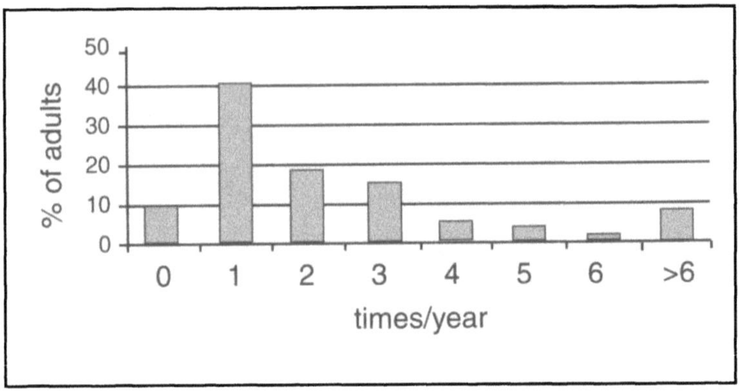

Figure 10: Internet usage of the adult segment

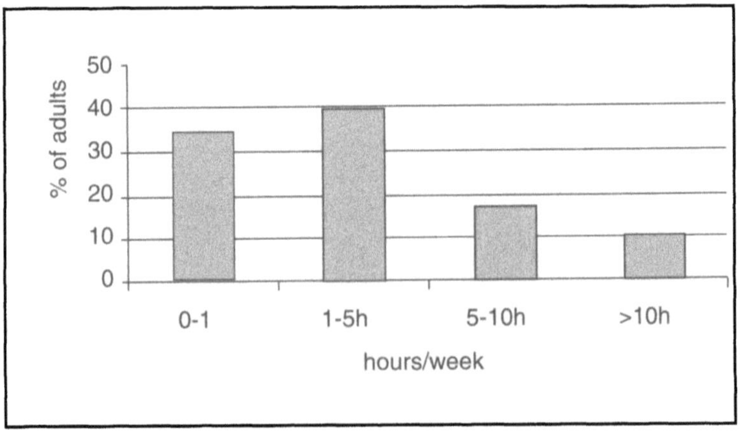

Figure 11: E-mail usage of the adult segment

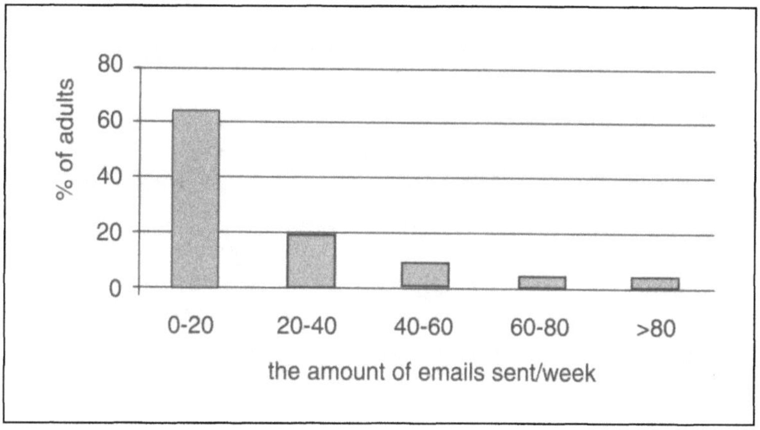

Figure 12: SMS usage of the adult segment

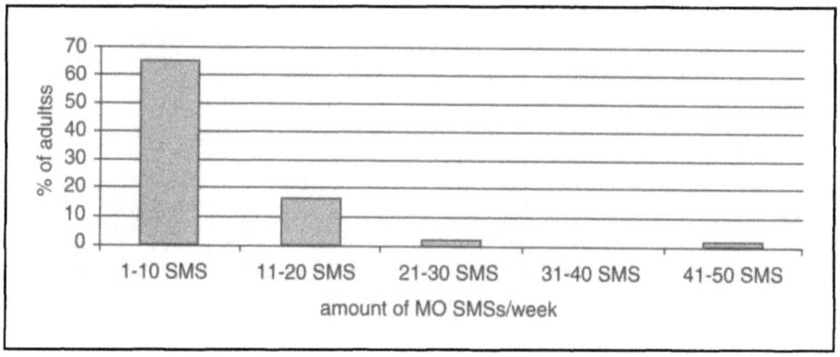

Figure 13: TV exposure of the adult segment

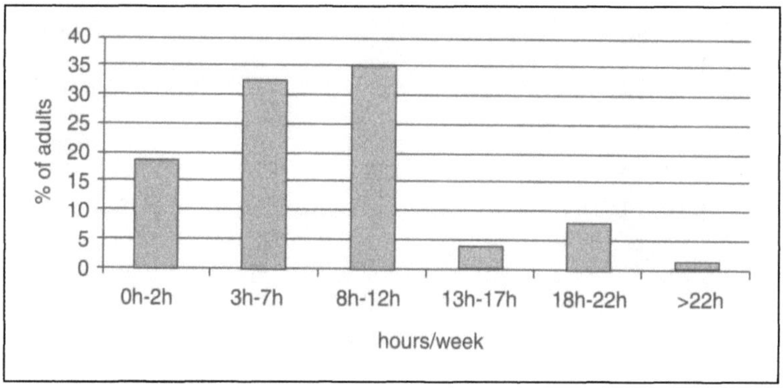

Mobile phone experience

96 % of the adult segment own a mobile phone. 19 % of the phones are Smart phones and 81 % legacy phones. The SMS penetration among adults is 86 %, and the average amount of mobile originated SMSs sent is 7. 78 % of the adult segment send 1-10 SMSs, and 19 % 11-20 SMSs a week (see Figure 12). Only 3 % of them send weekly more than 20 SMSs. 64 % pay their own mobile phone bills, others get it paid by their employer. Although only 19 % of the adult segment have a Smart phone, and 8 % have tried it at least once, 55 % of the adult segment are aware of Smart Messaging. 27 % of the adult segment, instead, are aware of Mobile Media Mode, and 11 % are interested in it. 24 % of adults use the ringing tone service.

Media habits

As many as 52 % of adults watch TV less than seven hours a week (one hour a day). However, TV is an efficient way to reach also the adult segment, because 82 % watch it at least two hours a week. The most typical time spent on watching TV is 8-12 hours a week (see Figure 13).

The mean time spent on newspaper reading is 3 hours and 40 minutes both among the whole innovator group and among the adult segment. Also, the most typical time spent on newspapers is 3-4 hours a week (see Figure 14). Because up to 88 % read newspapers at least one hour a week, it is a good channel for advertising MMS.

The most typical time spent on magazine reading is 0-1 hours and only 53 % of adults read magazines more than one hour a week (see Figure 15). The mean time spent on reading magazines is 2 hours and 10 minutes a week both among the innovators and the adult segment. That is, an adult reads magazines more than eight hours a month. Thus, advertising in magazines seems to be a good medium to reach the adult segment.

Personality characteristics

The adult segment tends to think that mobile phone is both useful (31) and fun (32). In addition, the adults are very outward (33) and interested in the latest technology (34). Likewise, they seem to be pleasure loving (35), sporty (36), impulsive (37), brand loyal (38), economical (39), and

Figure 14: Newspaper exposure of the adult segment

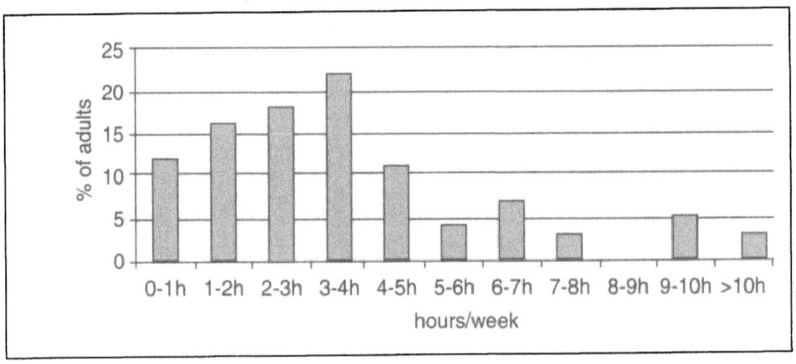

Figure 15: Magazine exposure of the adult segment

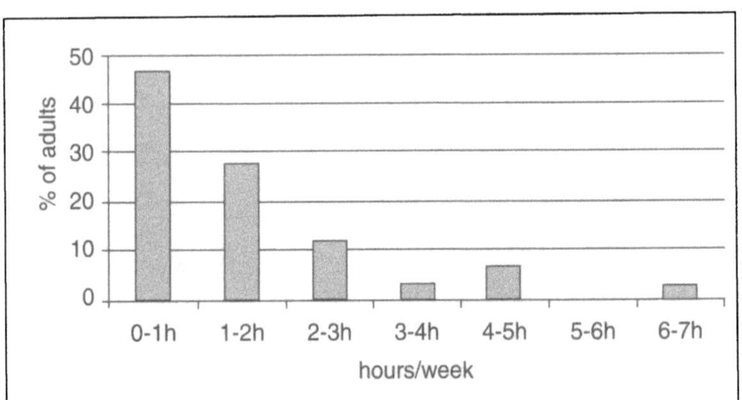

venturesome (40). They also tend to spend much time at home (41).
62 % of adults state that they have opinion leadership in the mobile phone product category in their reference group.

Demand for the different services

Next the nature and extent of the early demand for MMS are studied.

Usage situations
According to the study people would use SMS when they were abroad because sending a short message is cheaper than calling. Short messages were also found useful when trying to reach a busy person or when the recipient had turned his or her mobile phone off. Some women would send a shopping list to their husband by a short message.
Users would send an audio message when they were too lazy to write a short message, or when they needed to communicate with blind people. Some people would use it for sending a "Happy Birthday" – song to a friend, or for sending a love song to a girlfriend. Some people would use it for calling up a meeting. A recorder and a microphone in a mobile phone would be very popular.

According to users, Photo Message Service would be very useful both at work and at leisure time, especially when it contained a digital camera or a scanner. People would use Photo Message Service for sending a postcard, for getting to know someone (also with work-related situations), for getting a quick response from a colleague or a boss for something that has to be seen, or for a substitute for fax. Someone would take a photo of a city map in order to use it and find different places, for example, during vacation.
People who do visual work, for example, art directors and architects, are likely to find Photo Message Service most useful because it is a quick and easy way to get other peoples' opinions and agreements on something. Married people would use Photo Message Service when shopping alone; "How would you like this sofa, honey?"

If a mobile phone would contain a videocamera people would use Video Message Service when their child is taking his or her first steps and they wanted to show it to the child's grandparents. Some people would use a video message when they were on vacation and wanted to show someone what the place is like. Work-related potential usage situations were quite few; people might, foe instance, use Video Message Service for industrial spying.

MMS with all four elements would be popular in all the situations that were mentioned in relation to other messaging forms, for instance, when sending a postcard, an invitation, a shopping list, or when showing one's boss how the job is proceeding.

According to the study, people would use Group Message Service when they needed to call up a meeting, keep customers or members of a project up-to-date, or invite friends to a party. Some respondents would use it when they needed to inform many people of change in their address or phone number. Many respondents would make a multimedia Christmas postcard and send it to relatives and friends via Group Messaging.

very interesting, entertaining, and useful, whereas Group Message Service is found very useful, pragmatic, and work-related. Thus, the youngsters find added value in Group Message Service. Table 3 illustrates

Table 2: The image of different services according to the youngster segment (1)

	SMS	AMS	PMS	VMS	MMS	GMS
Interesting	1.57	2.02	1.67	2.06	1.84	2.02
Entertaining	2.16	1.73	1.67	1.65	1.76	2.55
Useful	1.63	2.28	1.89	2.26	1.91	1.66
Pragmatic	2.08	2.51	2.11	2.50	2.08	1.87
Work-related	2.62	2.62	2.36	2.68	2.22	1.91
Freetime-related	1.65	1.80	1.75	1.69	1.68	1.88

The following tables summarize the innovators' images of each service. The figures presented in the tables are mean values; 1 = very much, 2 = much, 3 = a little, and 4 = not at all. Table 2 illustrates the image of different services according to the youngster segment.
As Table 2 shows, the youngsters consider all services very freetime-related. SMS is also found very interesting and useful, but Audio and Video Message Services only entertaining. Photo and Multimedia Message Services, instead, are considered the respective figures for the adult segment.
The adult segment considers all services very freetime-related. The other figures, however, are not as positive as they were for the youngster segment. Only Short and Group Message Services are found very useful, whereas other services are considered mainly entertaining. Photo Message Service is found very interesting.

It can be concluded that MMS is a very freetime-related service. The difference

Table 3: The image of different services according to the adult segment (2)

	SMS	AMS	PMS	VMS	MMS	GMS
Interesting	2.03	2.61	1.83	2.39	2.03	2.04
Entertaining	2.53	1.94	1.79	1.74	1.84	2.49
Useful	1.89	2.83	2.19	2.57	2.06	1.75
Pragmatic	2.17	2.96	2.46	2.71	2.23	2.01
Work-related	2.60	3.10	2.39	2.85	2.39	2.35
Freetime-related	2.00	1.94	1.85	1.86	1.81	2.06

between the two segments seems to be that freetime-relatedness brings much added value to the youngster segment, but not necessarily to the adult segment. The adults tend to seek clear benefits from messaging in general, whereas the youngsters seek merely fun from it.

(1) AMS = Audio Message Service, PMS = Photo Message Service, VMS = Video Message Service, and GMS = Group Message Service

(2) AMS = Audio Message Service, PMS = Photo Message Service, VMS = Video Message Service, and GMS = Group Message Service

Preferences

The youngster segment preferred the combination of Short and Photo Message Service the most. The second most popular service was plain SMS, and the third most a packaged MMS. Within the adult segment the same three services were found the most popular but in a slightly different order; the adults preferred plain SMS the most, the combination of text and photo the second, and the packaged MMS the third most. Both segments agreed that SMS can only be combined with Photo Message Service. Audio and Video Message Services, however, were not popular by themselves, but offered much added value to consumers when combined together. One could conclude that both segments demand separate, non-packaged services. It seems like there still exists a huge demand for plain SMS. The market appears to be ready to adopt the Photo Message Service.

According to mean values, the youngsters seem to be most interested in Photo (42), Group (43), Multimedia (44) and Short Message Services (45). The adults, however, were most interested in Short (46), Photo (47), Multimedia (48) and Group Message Services (49). Audio and Video Message Services seemed to interest neither of the groups.

Price perceptions
There are no major differences in price perceptions between the youngster and adult segments (see Figure 16). It seems as the richer the content of the service, the more the innovators are ready to pay for it. Both segments perceive the price for SMS and Audio Message Service to be around 1 FIM. According to the innovators, one photo message could cost twice as much as a short message, for instance, FIM 2. A video message could be three times as expensive as a short message. Finally, the price of a multimedia message could be around FIM 5. That is, pricewise it is perceived five times more valuable than a short message.

A group message to seven people is worth FIM 3 according to both segments. A group audio message is worth FIM 5, a photo message FIM 7, a video message FIM 10, and a multimedia message FIM 11 according to both youngsters and adults.

Figure 16: Price perceptions for different services

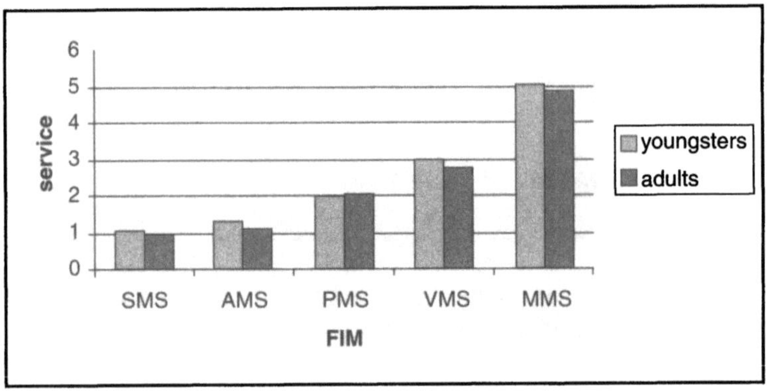

Figure 17: Price perceptions for group messaging to seven people

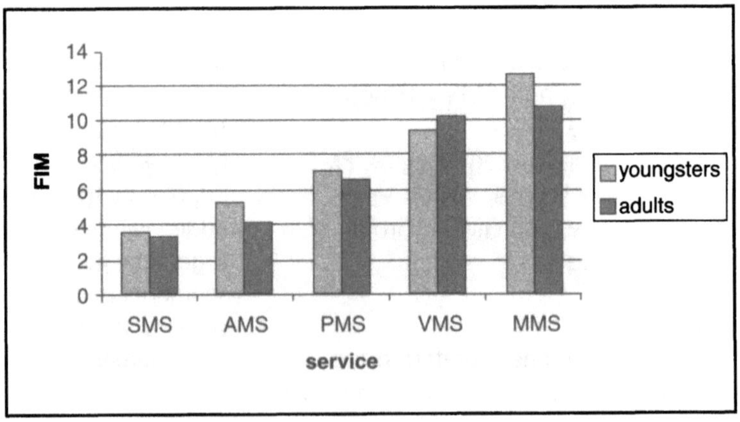

Storage

The youngsters would save a mobile terminated short message for one day, whereas the adults would save it only for a couple of hours. The youngsters would save an audio message for one day, and the adults from a couple of hours to one day. Both groups would save photo and multimedia messages for a few days. It seems that the youngsters demand longer storage times for mobile terminated short and audio messages (50).

The mobile originated messages would not be saved for as long as the mobile terminated messages. Both groups would save a short message or an audio message from a couple of hours to one day. Similarly, mobile originated photo, video, and multimedia messages would be saved for one day.

Figure 18: Storage of mobile terminated messages (1=would destroy immediately, 2=would save for a couple of hours, 3=would save for one day, 4=would save for a few days, 5=would save for a week or longer)

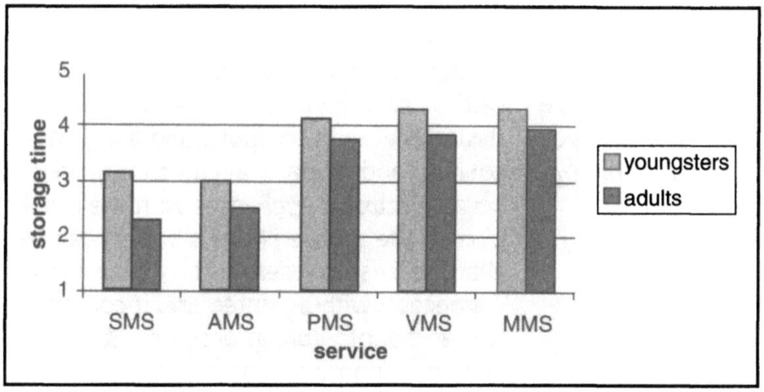

Figure 19: Storage of mobile originated messages (1=would destroy immediately, 2=would save for a couple of hours, 3=would save for one day, 4=would save for a few days, 5=would save for a week or longer)

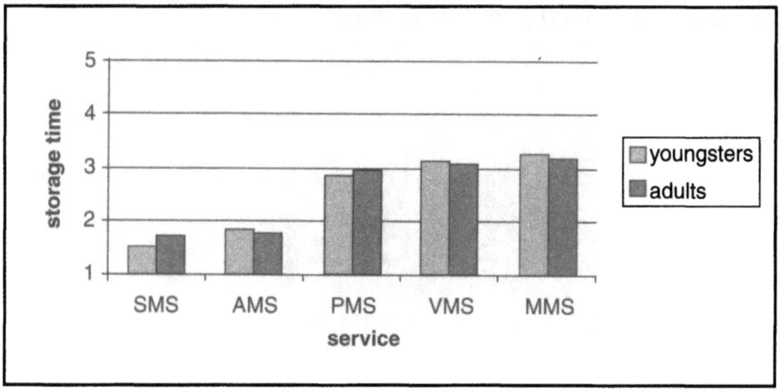

Group size

Both segments prefer the group sizes of 5-7 or 8-10 the most. Especially the adults find also bigger group sizes useful. This may be due to their most relevant usage situation for Group Message Service; communication between a project group, that usually includes more than ten people. Likewise, the adults often belong to a sports team, which might include more than ten players. Consequently, it would most likely be best to let the user customize the demanded group size him or herself.

Figure 17 showed the innovators' price perceptions for GMSs, when the group size was 7people. The prices for a group size of 5-10 people can be calculated accordingly;
Short Message Service: FIM 2.15-4.30
Audio Message Service: FIM 3.20-6.40
Photo Message Service: FIM 5.00-10.00
Video Message Service: FIM 7.15-14.30
Multimedia Message Service: FIM 7.85-15.70
Finally, in this study, the users were asked to tell their own hopes and ideas about the future mobile phone.

The ideal mobile phone

According to consumers, the future mobile phone is not just a phone anymore, it is also a digital camera, a videocamera, a tape recorder, a TV, a video recorder, a tamagotchi, a karaoke with a songbook, a ministereo with mobile loudspeakers, a playstation, and a GPS-navigator. It contains an electronic dictionary for twenty languages and an automatic word translator. It is very small, quick and light, and it can be totally controlled by voice.

The ideal phone contains maps of the world's most important cities, and you are able to scan more of them. The mobile phone tells your location if you are lost, or advises you about local services. It also informs you about traffic jams, bus timetables, menus of the nearest restaurants, and TV programs. The SIM-card is also a bank- and VISA-card, a driving license, a passport, and a key to your car and home. You can control your home electronic appliances with the mobile phone. The mobile phone's battery can be charged by solar energy. The phone has a connection with a printer, and it contains a text and photo-editing program. You are able to compose ringing tones. The phone includes a schedule, an address book, and a more advanced calculator, for example, for engineers. It is possible to receive phone calls only from certain numbers, for

Figure 20: The most preferred group size

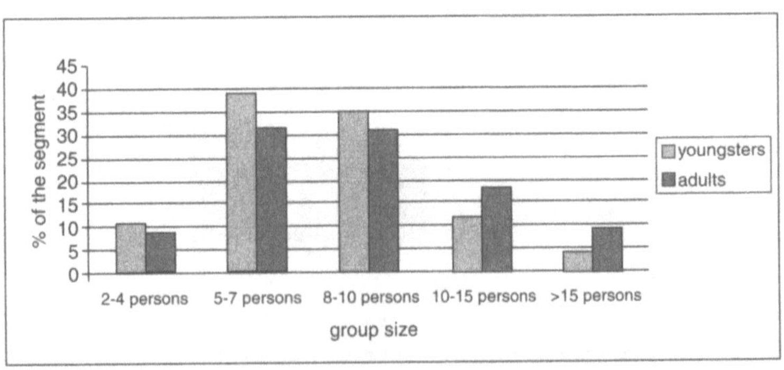

example, during a meeting. When it comes to Short Messaging, it is possible to write bold, italic, and underlined text, and one short message can contain at least 300 characters. The Short Messaging Service also includes the symbol of heart, and ready-made electronic postcards.

Conclusions

The innovators for MMS are people who are between 15 and 35 years old and live in an urban area, that is, the same people who are the heavy users of SMS. There were also innovators who did not use SMS at all or did not even own a mobile phone, which means that there are also other potential users than the heavy users of SMS. Users that have already adopted SMS find added value especially in Photo Message Service. They would use the combination of text and photo the same way people use E-mail nowadays, or as a substitute for postcards. A digital camera, a scanner, and a fax in a mobile phone would be very useful.

SMS is the only service that succeeds by itself, all the other services require each other to be successful. SMS works best with Photo Message Service, whereas Audio Message Service is most demanded when combined either with Video or Photo Message Service. A package with all four elements is the most demanded service after the combination of SMS and Photo Message Service and the plain SMS. Consequently, it might be best to sell the different services separately, that is, to charge according to the content of the originated message, and not to set a fixed price to include all kinds of multimedia messages. A roof price, however, should be set for a multimedia message, and to adjust the quality of the image or a video clip to it, and to make the usage of the services more easy and secure for the customer.

According to the innovators, sending a short message could cost FIM 1, and a photo message FIM 2. Sending a package of all four elements could cost about FIM 5 and a group multimedia message to seven people about FIM 11,50. Group Messaging is found very interesting and useful, and it is the only service in addition of MMS that is found pragmatic and work-related. The most demanded group size would be 5-10 people.
The richer the content of the service, the longer would users like to store it. Users would mostly like to save both mobile terminated and originated multimedia messages for a day to a few days.

In summary, a strong demand for Multimedia Message Service seems to exist. Some barriers might exist, however. Users probably do not realize how much more capacity is needed for transferring a photo from one terminal to another than for transferring plain text. FIM 5 is definitely not enough for such amount of data. In addition to this, users demand long storage times for mobile terminated multimedia messages, which require an enormous memory from the mobile phone. Other possible barriers may include the quality of services and the availability of multimedia mobile phones; people do not quite believe that, for instance, photo and video messages could be of high quality, and a fact is that users need multimedia mobile phones for Multimedia Messaging. Usability may also be a barrier, although not a very likely one, since Multimedia Messaging is as easy to use as traditional Short Messaging.

In addition to person-to-person and person-to-multiperson MMS, users seem to be very enthusiastic about person-to-application MMS. Together these multimedia services will enhance the usage of traditional messaging in the future.

Infinite InterChange WAP White Paper

Title: *Infinite InterChange WAP White Paper*
Author: *Infinite Technologies Inc.*
Abstract: *In the existing wired Internet, electronic mail is by far the most widely used application or technology, with over 95% of Internet users using some kind of electronic mail service. With WAP being billed as the "wireless Internet", it is expected that electronic mail is the first application that every new WAP device user will want to access. Infinite InterChange allows users to access the exact same mailbox from both the wireless and wired Internets. In today's world, the last thing users need is yet another mailbox.*

Biography: Since 1991, Infinite Technologies Inc. has worked to enrich basic network communication by offering a variety of enterprise messaging solutions. Infinite is aggressively expanding its presence in the WAP market.

Executive Summary

Infinite InterChange is the most powerful e-mail solution available for WAP (Wireless Application Protocol) compatible mobile phones and devices. Infinite InterChange extends the reach of both corporate e-mail systems and ISP (Internet Service Provider) e-mail systems to these mobile devices. In the existing wired internet, electronic mail is by far the most widely used application or technology, with over 95% of internet users using some type of electronic mail service. With WAP being billed as the "wireless internet", it is expected that electronic mail is the first application that every new WAP device user will want to access. Infinite InterChange allows users to access the exact same mailbox from both the wireless and wired internets. In today's world, the last thing most users need is yet another mailbox. The technology behind Infinite InterChange is a proven commodity in the marketplace, with initial versions of the product being introduced to the market in the fourth quarter of 1995. Since this time, Infinite InterChange, and its companion product WebMail, have been licensed for over 20 million mailboxes worldwide. Pre-WAP support, through implementation of Phone.com's HDML (Handheld Device Markup Language), has been available since the second quarter of 1997 giving Infinite InterChange a strong background in application considerations for the WAP environment.

Background on WAP

WAP is an open global standard for communication between a mobile handset and the Internet or other computer applications, defined by the WAP forum (http://www.wapforum.org). WAP-based technology enables the design of advanced, interactive and real-time mobile services, such as mobile banking or Internet based news services, which can be used in digital mobile phones or other mobile devices. The WAP specification enables solutions from various suppliers to work consistently for end-users on the digital networks. WAP creates new business opportunities for corporations by providing a new channel for existing services and the possibility for totally new services that can reach customers 24 hours a day wherever they are. Since WAP is an open protocol for wireless messaging, it provides the same technology to all vendors regardless of the network system. This means that there will be WAP compliant terminals from several manufacturers. Also the server technology is open, so operators and companies can select from a wide range of products. The common standard offers economies of scale, encouraging manufacturers, application developers and content providers to invest in developing WAP compatible products. Companies can benefit from WAP by offering end-users new mobile services that make everyday life much more convenient and are easily accessed by a mobile phone. The telecommunications industry avoids overlapping costs and investments thanks to the common, open platform and tool for wireless messaging. Operators can differentiate themselves by launching new and intriguing services. Developers and content providers are able to access all protocols and carriers with one product.

Key WAP application categories include:

- Wireless access to personal information: Wireless device users can access their e-mail, calendars, 'to do' lists, and even screen text headers for their voice mail messages.
- Wireless access to Internet content: Consumers benefit from immediate interactive access to the information they need at that moment.
- Wireless access to Corporate IT-systems (origin servers) and Extranets. Corporations can offer new channels for their services and also create totally new services for their mobile customers.
- Intelligent Telephony Services: Carriers can offer their customers secure access to their personal and other customer-related information.

The WAP protocol is being developed into WAP version 2.0, enhanced further by the increasing number of new WAP Forum members. WAP will be based on the latest existing technologies and the protocol aims to support them as quickly as possible. Market feedback will heavily influence future development.

Background on Infinite InterChange

Infinite InterChange is a server-based application that integrates seamlessly with any IMAP4 (Internet Messaging Access Protocol Version 4), POP3 (Post Office Protocol Version 3), SMTP (Simple Mail Transport Protocol), Microsoft Mail, Microsoft Exchange, Lotus cc:Mail, Lotus Notes, or Connect2 based e-mail system, or a mixed environment containing multiple protocols. Infinite InterChange also integrates with the underlying directory systems of these e-mail systems, including LDAP (Lightweight Directory Access Protocol). Remote wireless users connect through a WAP gateway to the InterChange server for direct access to their underlying mailbox, eliminating the synchronization problems and security risks inherent with other solutions that use a separate wireless mailbox. SSL (Secure Sockets Layer) and TLS (Transport Layer Security) protocols are used to ensure secure access. Infinite InterChange acts as an e-mail client proxy on behalf of the wireless user, providing the user with a view into their mailbox that is optimized for the small displays and bandwidth constraints of wireless environments. Through this proxy, the user interacts directly with their mailbox, performing all of the standard actions such as reading, replying to, forwarding, and deleting messages. All actions are carried out real time, with updates recorded immediately in the user's mailbox. If the underlying mail system tracks deleted messages in a deleted folder, then messages deleted from the wireless device are also moved into this folder. Similarly, if a sent messages folder exists, outbound messages from the wireless device are logged in the appropriate folder. This attention to detail, together with support for a wide range of messaging systems, distinguishes Infinite InterChange from other solutions. The Infinite InterChange solution is flexible in that it scales from small business environments supporting less than 50 users, to larger corporate environments with thousands of users, all the way to carrier-class and portal systems supporting hundreds of thousands or even millions of users in environments with fault tolerant requirements. Infinite InterChange is also flexible in that it can be installed as an add-on to an existing e-mail system, or it can be installed as a complete stand-alone SMTP mail server. In addition to supporting WAP clients, InterChange can also be configured as an IMAP4 and POP3 server to support conventional Internet e-mail clients. InterChange also integrates WebMail, which was the first (as evidenced by a "Best-of-Show" award at the Fall 1995 NetWorld+Interop trade show), and is arguably one of the best web-based e-mail interfaces available today.

Infinite Interchange in the WAP Marketplace

Limiting perspective to the WAP marketplace, Infinite InterChange provides solutions for the following environments:

1.) Corporate E-Mail Integration (figure 1) - Infinite InterChange can be installed on site in a corporate environment to provide corporate users with wireless access to their existing mailbox. Depending on customer security requirements and preferences, InterChange can be installed inside or outside of the corporate firewall. Transport layer security between the WAP gateway and the InterChange server is handled via SSL.
Corporate users can continue to use the e-mail system that they are most comfortable with. Direct access into the e-mail system

Figure 1

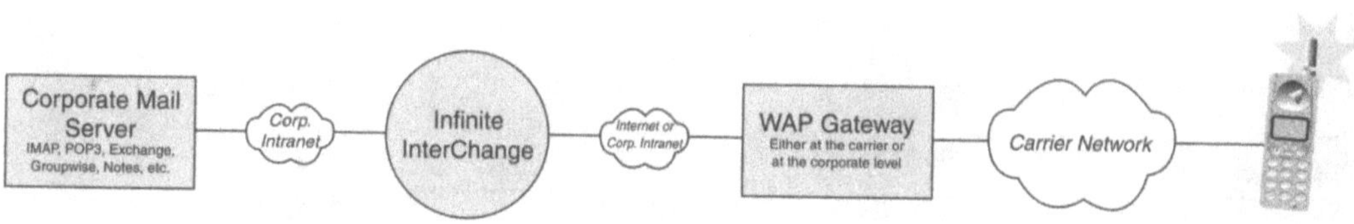

provides enhanced security in that users are not forwarding or replicating messages to an external non-secure mailbox. Forwarding solutions also present other dilemmas which do not exist in the Infinite InterChange solution. For example, users may be sometimes working in the office, and other times working outside of the office and requiring mobile access via a WAP device. If a user forgets to turn on forwarding, their mail will not be accessible from a WAP device. If a user forgets to turn off forwarding, they are often left with worse problems, such as the WAP specific mailbox filling up and having to be cleaned up, and possibly mysterious error messages being returned to the people sending them messages. Interestingly enough, network operators are often the first customers to encounter this requirement. As they begin to deploy phones into their organizations, they face the daunting task of how to access their corporate e-mail system from the WAP device. Infinite Technologies has worked with AT&T Wireless since the launch of their PocketNet service to provide access to AT&T's corporate Exchange system via mobile pre-WAP devices.

2.) ISP E-Mail Integration (figure 2) - Network operators often have related business units that provide ISP services. Integration between WAP e-mail services on the mobile phones and the e-mail services provided by the ISP can provide a compelling business advantage in the marketplace, as well as a way to draw users to sign up for both services. Infinite InterChange allows WAP devices to integrate with industry-standard IMAP4 and POP3 based e-mail systems, so that there is no need to install a separate e-mail system to support the WAP devices.

3.) Complete E-Mail System for Wireless and Wired Portals (figure 3) - E-Mail services are a core component of any portal strategy, whether the portal strategy is wired, wireless, or a combination of both. Infinite InterChange is available in configurations that support fault tolerant operations using clustering and load balancing technologies. In these environments, Infinite InterChange can provide a complete e-mail solution supporting hundreds of thousands or even millions of users. Users

Figure 2

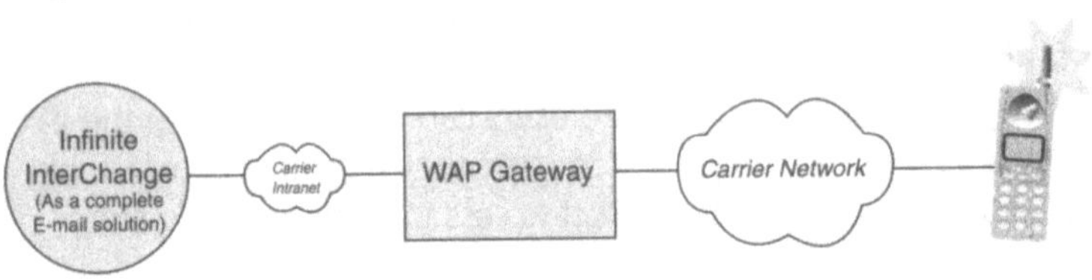

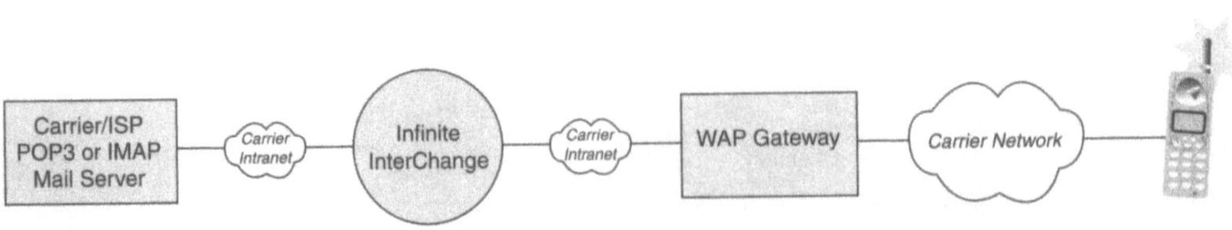

have the flexibility of connecting to their mailbox via a WAP device, via the WebMail interface, or via industry-standard POP3 and IMAP4 e-mail clients.
Infinite Technologies also provides outsourced mail hosting services for these environments.

4.) Individual Users and Small Offices (figure 4) - For individual users and small offices that have existing ISP e-mail accounts, Infinite Technologies provides the free MailAndNews.com service, which can allow these users to access their existing POP3 or IMAP4 mailboxes via a WAP device.

5.) (For UP.Link carriers) Web-based access to UP.Mail - Phone.com's UP.Link gateway includes an integrated WAP-based e-mail component. However, this solution will not be adequate for many environments. In addition to environments where the network operator wishes to integrate the WAP e-mail services with an existing e-mail system, there is also a strong need for users to be able to access their WAP mailbox from the wired world. Infinite InterChange's WebMail services can be used to provide world class web-based access into the UP.Mail system.

The Possibilities Are Infinite

Since 1991, the Owings Mills, Maryland based Infinite Technologies has worked to enrich basic network communication by offering a variety of enterprise messaging solutions. The company's product line includes message transport and gateway software, Internet and Intranet access solutions for networks, e-mail systems, and messaging utilities. Infinite continues to expand its connectivity solutions to include additional popular messaging platforms, bringing systems together to enable effective and efficient communication. Infinite Technologies can be reached at +1 410.363.1097, via e-mail at Sales@InfiniteMail.com, or via the Web at http://www.InfiniteMail.com.

Figure 4

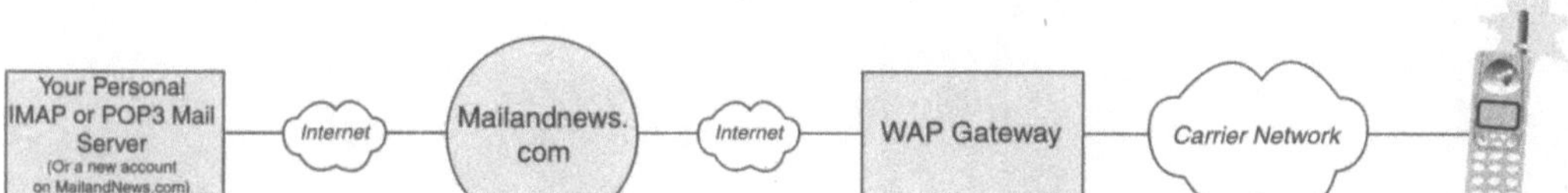

Figure 5

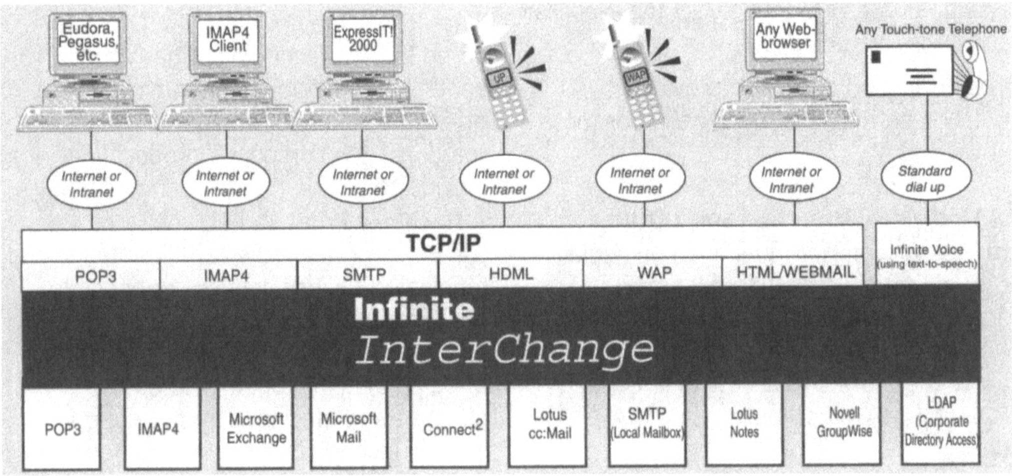
Eudora, Pegasus, etc.
IMAP4 Client
ExpressIT! 2000
UP
WAP
Any Web-browser
Any Touch-tone Telephone
Internet or Intranet
Internet or Intranet
Internet or Intranet
Internet or Intranet
Internet or Intranet
Internet or Intranet
Standard dial up
TCP/IP
POP3
IMAP4
SMTP
HDML
WAP
HTML/WEBMAIL
Infinite Voice (using text-to-speech)
Infinite InterChange
POP3
IMAP4
Microsoft Exchange
Microsoft Mail
Connect2
Lotus cc:Mail
SMTP (Local Mailbox)
Lotus Notes
Novell GroupWise
LDAP (Corporate Directory Access)

Is Mobile Data Really About to Take Off or are we seeing another false dawn?

Title:	*Is Mobile Data Really About to Take Off or are we seeing another false dawn?*
Author:	*David Balston, Adrian Golds, consultants with Intercai Mondiale*
Abstract:	*There have been many false dawns in the world of mobile data. Now however, so many critical factors are combining that many commentators believe that mobile data is about to have its day. In this paper we examine these factors and, while broadly accepting the premise that data use is going to grow at a dramatic rate, introduce a note of caution about the speed of its acceptance.*

Introduction

Data services have, to date, largely been the domain of specialist operators who offer packet radio access to a small market segment with a clear commercially justified need for such communication. No economy of scale, little competition, high equipment prices and a data-only capability have meant penetration has remained low, typically a few tens of thousand users in UK. Similarly, while data services have been available to the millions of cellular users for many years, the uptake has been very limited. The services are expensive (being connection rather than packet based), slow (9.6 kbps compared to the 56 kbps with a modern wireline modem) and, above all, inconvenient to use requiring special, expensive PC interface cards and a good signal location.

The first rays of the new day have already been seen however with the increasing popularity of SMS (Short Message Service). Although each SMS message is limited to 160 characters it is increasingly being used for a variety of information and alerting services. Despite some operators charging 10p a message, a rate that is more expensive than many off peak tariffs for a whole minute of talk time, it is estimated that there are more than 9 million regular users in Europe. It anticipated that this figure will grow to 25 million in 2000 [Dataquest Europe]. So what are the factors that will accelerate this move to data? We identify three critical developments.

Driver 1: The increasing ubiquity of IP and the Internet

Businesses, recognising the value of up-to-date data, have introduced Intranets that allow access to centrally held corporate information. In the US, and now increasingly in Europe, the Internet, and its associated technologies, are becoming an important tool in an information rich society. E-Commerce is beginning to become accepted, services are purchased over the 'net', immediate access to information is becoming the standard. In both the business world and the consumer market familiarity with, and reliance on, these services will create the pressure to have them available at all times, wherever the user is. We can easily envisage a world where scenarios such as those shown overleaf are quite common.

Driver 2:
Technology can make these scenarios realistic

The scenarios may seem a little far-fetched, but in fact they both use technology that is, if not widely available, is at least being trialed today. Headline email is available as an SMS based service from a number of service providers today. However, with the advent of packet data to the mobile terminal over the next year or two, as provided by GPRS in the GSM world (see panel), this sort of service will become a standard offering. Combine this with the trend toward larger mobile displays and intuitive keypad entry of text, then the service becomes practical on mobiles that look similar to the models available today.

The electronic purse applications (e-cash, e-ticket etc) are again examples of existing technology.
Various trials (such as Mondex in the UK) have already been carried out and features such as SIM Application Toolkit make for easy transference to the mobile sector. With the advent of dual slot phones, such as that from Motorola, the banking sector's

dream of an ATM in the pocket can become a reality.

Machine to machine communications has for some time been identified as a significant new market sector. One of the first examples of this is the soft drinks machine dispensing its wares and automatically calling for refills. Manufacturers are starting to produce low priced GSM units for exactly this type of application, and again, GPRS will make the operation economic. This is an area that has also had a significant profile in the various 3G debates – a significant traffic generator within a packet network, despite requiring extremely low bandwidths!

The virtual reality tour of houses for sale is an idea already being tested by some innovative web designers, and once perfected, the only thing stopping that being available on a mobile terminal is bandwidth. Once the bandwidth is made available, either from the forthcoming 3G networks or further enhancements of second generation technology (see panel describing EDGE), this sort of application is as easily viewed on a mobile PDA as a conventional computer screen.

Driver 3:
The entry of new players from the computer world

A most significant development is the entry of players such as Microsoft, Sun, 3Com and Cisco into the mobile data world. These companies have realised that mobile telecommunications has now become a mainstream business and their products must be capable of 'shrugging off the wired constraint'. This has caused the introduction of new concepts and

The Business Traveller

On arriving in Paris on a Monday morning, Mike checks his schedule on his PDA which has now been automatically updated since turning it back on after leaving the aircraft. He sees a new appointment in Frankfurt for the following day and headlines of a number of new emails. While scanning these new messages, he sees that several have large attachments, and so he decides to leave these for downloading later.

In the taxi on the way to his meeting, Mike uses the time to search the Internet for flights that evening to Frankfurt, and subsequently books a seat on the 17:55. The seat is confirmed and an e-ticket transmitted to the electronic purse smart card currently inserted into his PDA. He then searches the net for Hotels in Frankfurt, and books a room online.

Later at Charles de Gaulle, after checking in with his e-ticket and bypassing all the queues at the Lufthansa desks, Mike finds himself with a little free time in the lounge. He then checks his bank balance back in the UK and immediately transfers another £100 to his electronic purse. He can then carry out a bit of duty free shopping for his wife and children, all debited directly from his e-cash card, using the current Euro-Sterling exchange rate. The card is also useful to obtain a small snack from a vending machine, eliminating the requirement for coins in the local currency.

Once checked into his Frankfurt hotel, Mike sets up his laptop PC and is able to attend to the e-mails he deferred from earlier. One message, from colleague has a report attached, with a request for Mike to review it. Although his colleague has forgotten to zip up the 50-page report, this file is downloaded within only a few seconds and Mike starts his review straight away. Once finished, the modified report is sent back to its author and Mike accesses the company server to download a couple of documents he will need for the meeting tomorrow. Properly prepared, he can now relax and enjoy the rest of his evening.

The Consumer

Sarah leaves the house and plugs her smart card into the slot in the car's dashboard. It now automatically adjusts the seat to her preferred driving position and then dials the number for her husband's mobile in response to her speaking his name. Once arriving at her destination, she removes the smart card and plugs it into her handset. Following a call from a friend, she is reminded about a promise to meet for lunch in London. She uses the phone's browser to access the railway timetable, and downloads an e-ticket for the next train to Paddington. When she arrives at the station, she inserts her smart card into the machine on the platform, and her ticket is printed out.

She has some time before her train is due so she goes to a nearby soft drinks machine and buys a coke using her phone to pay for the can. In addition to handling the transaction the GSM chip in the vending machine can also alert the owner when the stock needs replacing or the cash box needs emptying (some people still use coins).

After meeting up and ordering lunch, Sarah's friend mentions that she had seen a house advertised in the area Sarah was currently looking. She brings out her PDA and connects to the web site of a national estate agent. She then selects a few parameters that cause the details of the property in question to be displayed on the PDA's screen. She shows Sarah, who is mildly interested at this stage, and then after selecting the picture of the property, they take a 'virtual tour'. This consists of selecting arrows with the PDA stylus, which changes the view of the property, giving the view of each elevation. They then select the front door, and are able to tour the house from the inside, moving around the rooms as if they were physically there. Finally, Sarah decides that she would like to find out more details, so presses the 'connect to agent' button and uses the PDA's voice facility to arrange an appointment for the next day.

Finally, after finishing their lunch, Sarah removes her smart card from her phone, and inserts it into the portable reader that the waiter brings to the table. The display already indicates the total from the lunch, to which she adds a tip and the total is immediately debited from the cash balance on the card. Thanking her friend for the recommendation of the house, she says goodbye and catches her train home.

supporting investment to develop new ideas.

These new entrants to the mobile world bring with them, not only a deep understanding of the direction that the data market, services and supporting protocols is taking, but, more importantly, the ability to influence the mainstream data world to facilitate service delivery via wireless networks.

This 'new blood' brings its own problems however. Not surprisingly, the approaches advocated by the new players differ from those developed to date by the more traditional mobile-centric companies. The danger is that a variety of solutions will be offered, the customer will become confused, scale efficiencies will not be achieved and the market will remain fragmented.

Such a cornucopia of solutions is not necessarily a barrier to growth however. In the fixed world there is a significant migration away from centralised network intelligence and a trend towards 'edge of

Figure 1: General Packet Radio Service (GPRS) shares radio resources with GSM, but is able to communicate with multiple users due to its packet nature. Within the network, routers or GPRS Support Nodes (GSNs) allow the packet communications to bepassed straight through to the Internet or other packet networks

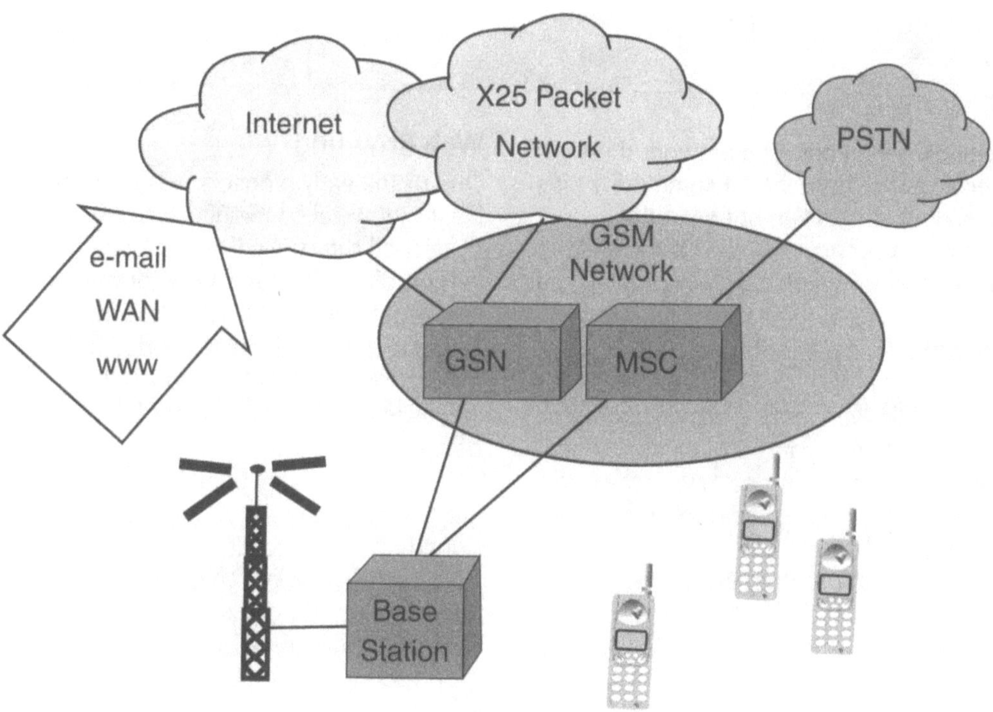

EDGE (Enhanced Data rates for GSM Evolution) When GSM was originally specified the main concern was to achieve good quality voice service even in poor signal quality areas. This meant that the receiver performance thresholds had to be set to cope with low Carrier to Interference (C/I) conditions. Now, 10 years on, most networks exceed these minimum standards and EDGE offers the GSM network operator the opportunity to take advantage of this to provide a better data service by using a less robust modulation scheme.
The chart shows that under exceptionally good conditions the user data rates could reach 48 kbps. At the 9dB C/I levels required for conventional services however the new scheme's performance is much poorer. The introduction of EDGE will require new radio equipment in the Base Station and upgraded handsets.

Figure 2: Typical Data Rates with and without EDGE (source: Ericsson).

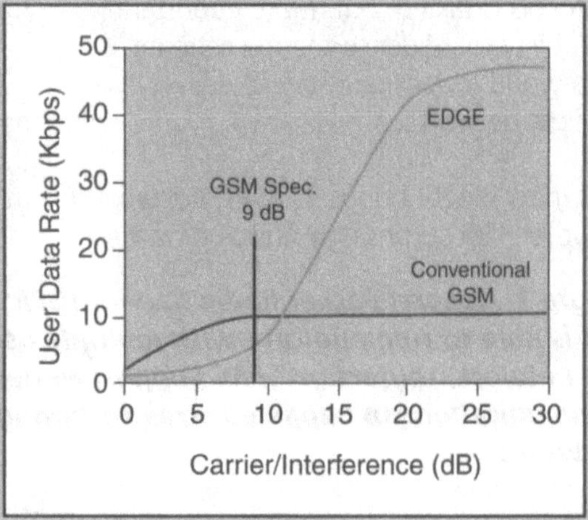

network' solutions. The network then becomes no more than a commodity pipe providing connection between the intelligent agents. Add to this the ability to download software, e.g. JAVA applets, and the capability of the intervening network becomes of less importance.

The mobile world has already recognised this and JAVA capable SIM cards with a SIM Toolkit facility that allows it to control aspects of the handset's display are now becoming available.
It is in the area of Web Browsing however that we can see the battle lines being drawn.

Web Browsing

One of the early pioneers was Unwired Planet (now called Phone.Com) who developed the concept of a mobile-friendly version of HTML (the WWW formatting language) called WML (Wireless Mark-up Language). In addition, a cut-down transmission protocol, called WAP (Wireless Access Protocol) has been defined.
This has proved very popular and all the major mobile handset manufacturers have signed up to the concept and are building WAP compliance into their products. In addition, over 80 companies (including Microsoft) have joined the WAP Forum, a group set up to promote the widespread adoption of WAP.

An important limitation of the approach however, is that it can only handle web pages that have been written in WML. Although many content providers, including big names like Reuters, have agreed to provide their content in WML form as well as the more conventional HTML, the fact remains that only a very small subset of the millions of pages on the Web are available in WML form.

An alternative approach advocated by Microsoft however concentrates on the ability of the display rather than on the limitations of the transmission medium. They propose a presentation-oriented protocol that mediates between the data available for presentation and the functionality of the display currently available for it. This clearly has much wider application than the mobile specific solution developed by the WAP Forum. It does however suffer from the disadvantage for the wireless community that all the data must still be downloaded to the presentation device.

An innovative approach, called Oasis, has been developed by yet another company, On-line Anywhere. This claims to be able to interpret the information provided in a conventional HTML page and automatically filter out that part which can be sent to, and displayed upon, a handhel device.

One of the less obvious consequences of the drive to simplify the data sent to a mobile device is that much of the data that is filtered out is going to be advertising copy. If this can no longer be sent to the customer there will be less incentive for the advertiser to subsidise the site and we can expect this trend to lead to a more rapid development of subscription sites and micro-payment facilities.

Operating systems

A second major area of contention is in the choice of operating system for handheld devices. Here, again, a comparatively small company has gained the high ground and Microsoft has arrived on the scene with guns blazing. The EPOC operating system,

Figure 3

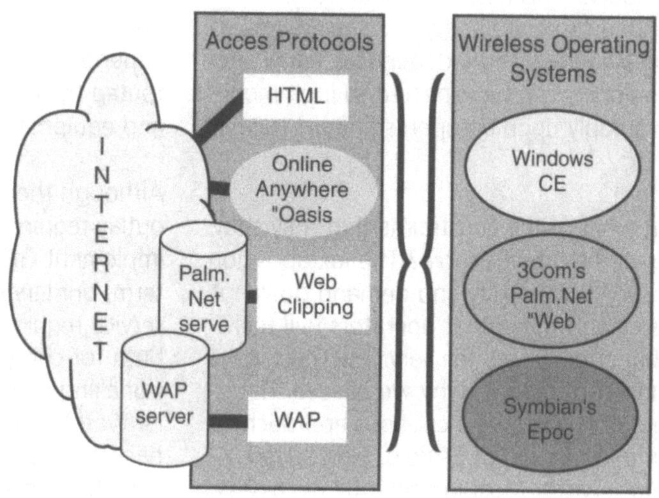

created by Psion, was transferred into to a new company called Symbian set up by them with Nokia, Ericsson and Motorola. Their intention is to make EPOC the universal operating system for handheld devices.

Microsoft, on the other hand is promoting its Windows CE operating system and, although it remains memory hungry, Microsoft's strength and the ubiquity of Windows and its associated applications, aided by advances in memory technology, place it in a powerful position. A position that is strengthened by the adoption of Windows CE by most of the PDA manufacturers, with the notable exceptions of 3Com and Psion.

A negative factor of significant import however derives from the very strength of Microsoft.
Many companies have, in the past, found to their cost that their niche products have been usurped by Microsoft which has subsequently built the functionality into Windows. This has, not unnaturally, created some antipathy towards Microsoft and Windows and independent application developers can be expected to look more favourably upon an open standard such as EPOC.

But what of the constraints that may slow down the growth curve? The juxtaposition of service capability and demand does not necessarily mean that operators will make a big effort to sell the services. There are a number of reasons why we believe that operators will take a cautious approach to the introduction of flexible, high speed data services. Most importantly we assess the potentially adverse impact that their introduction may have on the basic voice service.

This is compounded by the likelihood that the total revenue obtained from a fully loaded base station will be lower than if voice alone was offered. Finally, the momentum towards 3G is now gathering pace with a resolution of the air interface conflict and the start of licence award.

The Operators' GPRS Dilemma

Although announcements of new GPRS contracts seem to regularly appear in the telecoms press, the enthusiasm for this new technology amongst GSM operators is far from universal. For instance, although two of the four UK operators have announced deals for infrastructure, the world-wide total shows commitment from only around 10 out of the 300 plus GSM operators.
This is not an indication of lack of confidence in the technology, but a more realistically cautious approach to the impending 'data revolution' currently being touted by many industry commentators and equipment manufacturers.

Although there is a considerable capital outlay required for an operator to implement GPRS, this may not be the short term, primary cost of the service. The service requires dedicated radio resources, both for control channels (to allow signalling between the terminal and network) and for carrying the actual data packets. In particular, if the much-advertised speeds of up to 115kbit/s are to

Figure 4

GPRS Orders by Manufacturer	
Operator	**Country**
Ericsson	
T-Mobil (Also with Alcatel)	Germany
Omnipoint	USA
One-2-One	UK
Telfort	Netherlands
SmarTone	Hong Kong
Nokia	
Sonera	Finland
KG Telecom	Taiwan
Hong Kong Telecom (MoU for GPRS development only)	Hong Kong
Motorola	
Cellnet	UK
France Telecom (Trial only)	France

be offered to the customer, then a whole radio carrier must be dedicated to GPRS – a resource that can carry eight simultaneous voice calls. The manufacturers will claim, correctly, that because GPRS is packet based, this allows many more than 8 users to share the resource.
However, this will only be achieved when there are a significant number of GPRS users registered with the network.

The concern for the operator, therefore, is not only the cost of additional infrastructure, but also the requirement to allocate valuable radio resources to GPRS. This has a number of potential impacts:

- It effectively reduces the grade of service for voice users (by denying them access to some radio resources, so more 'network busy' failures at peak times)
- It reduces trunking efficiency, and therefore overall network capacity.
- It will, initially, result in lost revenue. This is because the idle GPRS resources could have been carrying, and charging for, speech calls that are now being blocked.
- In addition there is currently no telecommunications billing system commercially available to bill by amount of data transmitted rather than time. This will be required to enable tariff structures to be put in place to attract customers to the packet based GPRS service.

The lost revenue issue is particularly acute because the wireless operator is not, by itself, in a position to provide the end-to-end service that GPRS will facilitate. Traffic will typically be generated by 3 rd party services and applications –either entertainment portals or business specific solutions. We can expect some operators to forge liaisons with content providers but, in general, the advent of GPRS is expected to herald a change in the structure of the cellular service value chain. We do not foresee a 'killer application' such as voice for the data world. Popular solutions will be content and services dependent and thus will not be integrated wholly within the cellular operator's business.

These factors have probably led to the majority of operators adopting the 'wait-and-see' approach. Particularly in light of some of the other developments in mobile data already discussed, operators may wait for applications such as WAP to establish a significant data market using conventional circuit switched connections before re-deploying resources to a packet network. If they then find that mobile data really does take off, they can then introduce GPRS and have the resources almost immediately fully utilised, and take full advantage of the capacity gains that will then be available.

Conclusions

We have seen that there are many positive drivers accelerating the take-up of data services in both the fixed and mobile domains.

While there remain uncertainties about which standards will prevail and, crucially, how much traffic will be generated, we can expect the mobile operators to take a cautious stance in the enhancing of their networks to support the new services. With licences for 3G services now becoming available operators will have to choose their investment strategies with care.

In any one market however the first mover can be expected to gain credibility and become well-positioned to take a major share of the data traffic when it finally reaches an economic level. As the recent boom in subscriber numbers levels off however it will be a far-sighted operator who is prepared to take the lead.

Application of Smart Antenna Technology in Wireless Communication Systems

Title: *Application of Smart Antenna Technology in Wireless Communication Systems*

Author: *Richard H. Roy*

Abstract: *Wireless communications is widely recognized as one of the fastest growing industries in today's global marketplace. As with any nascent industry faced with unexpectedly large demand for its products, growing pains are inevitable. Consumers are continually pressing system providers to expand their suite of services, and to provide these services at ever-decreasing costs. Unfortunately for system providers, there is only so much product to sell and technological limitations are in many instances becoming significant barriers in the quest to meet consumers' insatiable demand. In this article, a brief description of some of the current technical challenges facing wireless service providers is given, historical attempts to address these challenges are summarized, and an overview of intelligent antenna technology as a solution thereto is presented.*

1. Introduction

In a majority of currently deployed wireless telecommunication systems, the objective is to sell a product at a fair price, the product being information transmission from one or more points to one or more points. Cellular and next generation Personal Communication Services (PCS) systems, in some sense the killer-apps of wireless, are classic examples where operators sell point-to-point information transfer to consumers largely in the form of circuit-switched (voice) links over which standard phone calls can be placed.

From a technical standpoint, information transmission requires resources in the form of power and bandwidth. Generally, increased transmission rates require increased power and/or bandwidth independent of the medium. While transmission over wired segments of the links can generally be performed independently for each link (ignoring cross-talk in land lines), this is not the case for wireless segments. While wires (fibers) are excellent at confining most of the useful information/energy to a small region in space (the wire), wireless transmission is much less efficient. Reliable transmission over relatively short distances in space requires a large amount of transmitted energy, spread over large regions of space, only a very small portion of which is actually received by the intended user. Most of the energy is considered interference to other potential users of the system (cf. frequency reuse).

Somewhat simplistically, the maximum range of such systems is determined by the amount of power that can be transmitted (and therefore received) and the capacity is determined by the amount of spectrum (bandwidth) available. For a given amount of power (constrained by regulation or practical considerations for example) and a fixed amount of bandwidth (the amount one can afford to buy at auction these days), there is a finite (small) amount of capacity (bits/sec/Hz/unit-area, really per unit-volume) that operators can sell to their customers, and a limited range over which customers can be served from any given location. Thus, the two basic problems that arise in such systems are:

1) how to acquire more capacity so that a larger number of customers can be served at lower costs in areas where demand is large, and
2) how to obtain greater coverage areas so as to reduce infrastructure and maintenance costs in areas where demand is relatively small.

II. The Quest for Capacity

In areas where demand for service exceeds the supply operators have to offer, the real game being played is the quest for capacity. Unfortunately, to date a universal definition of capacity has not evolved. Free to make their own definitions, operators and consumers have done so. To the consumer, it is quite clear that capacity is measured in the quality of each link he gets and the number of times he can successfully get such a link when he wants one. Consumers want the highest possible quality links at the lowest possible cost.

Operators, on the other hand, have their own definition of capacity in which great importance is placed on the number of links that can simultaneously be established. Since the quality and number of simultaneous links are inversely related in a resource constrained environment, operators lean toward providing the lowest possible quality links to the largest possible number of users. That the war wages on is evidenced by churn and the lack of acceptance of current digital systems in markets where competitive analog service is available. Consumers are wanting better links at lower cost, and operators are continually trying to maximize profitability providing an increasing number of lower quality links at the highest acceptable cost to the consumer. Until the quest for real capacity is successful, the battle between operators and their customers over capacity, the precious commodity that operators sell to consumers, will continue.

Attempts to obtain more capacity to sell have recently centered around transitions to digital systems. Unfortunately, this transition has involved giving less capacity to each customer (through voice compression), and having long associated digital technology with superior performance at lower cost, consumers are largely disappointed with digital wireless performance and cost (you can't fool the wireless consumer!). Attempts at increasing system capacity have involved the development of sectorized antennas for cellular systems, culminating with the more recent attempts to develop cost effective microcellular concepts. Such systems suffer from trunking inefficiencies, and higher costs per line, but can represent a solution to increasing capacity where the costs can be justified. As a long-term solution however, providing base stations everywhere for everyone may not be economically viable.

III. The Requirement for Range

There are many situations where coverage, not capacity, is a more important issue. Consider the roll-out of any new service such as PCS in the United States. Prior to initiating service, capacity is certainly not a problem; operators have no customers. Until a significant percentage of the service area is covered, service can not begin. Clearly coverage is an important issue during the initial phases of system deployment.

Consider also that in many instances only an extremely small percentage of the area to be served is heavily populated. Furthermore, the degree to which customers in those areas can be enticed to sign-on is a strong function of the coverage provided by the operator in the other 99.9% of the service area. The ability to cover the service area with a minimum amount of infrastructure investment is clearly an important factor in keeping costs down and customers happy.

As is often painfully obvious to operators, the two requirements, increased capacity and increased range, conflict in most instances. While current technology can provide for increased range in some cases, and up to a limit increased capacity in other cases, it rarely can provide both simultaneously. Operators who would like to keep infrastructure costs down by deploying fewer sites have to sacrifice revenue generating capacity, and those that opt for more capacity must pay the price. Intelligent antenna technology offers the potential to ease the pain by providing systems with the capability to both increase coverage and capacity at the same time, and more importantly, the exibility to adjust to the particular needs of the operator as the system requirements evolve.

IV. Space – "The Final Frontier"

Space is truly one of the final frontiers when it comes to next generation wireless communication systems. Spatially selective transmission and reception of RF energy promises substantial increases in wireless system capacity, coverage and quality. That this is certainly the case is attested to by the significant number of companies that have been recently formed to bring products based on such concepts to the wireless marketplace. The approaches range from switched-beam to fully-adaptive, uplink only to uplink and downlink, with the benefits provided by the various approaches differing accordingly. Clearly, as Qualcomm's founder Andrew Viterbi stated: "Spatial processing remains as the most promising, if not the last frontier, in the evolution of multiple access systems."

A. Switched-diversity Technology

One the earliest attempts at addressing the difficulties of the mobile RF environment was to employ two identical antennas separated by several wavelengths, each instrumented with conventional receivers. The basic principle underlying such designs is that in complex RF environments, there is sufficient scattering of the RF fields to practically decorrelate signals received from antennas sufficiently far apart. The import of this is that the probability of the signals in both of the antennas becoming extremely weak at the same time is very small, and selection of the strongest signal will always improve matters. While these techniques are still in widespread use today, they do not increase range or capacity, though they do address an important problem facing mobile systems today, that of fading (uplink signal quality) in complex RF (dense urban) environments.

B. Switched-beam Technology

As an extension of the microcellular concept, switched-beam technology is being investigated by several companies as a possible solution to the requirement for range and the capacity crunch. The design of such systems involves high-gain, narrow azimuthal beamwidth antenna elements (using conventional or Butler matrix array technology), and RF and/or baseband digital signal processing hardware and software to select which beam or sector to use in communicating with each user. In order to overcome the well-known trunking efficiency problem of small cells, pooling of radio resources is being investigated by several proponents of this technology. Additionally, many of the system related issues dealing with access and control channels require some special care, and there are certainly interesting challenges to be faced concerning the downlink in such systems. While previous unsuccessful attempts at such solutions date back to the 1970's when six-sectored systems were tested as improvements over three-sector technology, advances in DSP technology may provide solutions to the challenges faced by highly restricted fields-of-view in rapidly changing mobile environments.

C. Intelligent Antenna Technology

At the other end of the spectrum is Spatial Division Multiple Access (SDMA) technology, a technology currently being productized in ArrayComm's IntelliCellTM line of products. SDMA technology employs antenna arrays, standard RF and digital components, and multi-dimensional nonlinear signal processing techniques to provide significant increases in capacity and quality of many wireless communication systems. It is especially well-suited to the current and next generation cellular systems termed Personal Communication Networks (PCN) and Personal Communications Services (PCS) and wireless local loop (WLL) systems.

Antenna arrays coupled with adaptive digital signal processing techniques employed at base stations improve coverage, capacity and trunking efficiency allowing for lower cost deployments with reduced maintenance costs. In addition to these immediate benefits, the exibility of the technology also allows for the creation of new value-added services which can provide a significant competitive advantage to operators offering these services. SDMA is not restricted to any particular modulation format or air-interface protocol, and is compatible with all currently deployed air-interfaces.

SDMA is an adaptive array approach, and is realized as a combination of antennas (antenna arrays), analog radio frequency (RF) and digital electronics, and estimation and detection algorithms. Resource allocation algorithms to effect efficient use of system resources are also employed. A block diagram of a typical intelligent antenna system configuration is shown in Figure 1.

Figure 1: Typical intelligent antenna system configuration

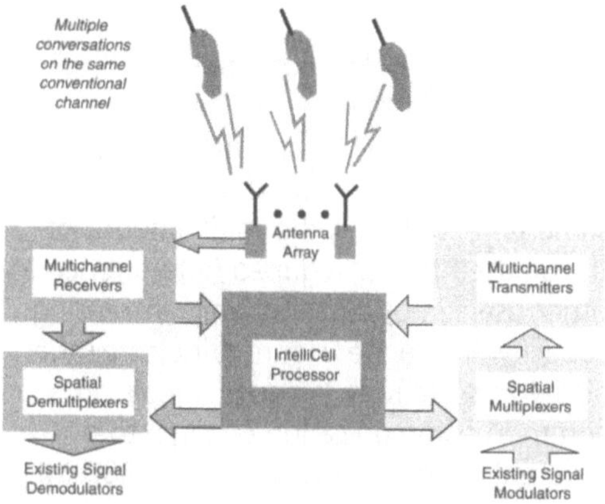

With this basic configuration, the issues of capacity and coverage can be addressed with a common architecture. Optimal processing of the outputs of multiple antennas at a base station yields significant improvements in signal quality through increasing desired signal strength and reducing levels of interference (i.e., increased SINR). While the details of the algorithms in a particular implementation certainly depend upon various factors including the temporal modulation format and the complexity of the RF environment, the objective of maximizing capacity and quality remains the same.

Shown in Figure 2 are field test results in a suburban environment with an 8-element intelligent antenna DCS-1800 system developed in cooperation with Alcatel. The cumulative histograms collected during 25 minutes of driving at ranges from 5 to 15 kilometers from the base station show the improvement obtained by using an intelligent antenna system compared to a single antenna system, a 2-element switched diversity system (antenna separation approximately 4 wavelengths), and a switched-beam system with 8 beams covering a 120° sector. These results clearly indicate the potential uplink improvement intelligent antenna systems provide over conventional single antenna and switched-diversity systems, as well as over proposed switched-beam systems, even those with extremely narrow beamwidths (nominally 15° in this case). The improvement over switched-beam is in accordance with that expected given the particular switched-beam parameters chosen. The improvement over a single antenna system is expected to be 9 dB which is indeed the case. Also as expected in such environments, switched-diversity performance is quite similar to that of a

Figure 2: Suburban environment uplink field test results comparing intelligent antenna technology to switched-beam, switched-diversity and single antenna conventional systems

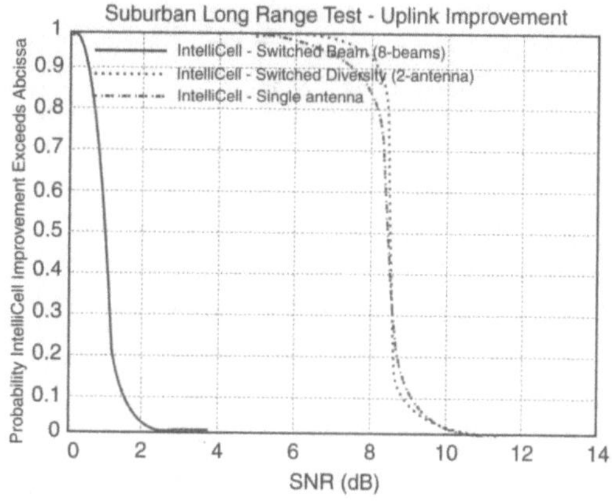

single antenna system. Note that the 9 dB improvement in uplink signal quality significantly increased the range at which the mobile unit's signal could be successfully demodulated (roughly a factor of 2 in these tests).

In Figure 3, the potential ability of intelligent antenna technology to handle multiple cochannel signals is clearly manifest. Shown are bit-error-rate (BER) measurements (using DCS-1800 signals) made for each of the aforementioned systems in the presence of an interfering signal of varying strength approximately 15° separated in angular position from the desired signal. The noted improvement in signal-to-interference-plus-noise ratio (SINR) not only means signals can be received from greater distances (increased uplink range), but more than one can be received simultaneously (increased uplink capacity). It is interesting to note that actually interference in mobile systems is more often than not the consequence of another customer attempting to use the system in another cell (spatial location).

In the downlink direction, multiple lowpower transmitters are used to selectively transmit information to one or more users on the same channel at the same time, reducing the total amount of transmitted power required (interference to other users in other cells) while increasing capacity and range. Figure 4 indicates the ability to selectively transmit signals on a common communication channel to different users in an experimental 800 MHz intelligent antenna system.
Shown are cumulative histograms from data collected during a 10 minute drive test in a suburban environment indicating the amount of signal versus interference and noise received at one of several mobile units being communicated with

Figure 3: Cochannel interference tolerance test results comparing intelligent antenna technology to switched-beam, switched-diversity and single antenna conventional systems

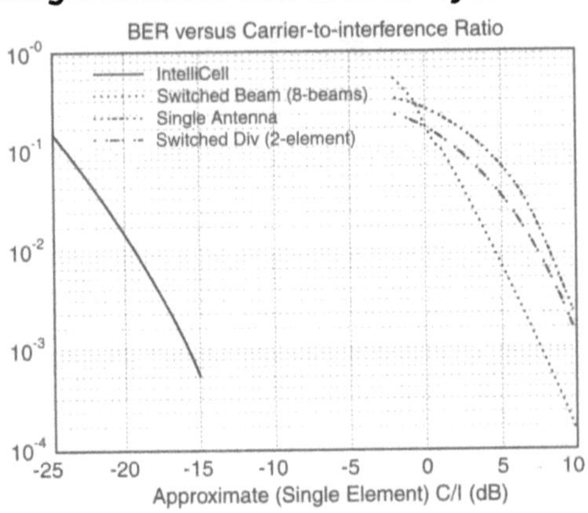

Figure 4: Suburban environment downlink spatial selectivity field test results using intelligent antenna technology

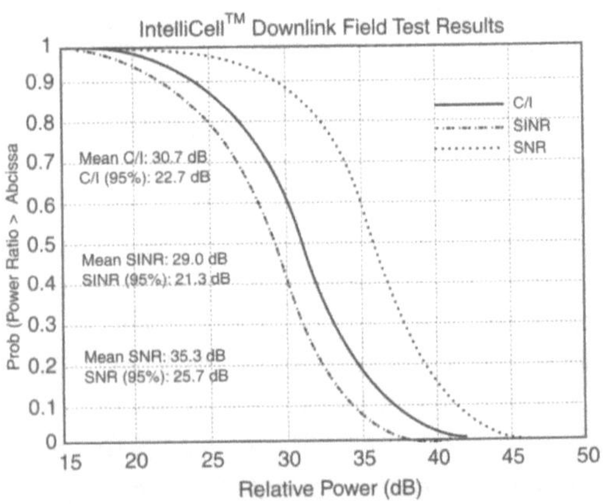

simultaneously on a single channel during the test. Similar results were obtained for the other mobiles. The results clearly indicate the potential of intelligent antenna systems to establish multiple links simultaneously and thereby increase capacity at a single cell site.

To quantify the system-level benefits of intelligent antenna technology, Table 1 provides numerical examples of the improvement that could be expected deploying a 10-element array and assuming R3.5 range-dependent attenuation. Note that for SINR improvements in excess of the stated SNR improvement, there must be the requisite amount of interference present. Also, the reduction in base station emissions relates to traffic channels and does not account for broadcast channel requirements associated with many mobile system protocols.

Table 1: Potential 10-element Intelligent Antenna System Performance Improvements (= 3:5)

Uplink Related	
▸ SNR Improvement	10 dB
▸ Range (Area) Increase	1.9 (3.7)
▸ SINR Improvement	> 25 dB
Uplink & Downlink Related	
▸ DL Tx Power Reduction	90%
▸ Real Capacity Increase	> 2

In summary, there is a growing need for improvements that increase coverage, capacity, and quality in many wireless communication systems. Optimal exploitation of the spatial dimension through intelligent antenna technology has the potential to meet these needs in addition to providing operators the ability to offer new value-added services to their customers.

References

[1] Stanford University, Second Workshop on Smart Antennas in Wireless Mobile Communications, Stanford University, Stanford, CA, July 1995.

[2] Stanford University, Third Workshop on Smart Antennas in Wireless Mobile Communications, Stanford University, Stanford, CA, July 1996.

WAP: the Key to Mobile Data?

Title: *WAP: the Key to Mobile Data?*
Author: *Henry Harrison*
Abstract: *Is WAP the key that will finally unlock the door to the market for mobile data services? Every one of the past few years has been declared 'the year of mobile data', and yet, barring a few notable exceptions, the market has equally consistently failed to materialise. Why is this, and where does wireless application protocol (WAP) fit into the equation?*

Biography: Henry Harrison is a mobile consultant at independent telecoms consultancy, Schema.

Looking at the success of the Internet in the fixed telecoms environment, there is good reason to believe that a similar 'data wave' in mobile could sustain growth rates in the mobile industry at their current dizzying levels, even as mobile phone penetrations begin to saturate.

In developing the WAP standard, members of the WAP Forum, predominantly cellular operators and manufacturers of mobile terminal devices, have taken aim at one of the main barriers to have impeded the growth of the mobile data market: the lack of a standardised IT environment for developing mobile data applications. While a number of proprietary IT environments have been available for some years from companies such as Racotek or IBM, the lack of standards has tended to put off a community of application developers growing increasingly used to open standards.

This is not the only barrier, of course. Data rates available for mobile services have historically been slow compared with fixed data network speeds. And terminals for mobile data applications, such as personal digital assistants (PDAs) have typically less power than desktop PCs, with restricted screens and input capabilities. However, the WAP Forum believes that network speed and terminal functionality are factors that will have to be coped with rather than overcome.

What does WAP provide?

The objective of WAP is to provide a development environment, which is optimised for narrow-band bearers with potentially high latency, and for efficient utilisation of device resources such as CPU usage and power consumption.

WAP defines a complete stack of protocols to be used to support mobile data applications. The protocols are to be implemented over a variety of mobile communications networks, making the development of mobile data applications network independent. The same application should be able to run over GSM, TETRA, CDMAOne and other networks.

Most attention so far has focused on the WAE or wireless application environment. This is a web-like environment providing wireless mark-up language (WML -- a lightweight version of HTML) and WMLScript (a lightweight version of Javascript). WML and WMLScript are optimised for wireless -- for example they are transmitted over the air in binary format in order to save on bandwidth -- but allow content similar to web pages to be developed for mobile devices.
But other elements of the protocol stack (Table 1) may also prove to be of great importance. The WAP Forum suggests that mobile data applications could be written either to WAE -- similar to developing for the web -- or to the wireless transaction protocol (WTP). This would be equivalent to developing direct to transmission control protocol/Internet protocol (TCP/IP) in the fixed environment, and is likely to prove of interest to those developing applications which require more client-side processing capabilities than WMLScript can offer. This could be the case, for instance, with applications developed to support particular business processes or business functions such as field maintenance forces.

Implementation of WAP-based services is via a WAP proxy, which is a gateway between the mobile network carrying WAP, and the Internet or another IP network. WAP services are stored on a server connected to an IP network: the proxy translates between this server and the WAP device. For an application written to the WAE, the set up is shown in Figure 1.

Figure1

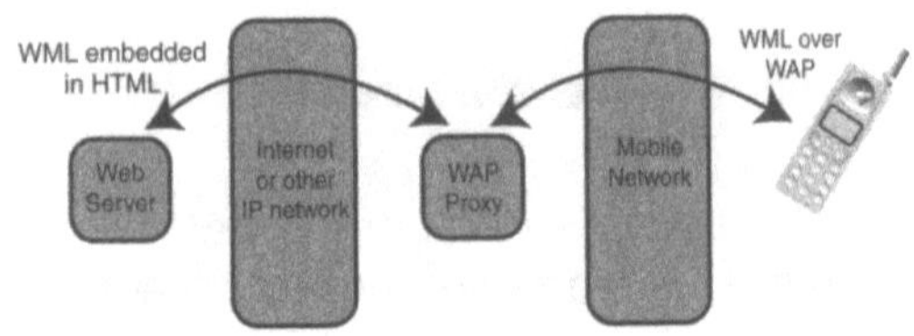

Will WAP succeed?

Not everyone is completely sold on WAP. Both Microsoft and 3Com, whilst agreeing to support WAP, are only lukewarm in their backing. Microsoft would prefer to see Internet standards used for mobile, while both Microsoft and 3Com are in favour of an approach whereby web pages from the fixed Internet are automatically cut down to remove graphics and other features before forwarding to the mobile device.

There are strong arguments to be made that mobile devices should have access to

Table 1: The WAP protocol stack

Applicant Layer	-	Wireless Application Environment (WAE)
Session Layer	-	Wireless Session Protocol (WSP)
Transaction Layer	-	Wireless Transaction Protocol (WTP)
Security Layer	-	Wireless Transport Layers (WTLS)
Transport Layer	-	Wireless Datagram Protocol (WDP)

the complete content of the Internet rather than accessing content designed for a mobile-specific technology such as WAP. There are also reasons to believe that mobile will continue having to be treated differently from fixed, specifically that:

- data rates available over mobile will continue in general to be lower than for fixed networks;
- user interfaces for mobile devices will continue to be different to fixed devices.

Many mobile operators are currently investing in technologies to improve the data rates available over their networks. For example, in GSM, the forthcoming high-speed circuit switched data (HSCSD) and general packet radio service (GPRS) technologies boast potential peak data rates of up to 115.2 kbps. While on the face of it, this should considerably level the playing field between fixed and mobile, it could be that the gap between fixed and mobile will actually get larger.

From the point of view of mobile networks, peak rates of 115.2 kbps might be technologically achievable, but may be economically impractical. In order to achieve such rates, eight voice channels must be taken out of service -- a considerable drain on network resources in areas where networks are already heavily loaded with voice traffic. In order to fund the added network capacity this would require, prices would have to be high -- and could prove prohibitive. Many implementations of HSCSD and GPRS will offer peak speeds lower than the technological maximum.

At the same time, speeds available in the fixed network are not standing still. ISDN is already popular in many European countries, while fixed operators around the world are already rolling out, or are about to roll out, broadband access solutions such as cable modem and ADSL which will deliver between 500 kbps and 2 Mbps.

Although the processing capabilities of mobile devices will continue to develop at a rate determined by Moore's law, screen size and input devices are less determined by technology than by ergonomics. Technologies such as voice recognition may eliminate some of the current disadvantages of mobile devices, but it is highly unlikely that these same technologies will prove ubiquitous in the fixed environment. Ultimately, people's behaviour and requirements are different when they are mobile to when they are seated at their desk or in their living room, and the designs of fixed and mobile devices are likely to reflect this.

Technologies such as web clipping are aimed at reformatting standard web content for small profile devices, and do in many cases address requirements. But experience shows that content providers are often not keen to have their content reformatted in ways they have no control over. An example can be found in the early days of the web, when a 'key feature' of web browsers was that the user could choose the fonts applied to display sites. Content providers soon started to clamour for ways to regain control over the way their content was displayed. The same development is expected to be seen in mobile: content providers who are seriously addressing the mobile environment will want their own control over exactly what is displayed on the mobile screen.

What will WAP be used for?

The potential market for mobile data services can be divided in terms of four main application areas:

- 'vertical' applications to support particular business processes -- for example, despatch or field maintenance;
- 'horizontal' applications, focusing mainly on connectivity and productivity -- for example email and group scheduling;
- information services -- the sort of services widely available on the web and moving constantly from the pure provision of information to more interactivity and ultimately ecommerce;
- support for unified telecom services*.

Vertical applications are often considered the poor relation of more glamorous uses for mobile data. But in practice, there is a much wider range of applications that could make use of mobile data than those that are currently in existence, and even these have often found it a tough job to get to where they are. A lack of development standards is never a favourite situation for IT departments, who foresee endless support and upgrade problems arising from the creation of a proprietary system.

WAP could provide the development standards that IT departments and software vendors require, either by providing a web-like user interface via WAE, or by supporting the communications infrastructure by providing application programme interfaces (APIs) to WTP or other elements of the protocol stack. A lot will depend on whether implementations of WAP are made available for a wide range of terminal devices -- vertical applications often require specialist terminal devices provided by companies such as Symbol or Telxon.

Horizontal applications such as email are a prime target for mobile operators, who already see some users dialling up over mobile links from their laptops, despite the slow connection speeds and often complex procedures to get the connections set up.

While WAP could form a natural infrastructure for this sort of application, the risk to its supporters is that these applications fall into Microsoft's backyard. Given that most users of these kinds of solutions in the fixed environment use Microsoft applications such as Outlook, Microsoft has a natural advantage in supplying these services to mobile users. If Microsoft can deliver good wireless Windows CE solutions which deliver these applications using its own technologies, an important section of the market will be using devices where WAP is the secondary, rather than the primary environment.

Information services provide the current main focus for backers of WAP. Nobody really knows whether this market will really take-off, but the potential in terms of numbers of users is certainly huge -- both looking at the number of mobile subscribers, and at the number of fixed Internet users.

Although the promise is there, getting the marketing right for these services will be critical. One big question facing operators looking to launch such services is how to charge for them, as the value proposition has to match the customer's expectation. It is not certain whether consumers will be prepared to pay for time online (as they do for Internet access in Europe) given that mobile access charges will be considerably higher than for fixed line. This situation will become even more difficult when WAP services are delivered over packet networks such as GPRS.

With packet networks, access charges have to be on a per-packet basis. This is not a concept easily understood by the mass market, and not a very customer-friendly way of charging, given that the user has no control over how many packets are likely to be delivered. One alternative is to bill on a content basis -- charging the user on the basis of the information they request while making the content provider responsible for the traffic costs incurred in delivering the information. This allows operators and content providers much more flexibility in meeting customer expectations.

However, a content-based approach to charging has a significant impact on the way services are defined, and the way they interact with the user interface. Consumers have to know how much they are going to be charged for any given information service before they select it. These requirements are not currently addressed by the WAP standards, and operators taking this approach would have to rely on proprietary extensions to WAP such as those provided by Finnish company, More Magic.

Even if the information services market proves less successful than operators hope, WAP is likely to find a place in mobile operators' networks as a way of supporting various kinds of intelligent network-based services. Using WAP, operators could rapidly deploy menu-based user interfaces to innovative new IN services, without the need to update SIM cards.

Ultimately, the potential for mobile data services is huge, but as ever, making the market happen will not be simple. The industry has been waiting for some time for standards such as WAP, but its arrival does not mean that all the pieces of the puzzle are in place -- yet.

THE EVOLUTION OF GSM DATA TOWARDS UMTS

Title: *The Evolution Of GSM Data Towards UMTS*
Authors: *Kevin Holley, Mobile Systems Design Manager, BT, United Kingdom, and Tim Costello, Cellular Systems Engineer, BT, United Kingdom*

Abstract: *GSM data services have launched a new era of mobile communications. The early analogue cellular modems were unattractive to the market as they were slow and unreliable. GSM now offers a high quality 9600 bits per second link which opens the way for useful data applications and the Short Message Service provides guaranteed delivery of small data packets even if the phone is switched off when the message is first sent. Now the market for data is moving onwards (more bursty) and upwards (more traffic), and the ETSI standards groups are working towards higher data rates but more significantly also towards packet data services, initially with GPRS and multislot data but eventually UMTS will provide much higher bandwidth.. This will certainly broaden the appeal to end users because data is routed more efficiently through the network and hence at lower cost, and also access times are reduced.*

Biography: see end article

Introduction

Cellular access to data services has been around for a long time. Even in the mid 1980s engineers were busy designing modems which could counter the effects of radio fading. Yet the data access services were not popular. Connections were unreliable and slow. Equipment was bulky, with the term "soap on a rope" in vogue as users struggled with many boxes to connect between the phone and the data terminal. GSM has turned this around, with high quality data solutions involving reliable connections and compact phones with only a single wire between phone and terminal.

At the same time as GSM networks became popular, the growth of the Internet started to reach out to the general public. The now-ubiquitous web browser has inspired a generation of data users and it is possible to find nearly everything somewhere on "the web". Large companies all over the world have embraced this technology and made it their own by providing in-house Intranets, firewalled from the outside world yet accessible to all in the company.

The mobile computer user has also seen dramatic changes over the last 10 years. The "laptop" computer is now virtually extinct, with high speed colour PCs now available. Much smaller handheld devices called Personal Digital Assistants or PDAs have emerged and become very popular.

With these developments in the role and presentation of data particularly in businesses the main demand for data services can be described as open access to public information on the Internet, or secure access to private information on company Intranets by a mobile population using handheld PDAs or notebook PCs. But this is only what the end user wants to see, so how can we achieve that in reality?

Figure 1: Businesses are very dependent upon corporate LANs

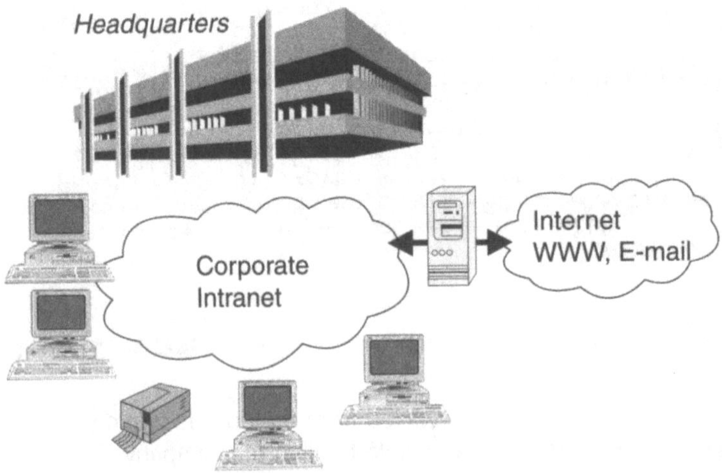

Today's GSM Data

GSM was conceived to be the "mobile arm" of ISDN, using signalling based on Q.931 and providing access to all the ISDN data services. The communications path is digital so the speech is converted to ones and zeros before transmission.. This made it quite easy to introduce data services because there was no need for any modem at the user end. GSM is a widely adopted standard with roaming (enabling subscribers from one operator's network to communicate via another operator's network), a key consideration of all developments. Thus it is possible to use the same application in many countries without the need to reprogram the dialled number or any other setup change.

GSM provides two basic data services, transparent and non-transparent.

The transparent service provides a consistent delay and a throughput of 9600 bits per second, with variable error rate according to the propagation conditions.

The non-transparent service provides a consistent low error rate with a throughput up to 9600 bits per second and variable delay according to the propagation conditions.

Whichever "bearer type" is chosen, there are many options for connection between the GSM network and the far end data network. The most commonly used ones are PSTN modem and ISDN data service. With either of these, the throughput is limited to 9600 bits per second and so flow control and/or rate adaptation are required. The rate adaptation as well as the basic data conversion from GSM protocols to the fixed network protocols is performed by an Interworking Function (IWF).

For the mass-market, the most obvious data service to offer is the non-transparent service, as this provides built-in error correction. However we have already seen devices on the market which allow end to end data compression by using the transparent data service. Claims of 36000 bits per second can be met and in some cases exceeded, depending on the type of data being transmitted. In addition GSM has now developed a standard data compression capability built on V.42bis, which can be applied between the mobile terminal and the network. This service will allow non-compressed data from a users terminal to be compressed in the mobile station and de-compressed in the network, thus overcoming the limitation that the reduced speed on the air interface gives.

Facsimile devices are very complex. They change modem speed during the call, have training to determine the acceptable data rate, cannot be flow controlled and sometimes even operate to a non standard protocol. This makes it very difficult to insert a digital bearer in the middle of a fax call. To get facsimile to work with GSM, the standards-makers had to terminate the fax protocol and re-create it, both at the terminal side and in the interworking function. Despite the complexity, GSM provides a very good facsimile group 3 capability.

The GSM standard incorporated a two-way paging function. Called Short Message Service, or SMS, this allows 160 characters of text to be sent to or from a GSM handset. Messages are sent to a Short Message Service Centre (SMSC), which stores SMS until it can be successfully delivered to the destination. Whilst in many cases the first attempt is successful, it may be that the destination handset is switched off or out of coverage, so the SMSC makes many attempts to send the message, possibly over several days. SMS has two distinct service classes: Mobile Terminated (messages sent to the mobile) and Mobile Originated (messages sent from the mobile).

Figure 2: Basic functions in the IWF

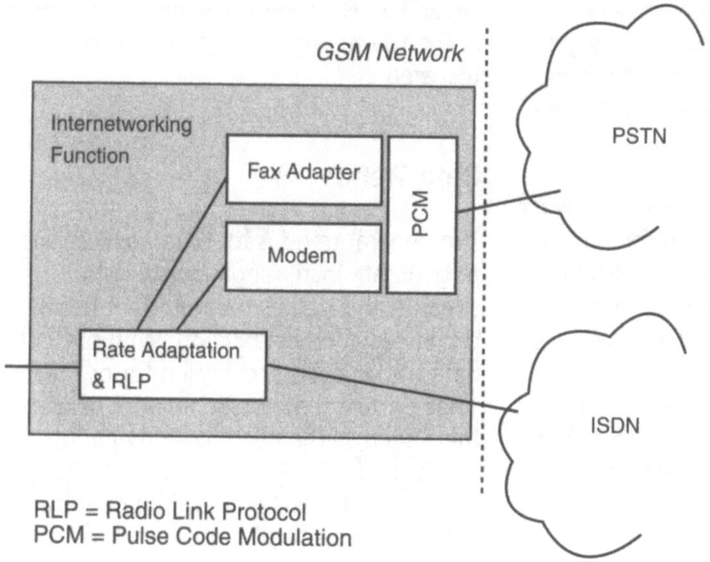

There is another short text-based messaging service called Cell Broadcast or CBS, which has the ability to broadcast messages in a given geographical area, making it suitable for information type applications available to all subscribers in a given area.

Figure 3: General access to SMS-SC and onward connection

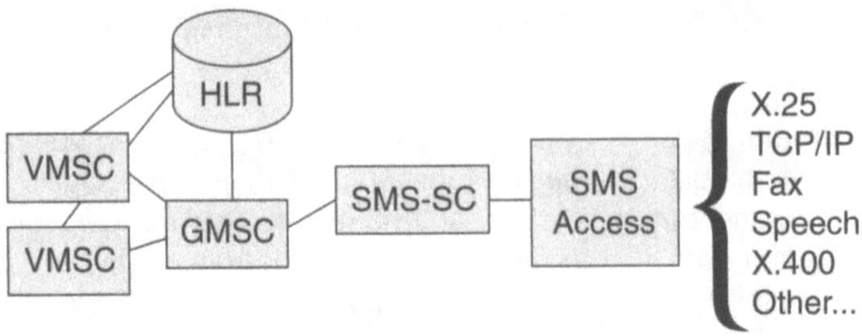

The PDA market is not served as well by the GSM handset makers. PDAs have difficulty in supporting a PC card slot because the power requirements are significant. They are also limited in terms of memory and CPU power and so complex communications drivers are a non-starter. The best solutions for PDAs are provided by manufactures who have a phone with direct data connectors (physical or other e.g. optical).

Some manufacturers have taken the step of integrating PDAs with phones. These are attractive because everything comes in one box and all services work together seamlessly. However, the offerings to date have been quite limited because they are large for a phone and do not have all the capability which people expect from a PDA. In the future we expect much more from this area.

Implementing GSM Data

All of these data facilities have been implemented for some years by the manufacturers. We have seen the emergence of the GSM data card, a PC card which looks like a modem to a PC, but instead of converting the PC data to modem tones it converts the PC data to a proprietary protocol which is then in turn converted by the handset into standard GSM data protocols. The GSM standards-makers, however, always envisaged that GSM handsets would provide a direct data connector, and this has been achieved by some manufacturers already. Some rely on additional PC software drivers but others effectively implement the entire GSM data stack in the mobile phone.

These PC interconnect solutions provide for data and fax calls to and from the mobile user. The Short Message Service / Mobile Terminated is now implemented in all mobile phones, and quite a few also have the Mobile Originated capability too. However mobile phone keypads are limited in terms of the speed of data entry and a PC interconnect solution for SMS makes the business of entering and sending messages much easier.

Data Trends

The general trend is for data applications to generate increasingly bursty data streams, this makes for inefficient use of a circuit switched connection. With present data traffic levels accounting for only a small part of the mobile network usage, there seems little need to worry about efficiency. However, fixed networks have seen an enormous growth in data traffic, not least because of the rise and rise of Internet access demand, and there is no reason to suppose that this will not spread to the mobile networks as technology and customer expectations move on.

The existing GSM network is optimised for circuit switched voice calls and therefore is inefficient at setting up and carrying very short data bursts.. Figure 4 shows a comparison made of the number of radio timeslots required to carry data traffic generated as the user base increases. It shows that with only a few subscribers the same number of channels are required for circuit and packet switching. As the user base (and traffic volume) increase then the number of channels required increase faster if circuit switching is used. Packet switching is more efficient at carrying bursty data, and mobile networks should employ packet switching right up to the end terminal if they are to cope with the expected demand for future data services. The network and radio capacity required to support a large amount of bursty data would make it uneconomic or impossible (ie limitations on the number of physical radio channels available) to implement.

A packet switched mobile network opens the way for a range of new applications. IP interconnectivity and Point-to-Multipoint (PTM) transfers become a reality with GPRS, applications currently restricted to fixed line connections will migrate to the mobile world.

Tomorrow's GSM Data

The very robust 9600 GSM data service works with good quality of service throughout existing GSM cells. With any design for high speed mobile data there is a trade-off between data rate and

Figure 4: Radio timeslots required to support growing data

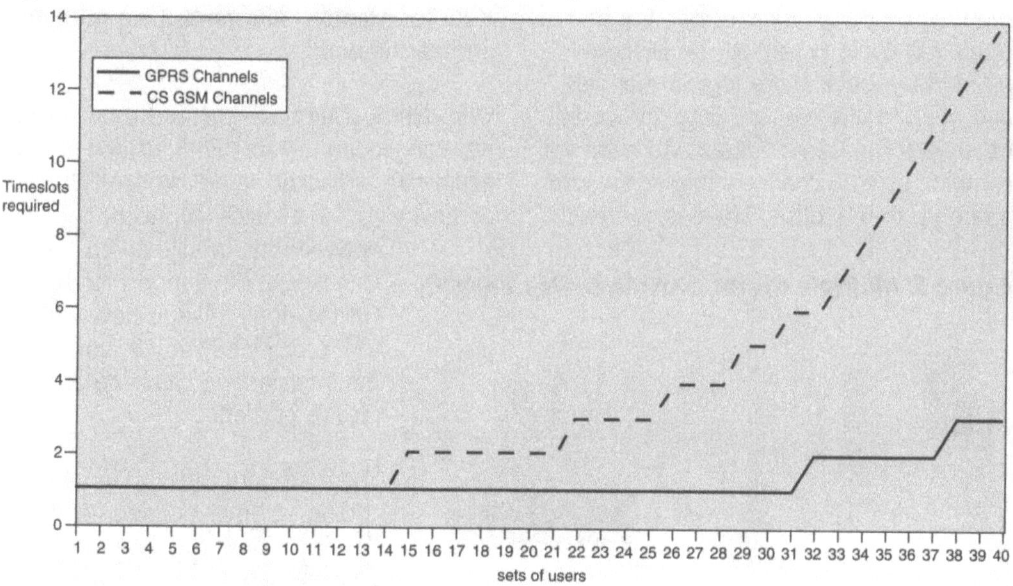

coverage area, and GSM has provided a good data speed with coverage wherever there is voice coverage. A strong requirement for a new faster service has emerged with a more relaxed requirement for total coverage. Thus the GSM standard will soon include a 14.4k bits/s data service as a basic enhancement to the existing bearer service classes. This will be achieved by reducing the error protection mechanism and therefore it will not be possible to use 14.4k bits/s over the whole of the GSM service area, instead requiring the user to stay in good to moderate coverage.

As well as increasing the rate for a normal call, the GSM standards will offer timeslot concatenation. To explain this it is necessary to go into a little more detail about the GSM radio interface. A single radio carrier on GSM is 200kHz wide and provides up to 8 simultaneous channels. Each logical channel is given one eighth of the time on the physical carrier. But a single call could potentially be allocated more than one of these logical channels and hence the aggregate data rate could be higher. The GSM multislot standard will combine up to 6 channels to give the user a rate up to 64k bits/s. There is no benefit in increasing the data rate beyond this because the switch network is based on narrowband ISDN 64kbits/s circuits, so the rate limitation starts to be the core network rather than the access.

But the biggest change is that of providing packet radio access. The new Generalised Packet Radio Service (GPRS) will do just that, it will offer operators the ability to charge by the packet, support data transfer across a high speed network and up to 8 timeslot radio interface capacity.

GPRS introduces two new nodes into the GSM network - the Serving GPRS Support Node (SGSN) and Gateway GPRS Support Node (GGSN). The job of the SGSN is to keep track of the location of the mobile within its service area and to send and receive packets from the mobile, passing them on, or receiving them from the GGSN. The GGSN then converts the GSM packets into other packet protocols (IP or X.25 for example) and sends them out into another network.

With GPRS, a terminal logs on to the network separately for GPRS. It then establishes a Packet Data Protocol (PDP) context which is a logical connection between the mobile and a Gateway GPRS Support Node. Having done this it is now visible to the outside fixed network, e.g. the Internet and can send and receive packets.

There are four radio channel coding schemes defined for GPRS, to allow the data rate to be increased when coverage is

Figure 5: Multiple routes provide higher capacity

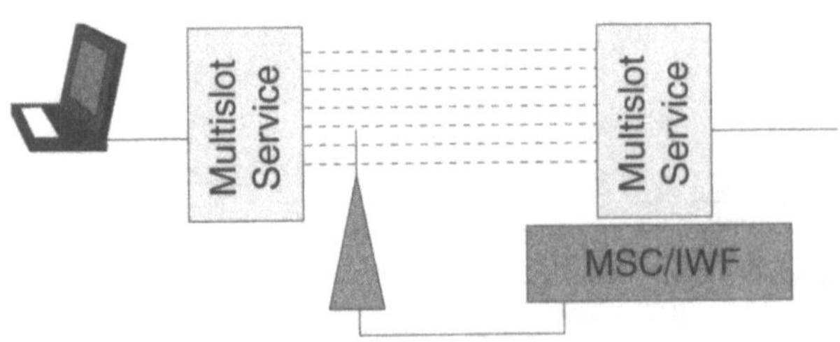

Figure 6: GPRS network architecture

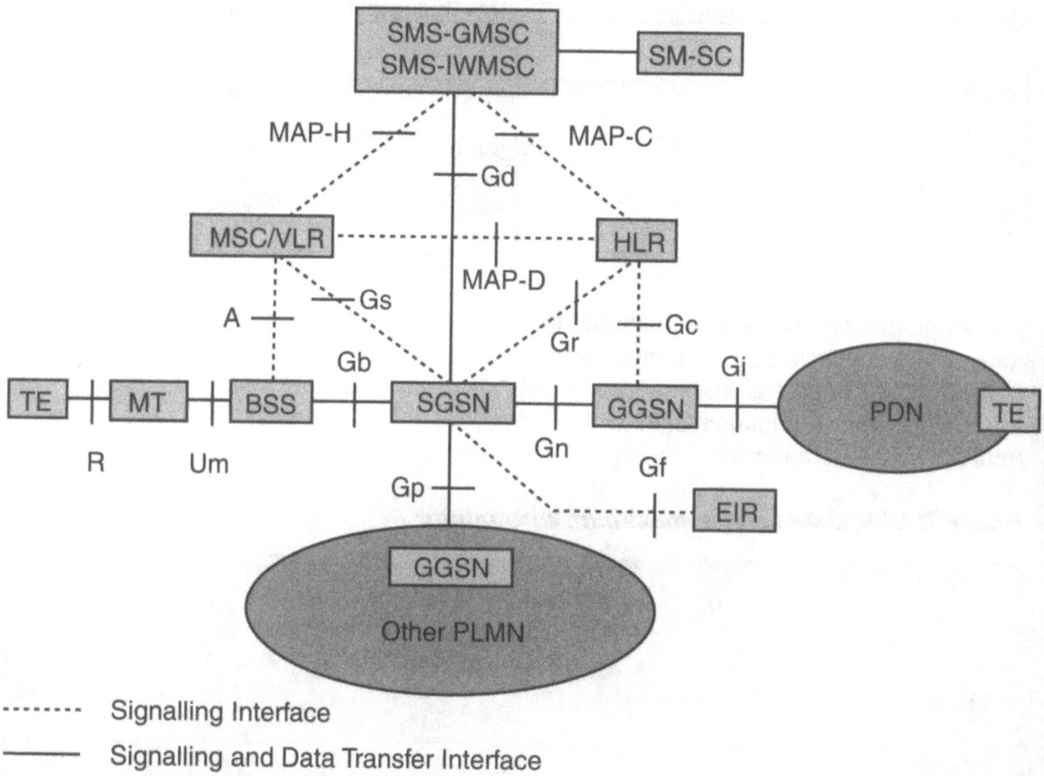

good, and ensure packet loss is never too great when coverage is not so good. These coding schemes allow between 9.05kbit/s and 21.4kbit/s per GSM timeslot in use. Thus the maximum data rate for GPRS is 8 times 21.4kbit/s or approximately 164kbit/s.

GSM Data Service Performance
The performance of all GSM data services will vary according to radio conditions. However it is possible to categorise GSM data services according to a few key elements:

- access time - how long it takes to get into a ready-to-transfer-data state
- throughput - users data transfer rate through the GSM system
- packet size - Size of data that is managed as one entity through the GSM system

Table 1

GSM Service	Acces Time	Max Throughput	Packet Size
Circuit Switched Transparant	3s	9600	N/A (bit oriented)
Circuit Switched Non Transparant	3s	9600	192 bits
Multislot T	4s	64000	N/A (bit oriented)
Multislot NT	4s	64000	192 bits
GPRS	<1s	164000	2k bits
SMS	3s	500	1120 bits

Shown graphically, this gives a much better idea of how the limitations of narrowband ISDN switching impact on the maximum throughput when comparing GPRS with multislot data (see figure 7).

Figure 7: GSM data service maximum throughput

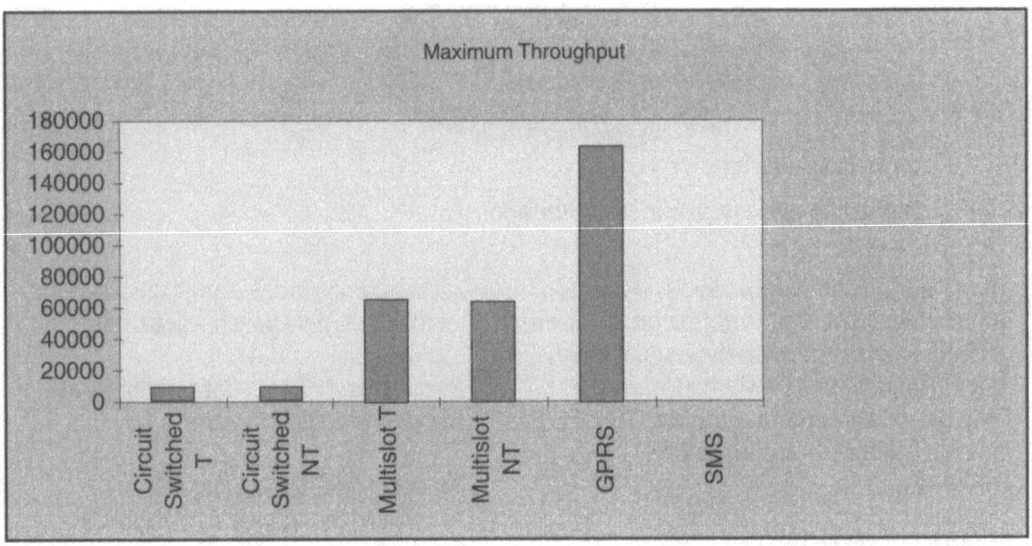

How to Fit an Application

All data services can be broken down into a few key components as follows:

Table 2

Data Volume	Transmission Mode	Interconnect Capability	Terminal size	Data Storage Need
File transfer	Unicast	Fax	Phone only	A few small messages
Packet transfer	Multicast	WWW	PDA size	Kilobytes
		FTP	Notebook size	Megabytes
		Video	In-vehicle terminal	
		Internet email		
		X.400 email		

All of the GSM data services can be used with any of the interconnect options in the centre column, however clearly the bit rate will be very important to Video in particular. The main element in choosing the GSM service is the size of "packet". For very small packets sent unicast or small-scale multicast, then SMS would provide an ideal transfer mechanism over GSM, whatever the end interconnect requirements. For larger packets then GPRS becomes the most attractive bearer, but this will not be available for a while so circuit switched data may have to be used. The terminal in use will also be important, not only from its size but also how much data is required to be stored. A despatch service would only need a few messages, whereas an email application would require more storage.

Whilst this way of studying the requirements is very useful in terms of defining the service to use, it is of paramount importance to consider the service required by the user rather than looking at the technology available and making mobile access possible. The following examples show what a difference this can make to the user perception:

World Wide Web access

The user wants to be able to view web pages and hyperlink from one page to another. One way to provide this is to set up a stack consisting of HTTP, PPP, TCP, IP and a bearer connection based on GSM with its own inbuilt error control. However all of this is rather inefficient for the transfer of HTML text and graphics. A really big gain could be made by sending the page requested as a Mobile Originated packet and then assembling the required data at an intermediate point before sending an alert to the mobile which then dials up and downloads the information all in one go. This allows pre-compression of the data and removes any Internet delay from the user's airtime. If graphics are not required then in many cases the Short Message Service alone can provide a transport mechanism! Solutions along these lines are being developed one called MASE (Mobile Application Service Environment), which is an API lying between the application and the communications system. Another is Mowgli which is a server positioned between two networks which assists in the communications between two users by modifying the data to be suitable/optimised for that network/user capabilities.

Email access

The user wants to be able to send and receive email. This can be achieved with POP3/SMTP on top of PPP, TCP, IP and a GSM bearer. But again this is inefficient for text transfer, which can be achieved using compression and file transfer techniques like zmodem, rather than an open Internet stack.

The difficulty with moving forwards from the present position is that there is a need for more than the traditional systems integration tasks. As well as providing user software and drivers, the systems integrators need to get additional functionality built into the Office Automation, Email and Internet Service Provider business so that online times can be reduced. There is some scope for additional value added service businesses in taking the Internet data and converting to a more mobile-friendly format before passing on to the mobile user, however for the Corporate customer this is unlikely to be popular in the long term as concerns over the security of intranet data are very high.

Services like multislot data and GRPS are very useful in moving the base technology forwards, but if the same goals can be achieved with the existing data services, we should be looking at prototyping services now on the current networks. We need to produce reliable user solutions which can cope with dropped calls without restarting a session, perhaps automatically re-establishing should this be needed. We should avoid using inefficient protocols for mobile just because they are existing and in common usage. People need access to common services, not necessarily using the same network protocols as are used in the higher bit-rate, lower error-rate fixed networks. The time is now right to look at producing a standardised mobile access mechanism for fixed network services, focusing on increasing the effective throughput and immunity to dropped calls and thus reducing the needed airtime.

Where do we go from here?

There are many goals yet to be achieved with wireless data. The whole data industry is moving away from X.25 based services towards TCP/IP based services. Data transport is becoming cheaper and cheaper as economies of scale drive down the cost of routing equipment. User devices are becoming more sophisticated and smaller. The idea of email and web browsing in your pocket can be achieved today yet it is expensive and in some cases awkward to use. But the mass market for data devices is the key to future success. Defining that mass market and the most useful services is the key. We must make the best use of the available data services to avoid excess cost.

The long-term goal is the UMTS vision, quoted from the UMTS Forum Regulatory report as:

"The Universal Mobile Telecommunications System, UMTS, will take the personal communications user into the Information Society of the 21st century. It will deliver advanced information directly to people and provide them with access to new and innovative services. It will offer mobile personalised communications to the mass market regardless of location, network or terminal used."

Basically the implications of this are that all telecommunications networks are integrated and users really can get fast data rates of 2Mb/s in some areas. But 2Mb/s is only useful for the right applications and those applications need some market experience before they can be well defined. There are bound to be some winners and some losers but at least some of the battles must be fought with the existing second generation cellular capability of up to around 64kbits/s. The future expectation of growth in the mobile market is driven by data and multi-media. This expectation is now being realised on fixed networks, mobile networks through the enhancement on GSM and introduction of new capabilities such as GPRS will start to build that market. UMTS is the "utopia" for meeting the market expectation, however, it needs to be technologically advanced sufficiently to meet that need.

Conclusion

GSM data provides a wealth of opportunity for transferring a variety of information to the user on the move. From simple email to video and multimedia, almost anything is possible. The future will bring even higher bandwidths, larger markets and with them lower costs. However in order to achieve the goal of mass market data use the applications must be developed with mobile in mind. Users will not want to bother with redialling just because of a dropped call, and they will not be happy to pay for two timeslots when one would suffice if only the data was pre-compressed and used an efficient transfer protocol. We need to look at the system as a whole, starting with the users needs and ending with the data transport mechanism.

Terminals will be very important to the success of mobile data. Whilst there will be roles for all types of data terminal, the most successful ones will provide unified mailboxes, not one button for fax, another for internet email, yet another for X.400 and a fourth for SMS.

The future is bright for data, if we get it right!

References

1. Global System for Mobile communications - what's in store?, K A Holley, BT Technology Journal Vol. 14 No. 3 July 1996
2. Mobile Data Services, C J Fenton, W Johnston and J D Gilliland, BT Technology Journal Vol. 14 No. 3 July 1996
3. GPRS standards
   GSM 02.60 GPRS Service Description: Stage 1
   GSM 03.60 GPRS Service Description: Stage 2
4. Mowgli Developments - http://www.cs.helsinki.fi/research/mowgli/
5. 5. A Regulatory Framework for UMTS, June 1997, UMTS Forum

Acknowledgements

The authors would like to acknowledge the support provided by colleagues at BT Laboratories, in particular Alan Clapton and Chris Fenton.

Biographies:
Kevin Holley is currently working as a Mobile Systems Design Manager with BT. He has been involved in the development of the GSM standard since 1988, particularly on the data services and Short Message Service. He also works closely with the UK operator Cellnet in defining their GSM services. He is now chairman of SMG4, the ETSI data development group for GSM and UMTS.

Tim Costello is currently working as a Cellular Systems Engineer with BT. He is working on future mobile data applications and particularly GPRS.

Wireless Access Protocol set to take over – WAP addresses the shortcomings of other protocols

Title: *Wireless Access Protocol set to take over – WAP addresses the shortcomings of other protocols*

Author: *Rawn Shah*

Abstract: *A new protocol family for mobile devices, Wireless Application Protocol (WAP), is on the horizon, and analysts are predicting it will become the standard for supporting smart cell phones, pagers, wireless personal digital assistants, and mobile computers by 2001. Rawn Shah investigates WAP and finds potential for greatness.*

Biography: Rawn Shah is an independent consultant in Tucson, Ariz. He has written for years on Unix-to-PC connectivity and has watched many of today's systems come into being. He has worked as a systems and network administrator in heterogeneous computing environments since 1990.

Unlike most computers, which use protocols such as HTTP over TCP over IP to access the Web, Wireless Application Protocol (WAP) has its own protocols that operate in a format more natural for wireless delivery. All WAP Internet connections still have to go through a gateway that connects the two protocols, but the gateway itself can add other services that optimize communications.

WAP is designed to handle higher latencies, unpredictable service availability, unpredictable connection stability, and lower bandwidth, all of which present problems for wireless communication. In particular, instability, service availability, and latency problems make it hard to maintain connection-oriented services like TCP, which means most Internet applications that use TCP as their delivery mechanism won't work reliably on a wireless network. Thus the need for WAP.

How to WAP

As Figure 1 indicates, WAP is actually a family of protocols. The network layer at the bottom represents a combination of

Figure 1: Wap contains layers of protocols

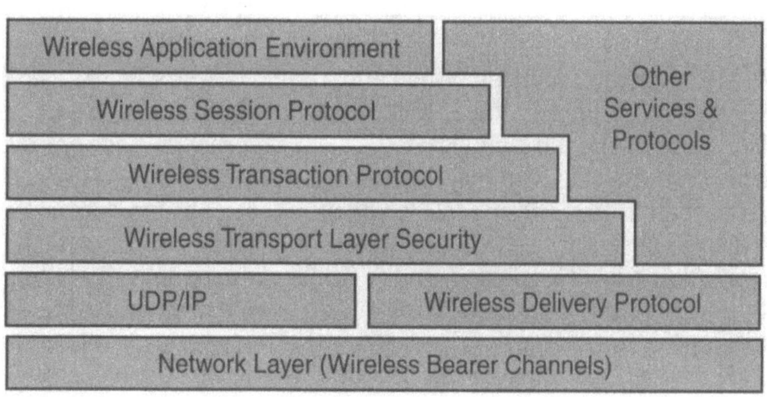

layers, including the datalink and physical layers. The data are placed in network packets, encapsulated in a hardware format, and sent out as a signal. In WAP, the network layer includes protocols like CDMA (Code Division Multiple Access), Short Messaging Services, and Cellular Data Packet Delivery, all of which combine the functions of the three layers into other protocol families. Individual vendors dictate the type of delivery mechanism that WAP runs on their products.

WAP also supports the User Datagram and Internet Protocols, so applications can use the datagram service to communicate with each other. Alongside is the native WAP datagram service known as Wireless Delivery Protocol (WDP), which handles primary communications between WAP devices. The successive layers above handle encapsulation for security and transactions, and there is even a session layer on top of the datagram service. At the very top is the Wireless Application Environment (WAE), which defines application layer protocols for communication.

At the WAE level are application protocols for document transfer, which replace raw HTTP and HTML. Instead WAP has its own Wireless Markup Language (WML) and a specific XML definition. WML, together with its own scripting language, WMLScript, handles documents as cards rather than as free-scrolling HTML pages. In a concept similar to the HyperCard application originally released for the Apple Macintosh, a user can create WML applications by moving through a stack of cards. The types of tags are also fewer and specific to telephony and mobile applications.

The WAP specification doesn't include Java or JavaScript, specifically because Java is too heavyweight for most mobile devices, even in its most compact form. JavaScript is also too focused on the concept of a full desktop browser and too heavy for phones with two-line LED text screens. However, since WAP does not limit the kinds of applications that can be added to the system, vendors can create their own extensions on top of the protocol.

The typical WAP-capable device will include a full implementation of the WAP protocol stack, the additional implementations of WML and WMLScript, and a small microbrowser that can display information and receive input for the device's given physical format. This device then communicates over the wireless network with a WAP server or gateway and sends a URL request to the WAP server that downloads the information from a Web server, encodes the data into a WML document, and sends it to the device.

The WAP server must be smart enough to break interactive Web page forms into components that can be sent to the device in pieces, collated, encoded into an HTML response, and sent back to the Web site. The WAP server handles Internet services such as the domain name service, IP address management (in cases where the device uses UDP), and Web requests as a proxy server for the device.

To operate HTTP, WAP converts ASCII descriptions of headers into a compressed binary format and then uses the Wireless Session Protocol to rapidly suspend and resume sessions in order to save on power or handle disconnected operations. The Wireless Transaction Protocol handles higher transport level services similar to the way TCP handles them in TCP/IP. For HTTP traffic, WAP uses UDP as the low-level delivery protocol.

The WAP method opens Web documents in 10 fewer steps than does HTTP, TCP, or IP, and it reduces the protocol overhead from 65 percent to 14 percent. Thus, on each communication session more data than protocol headers are being sent back and forth. Overall the system is faster than systems with a full TCP/IP stack -- such as the QUALCOMM pdQ combination mobile phone and Palm Pilot.

Who's wapping?

Ericsson, Nokia, and Motorola, together with Phone.com (formerly Unwired Planet), were the original promulgators of the WAP standard. Today the WAP Forum has more than 100 members, including such major communications and software companies as Alcatel, AT&T Wireless Services, BT (British Telecom) Cellnet, Cable & Wireless, Deutsche Telekom, France Telecom, Hewlett-Packard, Hitachi, IBM, Intel, Lucent Technologies, Nortel Networks, NTT Mobile Communications Network, Oracle, Psion, QUALCOMM, SBC Communications, Star Media, and Sun Microsystems. Microsoft was originally against the WAP concept, but it has now joined the forum and plans to integrate WAP with Mobile Explorer, a microversion of its popular browser, for inclusion in future wireless products. WAP devices are now available from Nokia and Ericsson and will soon be available from Motorola. Nokia has also released a WAP server for backend services; IBM will add middleware and resell the package in its Netfinity PC servers.

The bandwidth issue might become moot several years from now. Imagine 2 Mbps connections from any mobile device anywhere. That may be one-fifth the speed of ancient Ethernet, but it's more than 200 times the speed of most wide-area wireless devices today (9.6 to 14.4 Kbps). The amount of research going into these third-generation mobile devices has kept their release on schedule for arrival in 2002-2003.

The only real problem lies in the deployment issues that surround WAP's coding scheme. Most of the world uses Global System for Mobile Communications (GSM), but the United States has yet to agree on a standard. Although GSM is also available in some parts of the United States, it isn't widely deployed, in particular because vendors are keeping to their own technologies. Even if bandwidth problems lessen, the need for a protocol specific to wireless communications will continue, whether it's WAP or some future development. Bandwidth by itself doesn't solve the other problems associated with wireless access.

Wireless Application Protocol addresses many issues that surround the Internet Protocol for wireless devices, but it does have its faults. For example, even though some WAP servers can translate Web pages instantly, information can be lost or misrepresented on some devices. However, implementing WAP in a mobile device doesn't exclude the device from running other protocols, so we may see WAP running on higher-end devices (for example, laptops) that support both WAP and IP in the not-too-distant future.

The WAP Vision

Title: *The WAP Vision*
Author: *Josh Smith*
Abstract: *The WAP vision promises compatibility and accessibility. But it also means that we must design our sites so that the presentation is separate from the information. Josh Smith points out that we need to make changes in our approach to design not only for handheld devices, but for those who have disabilities too.*

Biography: Josh Smith is a freelance writer who has worked for Trinity Publications, Practical Internet, the Writers' & Artists' Yearbook, Amazon, Wired, Miller Freeman Inc, the Daily Telegraph, the Times and many others. He is the author of two highly acclaimed books, Internet Culture in Easy Steps and Business on the Internet. His pet hates, in ascending order, include telephones, PR reps, cranky editors, and computers--which is why he spends most of his day, working with all of them.

PART 1

In the beginning, Berners-Lee created HTML and the Web. And the Web was without form, and void; and darkness was upon the face of the deep. And the Spirit of Tim moved upon the face of the routers. And Tim said, "Let there be tags," and there were tags. And Tim said, "Let there be design in the midst of the content, and let it divide the content from the content." And in the beginning there was universality, and everyone could parse the HTML, read the web pages, and Berners-Lee saw everything that he had made, and it was good. And the night followed the day, and the years passed ...

In time, it became clear that not everyone was playing by the creator's rules. Compatibility and accessibility issues revealed that the Web wasn't really the Eden-esque garden of information that new users expected it to be. For many, the appropriate resources were difficult to find and access—especially with non-standard interfaces. Software manufacturers and designers alike had eaten the forbidden fruit and had developed features that prevented many people, particularly the disabled and those using mobile devices, from viewing their web pages.

The time for penance is now, explains Greg Heumann, a Phone.com executive. "It's estimated that by 2003, there'll be 2 billion people with mobile phones. Chances are that a large proportion of these people will want to use these devices to access the huge amount of information that is available online." Heumann believes that people will embrace these itty-bitty two-inch screens to view content on the Web.

The real change, according to Heumann, is that designers need to provide for the variety of different interfaces that users might employ to access information. He believes this is a reasonable task. "The extra work involved in providing mobile connectivity is negligible. The majority of this is with regard to UI (user interface) design. Developers must think about that because it's not something that can be easily automated."

The XML answer

To accomplish this, it's essential that developers design their sites so that the presentation is separate from the information. The answer to this is XML (the eXtensible Markup Language), a meta-language that enables developers to define mark-up languages according to set specifications and syntax. WML (Wireless Mark-up Language) is an XML application, developed by the WAP (Wireless Application Protocol) Forum, used to create pages intended for consumption on mobile devices.

Another piece of this jargon jigsaw is the Composite Capability/Preference Profiles (CC/PP) exchange protocol, a creation of the W3C's Mobile Activity group. It allows a web server with CC/PP functionality to deliver the appropriate content for users, depending on the sort of browser and device they are using, and the limitations of their interface.

Johan Hjelm, a W3C fellow from Ericcson and Chair of the W3C's CC/PP Working Group, believes that in order for mobile access to really work, a new approach needs to be taken towards web development. "Designers have been doing the wrong thing for five years. The original intention of Tim Berners-Lee was never to create a new printing press, but a heterogeneous environment. The important thing about a web-based presentation isn't the graphics and design, but the interaction."

He describes the impetus of widespread Internet access as "a return to the roots for the Web, with mobile devices as the catalyst." Hjelm sees the concept behind the W3C, both with regard to mobile access and in general, as a drive to "unify the information space. A space where the presentations are different, but the information is the same."

But is this enough? Heumann isn't so sure. "It's not uncommon for many sites to have several hundred links on a single page, but this obviously isn't feasible on the screen of a mobile phone or even a PDA. The way we like to work it is that on each screen you can choose from nine different links, and a tenth option that will move you back up the site tree. These options can all be accessed using the ten numbers on every phone's dial pad."

It seems inconceivable that the paragraphs and paragraphs of information written for purveyance on a WebTV, a PC, a Mac, or a similar platform could be appropriate for a mobile user with a limited interface. "The requirements of a mobile user are different from those of users working through a standard web browser. They tend to want very specific information, or to perform very specific tasks. That might mean buying a CD or finding a bus timetable. Usually that won't extend to general browsing."

He warns us, though, not to dismiss human ingenuity, despite the massive hurdles that stand in the way of mobile connectivity. "The applications that will be available in a few times, I believe, will be absolutely mind blowing," prophesizes Heumann. "But the technology to make this become a reality already exists. One

Table 1: Screen comparisons of hand-held devices

Product	Resolution	Colordepth	Screensize (")	Lines per screen	Chars per line
Psion 3 (a/c/mx)	480 x 160	Mono guess 8 g/s	5 x 1.73	80	26
Psion Revo	480 x 160	Mono guess 8 g/s	Same?	Same?	Same?
Psion 5 (+mx)	640 x 240	Mono 16g scales	5.25 x 2.0	100	26
Psion 7	640 x 480	STN Colour	7.7*	VGA	VGA
Motorola Timeport P7389	-	Mono	1.3 x 0.8	4	16
Motorola Timeport P8167	-	Mono	1.3 x 0.7	3	15
Qualcomm pdQ Smart Phone	-	Mono	2.4 x 1.9	12	36
Qualcomm QCP-2760	-	Mono	1.4 x .09	3	12
Samsung SCH-3500	-	Mono	1.2 x 0.8	3	16
Sanyo SCP-400	-	Mono	1.1 x 0.9	4	15
Sprint PCS NP-100	-	Mono	1.5 x 2.0	9	17
Sprint PCS Touchpoint	-	Mono	1.3 x 1.3	5	15
Palm III **	-	Mono	2 3/8 x 3.25	15	40
Palm V **	-	Mono	2.25 x 3	13	40

* = Length of the diagonal
** = Measurements done by Web Review editorial staff

application that exists today allows you to enter search criteria for a CD or a book, and then purchase that product on the Amazon web site using the 1-click settings configured while on your desktop computer."

PART 2

Access for all

The problems of providing for non-standard user interfaces aren't just confined to cell phones and PDAs. Over 20 percent of the US population has some sort of disability—cognitive, visual, auditory, physical, or otherwise. In addition to the social incentives for creating accessible interfaces, there's also a strong financial incentive, given the preponderance of anti-discriminatory legislation present both in the US and around the world.

But it's not quite that simple, as Judy Brewer, Director of the W3C's Web Accessibility Initiative (WAI) explains. "In order to provide widespread accessibility for disabled individuals, significant change needs to occur in five main areas. This is how the work of the WAI [a project funded by the European Commission, the US government, the Canadian government, and industry giants such as IBM, Bell Atlantic, and Microsoft] is divided." These five areas cover the development of web content, user agents (or web browsers as everyone else seems to call them), and authoring and validation tools. In addition, the WAI is involved with education and outreach as well as with the funding and execution of research into pertinent issues.

The accessibility problems associated with providing for this small but burgeoning base of Internet users are truly gargantuan, particularly given the range of interface requirements.

The seemingly disparate issues of accessibility for the disabled and mobile connectivity have more in common than is initially apparent. These two sectors of the market are becoming increasingly important to cater to, albeit for different reasons. Solutions to the accessibility problems faced by page designers looking to the disabled and mobile markets are similar in nature. So let's take a peek at a few of these solutions, starting with XHTML.

Solutions to accessibility problems

Although WML may be considered the way ahead for mobile connectivity, most web pages are written solely in HTML and are likely to stay that way for a long while yet. The advantages of conforming to the XHTML specification are myriad, but in terms of providing for those with limited functionality interfaces, the fact that XHTML 1.0 requires that your code is compliant with XML syntax means that it will be easier for these interfaces to correctly interpret it, and easier for you to convert the code into WML or other mark-up languages at a later date.

It's important to use mark-up appropriately. This means abandoning devices such as the one-pixel GIF and designing layout properly, using CSS. W3C fellow, Johan Hjelm, advises that tables should be used solely for tabulating linearized information, not for dividing up the page. Frames, meanwhile, should be avoided. The Web Content Accessibility Guidelines suggest that you should also use CSS properties in preference to HTML elements when designing your page. Using relative rather than absolute values in layout definition is also a good idea so the design can scale according to the size of the screen on which the information is presented.

Another important point is to provide contextual information to the client explaining what the page is about, the function of the page's various elements, and how it all relates to the site as a whole. A good start is to ensure that you include descriptive page and frame titles. The hyperlink text should be descriptive and take into account the fact that not everyone will be using a mouse. Use meaningful text, not the inane and heinous "click here." You might also want to provide a site map or table of contents, and take particular care in ensuring that these are simply designed and widely accessible. Your navigation should be intuitive.

Beware of using client-side scripts on web pages. Not all devices will have the necessary processor power, or feature set, to correctly parse and execute scripts produced by the developer. These scripts most likely aren't appropriate for users on devices with limited interfaces. Developers shouldn't depend on the correct execution of a script to display a page's contents. If using such a script is unavoidable, then a server-side implementation might be a better choice. Conversely, client-side image maps tend to be preferable to server-side creations since the necessary data is available to the browser application. Client-side image maps can adapt the information therein to the requirements of the interface. On the other hand, if the data is all locked away on the server, then it's in the lurch.

Equivalent alternatives

Although the production of separate pages for different devices is rarely a good idea, it's important for data to be deliverable in a variety of forms. This is where, to use more of the W3C's jargon, "equivalent alternatives" come in. These come in two types: discrete and continuous. Discrete equivalents contain no time references and have no defined duration, while continuous equivalents may be linked to a time code.

Examples serve better to illustrate this distinction. Discrete equivalents are nothing new, the alt attribute of the <img> tag is probably the most basic example of this. Continuous equivalents are somewhat more complicated, the most likely example being synchronized captioning that might accompany a video stream so that people with hearing difficulties could still follow the dialogue.

To do this you'll need to use the Synchronized Multimedia Language (SMIL), a creation from those boffins at the W3C. While we can't go into all of them in this article, SMIL does have a number of interesting features to aid accessibility. SMIL elements support three discrete text equivalents in the form of the alt, title, and longdesc attributes. This last one allows designers to provide more detailed description of the element and its function. It is, however, fair to say that the first two attributes aren't exactly cutting-edge developments.

Continuous equivalents are a little more interesting, as this piece of SMIL code shows:

```
<audio alt="CNN News, recorded in
American English"
   src="cnn-aud-useng.rm"/>

<video alt="CNN News, standard NY
studio"
   src="cnn-vid-ny.rm"/>

<textstream alt="American English for
CNN News"
    src="cnn-useng-caps.rt"/>
```

The idea is that there are three separate streams of information that can be provided to the user: the audio stream, the video stream, and the text stream. Exactly which of these streams are delivered depends on the configuration of the user's browser. But most importantly, they all run on the same time track. Clever stuff.

Many people support the misconception that, with the advent of more consistent feature implementation across the spectrum of desktop web browser applications, the issues of compatibility and accessibility are going to fade into obscurity. Nothing is further from the truth. The growing complexity and range of information available for presentation on web sites, and the greater range of devices being developed that may wish to access those sites, mean that the difficulties of providing for a variety of different users and interfaces will continue to increase. We must, however, tackle these difficulties if we are to take the Web from something you merely access through your desktop PC to a system of information dissemination that is part of everyday life.

In order to face these challenges effectively, developers, designers, and clients need to not only learn about the new technologies and tools with which they can battle them, but develop a new approach to web development in general. They cannot expect the precision control of the print medium, because on the World Wide Web of the future, users and their devices will play a much greater role in determining how content is displayed. Many regard this period as a return to the roots of the philosophy of the Web, a re-creation of the universality of Tim Berners-Lee's original vision—a Noah's Ark of sorts. Amen to that.

WAP Overview - "The Internet In Your Hands"

Title: *WAP Overview – "The Internet In Your Hands"*

Author: *Surrey & City Consulting*

Abstract: *As the world becomes a smaller place the need to communicate with others and access information irrespective of one's location, becomes more than a luxury it becomes a neccessity.*

Mobile communication has been made possible by the invention of celluar phones. Access to information and knowledge held on the World Wide Web, private Intranets and Extanets, has until now, been available to those indiviuals who have a PC, Laptop or a PDA (with a connection to a machine connected to the Internet/Intranet).

Wireless Application Protocol (WAP) is a worldwide protocol for Internet and Intranet applications to be used in wireless devices. Unwired Planet, Ericsson, Nokia and Motorola founded Wireless Application Protocol Forum in June 1997. It goal was to offer license free standard to whole wireless industry so that everyone who wants can start developing WAP based services.

Detailed information presented in a very graphical way, however requires time to digest and if rushed through can lead to information overload. At many times the need is for only one or two lines of data; e.g. Someone's name and phone number, flight numbers and times, Currency and exchange rate etc. It is this information which people need to be able to have quick and easy access to, it is this information which WAP addresses. It's a

Figure 1

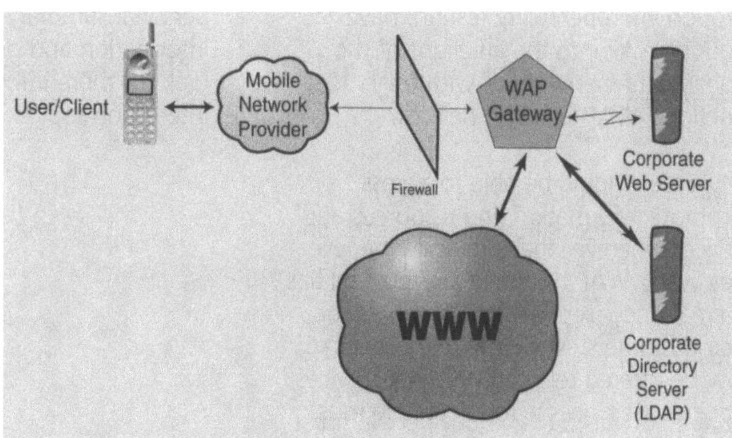

good point to bear in mind that WAP is NOT FOR EVERYONE. It will benefit people with a high demand for mobility the most.

Why WAP?

WAP provides easy access to information anywhere, anytime. For businesses it saves time, increases productivity improves business performance through continuous mobile access to corporate intra/extranets. It can also provide an expanding global market for new products and services.

As a global standard, WAP's popularity and adoption is being instigated by the leading Mobile telecommunications companies world-wide.The WAP forum today, define a set of protocols in application, session, transaction, security, and transport layers, which enable operators, manufacturers, and applications providers to meet the challenges and needs of the advanced wireless services into the millenium. For more information, visit the WAP Forum at http://www.wapforum.org.

The WAP contributing members, who, through interoperability testing, have brought WAP into the limelight of the mobile data marketplace with fully functional WAP–enabled devices.

Employees should be able to access corporate information and resources and services wherever they are and whenever they want. WAP has unlimited possibilities to create new services and products for their customers. Mobile Phone users can now be offered relevant personal services that suit their needs. Today people must carry with them at least a wallet, a calendar and a phone. Soon it will be only a phone – a media phone to make calls, pay bills, buy tickets, pay bills, buy tickets, check e-mails and manage your calendar at work.

- ***Ease of use:***
  Wireless devices have to be easier to use than even the simplest computer.
- ***Market size:***
  There are now about 200 million mobile users in the world and by 2005 that number will be over 1 billion.
- ***Price sensitivity:***
  Since in wireless phones small price change is more meaningful than in computers, should this solution add value with low price effects to be successful in these markets.
- ***Essential tasks:***
  Users want to scan their e-mail rather than read it all. He wants to see five most important stock quotes rather than see all the details of all stocks. Good application gives users extensive personal summary of important information and gives them the ability to look for more information about interesting topics.

WAP Products - Current Status

There are currently 4 products which support WAP :

- Nokia 7110e
- Siemens S25
- Ericsson R320S
- Ericsson MC218 - This is a clone of Psion's Series 5mx running the EPOC operating system with Ericsson's extensions and applications, and is now on sale.

The Future - Conclusion

Everyone in the mobile data business - including end users - seem to benefit from WAP:

- hardware and software manufacturers since users and network operators will be demanding new equipment and applications.
- network operators since they will be able to sell their customers new services and all services will work on all devices.
- users because they will have new portable services and they don't have to worry if a particular service will work with their device or not.

The currently introduced devices will cover a wide base of customer needs when they hit markets and many more are following since most mobile phone manufacturers are developing new ones to keep up with others. According to mobile phone manufacturers committed to WAP, most mobile phones will be WAP-enabled in a couple of years and since all major mobile phone manufacturers do support WAP, there doesn't seem to be anything in the way of WAP success.

Architecture

WAP requires 3 parts - a WAP-enabled device with the user, a WAP Gateway, and a conventional web server. The WAP Gateway processes the requests from the user's device, and performs one of three actions:

1. Supplies a WML file from the local hard disk.
2. Runs a CGI-program located on the

Figure 2: Architecture of the WAP Gateway

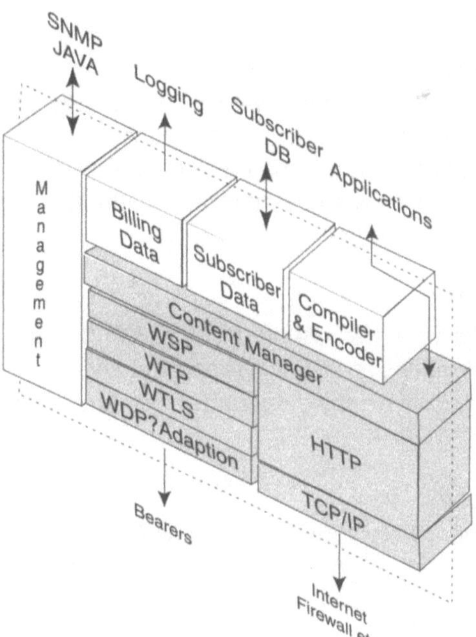

WAP Gateway, which would most probably be written in Java.

3. Performs an HTTP query of a seperate web-server to gather the WML content.

The Gateway then send the WML deck back over the mobile connection.

Mobile-Originated Example of WAP Architecture

The request from the mobile device is sent as a URL through the operator's network to the WAP gateway.

WDP
The WAP datagram protocol (WDP) is the transport layer that sends and receives messages via any available bearer network, including SMS, USSD, CSD, CDPD, IS–136 packet data, and GPRS.

WTLS
Wireless transport layer security (WTLS), an optional security layer, has encryption facilities that provide the secure transport service required by many applications, such as e-commerce.

WTP
The WAP transaction protocol (WTP) layer provides transaction support, adding reliability to the datagram service provided by WDP.

WSP
The WAP session protocol (WSP) layer provides a lightweight session layer to allow efficient exchange of data between applications.

HTTP Interface
The HTTP interface serves to retrieve WAP content from the Internet requested by the mobile device.
WAP content (WML and WMLScript) is converted into a compact binary form for transmission over the air

The WAP microbrowser software within the mobile device interprets the byte code and displays the interactive WAP content.

Chapter 2: How to Benefit

Applications of Wireless Networks

Title: *Applications of Wireless Networks*

Author: *Jim Geier, author of the Wireless LANs book.*

Abstract: *Wireless networking makes it possible to place portable computers in the hands of mobile "front line" workers, such as doctors, nurses, warehouse clerks, inspectors, claims adjusters, real estate agents, and insurance salespeople. The coupling of portable devices with wireless connectivity to a common database and specific applications meets mobility needs, eliminates paperwork, decreases errors, reduces process costs, and improves efficiency. The alternative to this, which many companies still employ today, is utilizing paperwork to update records, process inventories, and file claims. This manual method processes information slowly, produces redundant data, and is subject to errors caused by illegible handwriting. The wireless computer approach using a centralized database is clearly superior.*

 685 North Enon Road, Yellow Springs, Ohio 45387 USA Email: jimgeier@aol.com
Phone: +1 937-609-8370 Fax: +1 425-940-3019
Biography: Jim Geier is a management consultant specializing in mobile computing applications, technologies and implementation. He is author of several books, inculding Wireless LANs, Wireless Networking Handbook and Reengineering.

Retail

Retail organizations need to order, price, sell, and keep inventories of merchandise. A wireless network in a retail environment enables clerks and storeroom personnel to perform their functions directly from the sales floor. Salespeople are equipped with a pen-based computer or a small computing device with bar code reading and printing capability, with the wireless link to the store's database. They are then able to complete transactions, such as pricing, bin labeling, placing special orders, and taking inventory from anywhere within the store.

When printing price labels that will be affixed to the item or shelves, retailers often utilize a handheld bar code scanner and printer to produce bar coded and/or human readable labels. A database or file contains the price information either located on the handheld device, often called a batch device, or a server located somewhere in the store. In batch mode, the price clerk scans the bar code (typically the product code) located on the item or shelf edge, the application software uses the product code to look up the new price, then the printer produces a new label the clerk affixes to the item.

In some cases, the batch-based scanner/printer has enough memory to store all the price information needed to effectively perform the pricing function throughout a shift or entire day. This situation makes sense if you update price information in the database once a day, typically during the evening. The clerks load the data onto the device at the beginning of their shifts, then walk throughout the store continuously pricing items within the store. If the memory in the device is not large enough to store all the data, however, a wireless network is probably necessary. If the handheld unit is equipped with a wireless network connection, then the data can be stored in the much larger memory capabilities of a centralized PC server or mainframe and accessed each time the item's bar code is scanned. In addition, a wireless network-based solution has merits if it is too time consuming to download information to a batch device.

Warehouses

Warehouse staff must manage the receiving, putting away, inventory, and picking and shipping of goods. These responsibilities require the staff to be mobile. Warehouse operations have traditionally been a paper-intensive and time-consuming environment. An organization, however, can eliminate paper, reduce errors, and decrease the time necessary to move items in and out by giving each warehouse employee a handheld computing device with a bar code scanner interfaced via a wireless network to a warehouse inventory system.

Upon receiving an item for storage within the warehouse, a clerk can scan the item's bar coded item number and enter other information from a small keypad into the database via the handheld device. The system can respond with a location by printing a put-away label. A forklift operator can then move the item to a storage place and account for the procedure by scanning the item's bar code. The inventory system keeps track of all transactions, making it very easy to produce accurate inventory reports.

As shipping orders enter the warehouse, the inventory system produces a list of the items and their locations. A clerk can view this list from the database via a handheld device and locate the items needed to assemble a shipment. As the clerk removes the items from the storage bins, the database can be updated via the handheld device. All of these functions depend heavily on wireless networks to maintain real-time access to data stored in a central database.

Healthcare

Healthcare centers, such as hospitals and doctors' offices, must maintain accurate records to ensure effective patient care. A simple mistake can cost someone's life. As a result, doctors and nurses must carefully record test results, physical data, pharmaceutical orders, and surgical procedures. This paperwork often overwhelms healthcare staff, taking 50-70 percent of their time.

Doctors and nurses are also extremely mobile, going from room to room caring for patients. The use of electronic patient records, with the ability to input, view, and update patient data from anywhere in the hospital, increases the accuracy and speed of healthcare. This improvement is possible by providing each nurse and doctor with a wireless pen-based computer, coupled with a wireless network to databases that store critical medical information about the patients.

A doctor caring for someone in the hospital, for example, can place an order for a blood test by keying the request into a handheld computer. The laboratory will receive the order electronically and dispatch a lab technician to draw blood from the patient. The laboratory will run the tests requested by the doctor and enter the results into the patient's electronic medical record. The doctor can then check the results via the handheld appliance from anywhere in the hospital.

Another application for wireless networks in hospitals is the tracking of pharmaceuticals. The use of mobile handheld bar code printing and scanning devices dramatically increases the efficiency and accuracy of all drug transactions, such as receiving, picking, dispensing, inventory taking, and the tracking of drug expiration dates. Most importantly, though, it ensures that hospital staff is able to administer the right drug to the right person in a timely fashion. This would not be possible without the use of wireless networks to support a centralized database and mobile data collection devices.

Real Estate

Real estate salespeople perform a great deal of their work away from the office, usually talking with customers at the property being sold or rented. Before leaving the office, salespeople normally identify a few sites to show a customer, print the MLS (Multiple Listing Service) information that describes the property, and then drive to each location with the potential buyer. If the customer is unhappy with that round of sites, the real estate agent must drive back to the office and run more listings. Even if the customer decides to purchase the property, they must both go back to the real estate office to finish paperwork that completes the sale.

Wireless networking makes the sale of real estate much more efficient. The real estate agent can use a computer away from the office to access a wireless MLS record. IBM's Mobile Networking Group and Software Cooperation of America, for example, make wireless MLS information available that enables real estate agents to access information about properties, such as descriptions, showing instructions, outstanding loans, and pricing. An agent can also use a portable computer and printer to produce contracts and loan applications for signing at the point of sale.

Hospitality

Hospitality establishments check customers in and out and keep track of needs, such as room service orders and laundry requests. Restaurants need to keep track of the names and numbers of people waiting for entry, table status, and drink and food orders. Restaurant staff must perform these activities quickly and accurately to avoid making patrons unhappy. Wireless networking satisfies these needs very well.

Wireless computers are very useful in the situations where there is a large crowd, such as a restaurant. For example, someone can greet restaurant patrons at the door and enter their names, the size of the party, and smoking preferences into a common database via a wireless device. The greeter can then query the database and determine the availability of an appropriate table. Those who oversee the tables also would have a wireless device used to update the database to show whether the table is occupied, being cleaned, or available. After obtaining a table, the waiter transmits the order to the kitchen via the wireless device, eliminating the need for paper order tickets.

Utilities

Utility companies operate and maintain a highly distributed system that delivers power and natural gas to industries and residences. Utility companies must continually monitor the operation of the electrical distribution system and gas lines, and must check usage meters at least monthly to calculate bills. Traditionally, this means a person must travel from location to location, enter residences and company facilities, record information, and then enter the data at a service or computing center. Today, utility companies employ wireless networks to support the automation of meter reading and system monitoring, saving time and reducing overhead costs.

Kansas City Power & Light operates one of the largest wireless metering systems, serving more than 150,000 customers in eastern Kansas and western Missouri. This system employs a monitoring device at each customer site that takes periodic meter readings and sends the information back to a database that tracks usage levels and calculates bills, avoiding the need for a staff of meter readers.

Field Service

Field service personnel spend most of their time on the road installing and maintaining systems or inspecting facilities under construction. In order to complete their jobs, these individuals need access to product documentation and procedures. Traditionally, field service employees have had to carry several binders of documentation with them to sites that often lack a phone and even electricity.

In some cases, the field person might not be able to take all the documents with him to a job site, causing him to delay the work while obtaining the proper information. On long trips this information may also become outdated. Updates require delivery that may take days to reach the person in the field. Wireless access to documentation can definitely enhance field service. A field service employee, for example, can carry a portable computer connected via a wireless network to the office LAN containing accurate documentation of all applicable information.

Field Sales

Sales professionals are always on the move meeting with customers. While on site with a customer, a salesperson needs access to vast information that describes products and services. Salespeople must also place orders, provide status, such as meeting schedules, to the home office, and maintain inventories.

With wireless access to the home office network, a salesperson can view centralized contact information, retrieve product information, produce proposals, create contracts, and stay in touch with home office staff and other salespeople. This contact permits salespeople to complete the entire sale directly from the customer site, which increases the potential for a successful sale and shortens the sales cycle.

Vending

Beverage and snack companies place vending machines in hotels, airports, and office buildings to enhance the sales of their products. Vending machines eliminate the need for a human salesclerk. These companies, however, must send employees around to stock the machines periodically. In some cases, machines might become empty before the restocking occurs because the company has no way of knowing when the machine runs out of a particular product.

A wireless network can support the monitoring of stock levels by transporting applicable data from each of the vending machines to a central database that can be easily viewed by company personnel from a single location. Such monitoring allows companies to be proactive in stocking their machines because they will always know the stock levels at each machine. Comverse Technology's DGM&S subsidiary licensed software to BellSouth to support a vending machine monitoring service called Cellemetry, which uses the data channels of existing cellular networks.

WAP White Paper ...when time is of the essence

Title: *WAP White Paper ...when time is of the essence*

Author: *AU-system radio*

Abstract: *In 1997, Ericsson, Motorola, Nokia and Unwired Planet took the initiative to found WAP Forum, which mission is to bring the convenience of the Internet into the wireless community as well. By addressing the constraints of a wireless environment, and adapt existing Internet technology to meet these constraints, the WAP Forum has succeeded in developping a standard that scales across a wide range of wireless devices and networks. This paper gives some technical background.*

Biography: AU-System Radio has been an active participant in the WAP Forum standardization work since it started in June 1997. We offer highly skilled specialist consultants within the following WAP related areas: WAP business and service development; WAP infrastructure development; WAP service creation/development; and WAP client application development

Executive Summary

Mobile networks of today do often not provide desired flexibility when value added services are about to be introduced. This since it often is a rather complicated and lengthy task to launch such services. The Wireless Application Protocol (WAP) addresses this issue by introducing the concept of the Internet as a wireless service platform.

The Internet has proven to be an easy and efficient way of delivering services to millions of "wired" users. In 1997, Ericsson, Motorola, Nokia, and Unwired Planet took the initiative to found WAP Forum, which mission is to bring the convenience of the Internet into the wireless community as well. WAP Forum has today gained great credence in the wireless industry all over the world, more than 90 of the world's leading companies in the business of wireless telecommunication are members as of February 1999.

By addressing the constraints of a wireless environment, and adapt existing Internet technology to meet these constraints, the WAP Forum has succeeded in developing a standard that scales across a wide range of wireless devices and networks. Key features offered by WAP are:

- **A programming model similar to the Internet's**
  Re-use of concepts found on the Internet enables a quick introduction of WAP based services since both service developers and manufacturers are familiar with these concepts today.

- **Wireless Markup Language (WML)**
  A markup language used for authoring services, fulfilling the same purpose as HyperText Markup Language (HTML) does on the World Wide Web (WWW). In contrast to HTML, WML is designed to fit small handheld devices.

- **WMLScript**
  WMLScript can be used to enhance the functionality of a service, just as for

example JavaScript may be utilised in HTML. It makes it possible to add e.g. procedural logic and computational functions to WAP based services.

- **Wireless Telephony Application (WTA)**
  The WTA framework defines a set of features that provides a means to create telephony services. This is accomplished by introducing an in-client interface to the mobile network, handling of network events, a repository that allows real-time handling of services, and a mechanism supporting server initiated services.

- **Optimised protocol stack**
  The protocols used in WAP are based on well-known Internet protocols such as HyperText Transport Protocol (HTTP) and Transmission Control Protocol (TCP), but have been optimised to address the constraints of a wireless environment, such as low bandwidth and high latency.

The opportunity of creating wireless services on a global basis will attract operators as well as third party service providers, resulting in both co-operations and competition that do not exist today. WAP provides a means to create not only services that we are used to from the World Wide Web today, but also telephony services.

Introduction

The Wireless Application Protocol (WAP) is a result of joint efforts taken by companies teaming up in an industry group called WAP Forum. The objective of the forum is to create a license-free standard that brings information and telephony services to wireless devices. To access these services WAP utilises the Internet and the World Wide Web (WWW) paradigm.

WAP scales across a broad range of wireless networks, implying that it has the potential to become a global standard and that economies of scale thus can be achieved.

In order to provide wireless access to the information space offered by the WWW, WAP is based on well-known Internet technology that has been optimised to meet the constraints of a wireless environment.

Background
WAP could roughly be described as a set o protocols that has inherited its characteristics and functionality from Internet standards and standards for wireless services developed by some of the world's leading companies in the business of wireless telecommunications.

In 1995 Ericsson initiated a project which purpose was to develop a general protocol or rather a concept, for value added services on mobile networks. The protocol was named Intelligent Terminal Transfer Protocol (ITTP), and handles the communication between a service node, where the service application is implemented, and an intelligent mobile telephone. The ambition was to make ITTP a standard for value added services in mobile networks.

During 1996 and 1997 Unwired Planet, Nokia, and others, launched additional concepts in the area of value added services on mobile networks.
Unwired Planet presented Handheld Device Markup Language (HDML) and Handheld Device Transport Protocol (HDTP). Just as HyperText Markup Language (HTML) used on the WWW, HDML is used for describing content and user interface, but optimised for wireless Internet access from handheld devices with small displays and limited input facilities. In the same manner HDTP could be considered to be a wireless equivalent of the standard Internet HyperText Transport Protocol (HTTP), i.e. a lightweight protocol to perform client/server transactions.

In March 1997 Nokia officially presented the Smart Messaging concept, an Internet access service technology specially designed for handheld GSM devices. The communication between the mobile user and the server containing Internet information uses Short Message Service (SMS) and a markup language called Tagged Text Markup Language (TTML). Just like HDML, this language is adapted for wireless communication, i.e. narrowband connections. With a multitude of concepts there was a substantial risk that the market could become fragmented, a development that neither of the involved companies would benefit from. Therefore, the companies agreed upon bringing forth a joint solution. WAP was born...

WAP Forum
On June 26 1997 Ericsson, Motorola, Nokia, and Unwired Planet took the initiative to start a rapid creation of a standard for making advanced services within the wireless domain a reality. In December 1997 WAP Forum was formally created, and after the release of the WAP 1.0 specifications in April 1998, WAP Forum membership was opened to all. Today (February 1999) over 90 companies are members of WAP Forum. Among these companies many of the world's leading terminal and infrastructure manufacturers, software companies, operators, and service providers are found. The handset manufacturers in WAP Forum represent over 90% of the world market across all technologies, and the network operators are representing about 100 million subscribers.

The main objectives of the WAP Forum are:
- Independent of wireless network standard
- Open to all
- Will be proposed to the appropriate standards bodies
- Applications scale across transport options (GSM, IS-95, IS-136, PDC, etc)
- Applications scale across device types (mobile phones, PDAs, etc)
- Extensible over time to new networks and transports (e.g. 3G systems)

WAP – The Wireless Service Enabler

The following sections discuss why the Internet is a suitable platform for wireless value added services and the tight coupling between the Internet and WAP programming models.

Why Using the Internet?

During the last couple of years we have become used to the wide variety of services offered by the Internet and the WWW. Not only the services themselves attract us, but also the convenient way of accessing them via an Internet browser. We can access the same services all over the world as long as we have access to a computer and the Internet. Nobody can deny that the Internet has scored a tremendous success, nearly a 150 million users (end of 1998, source: Computer Industry Almanac) can not be wrong. Service providers do also benefit from the WWW paradigm since their services can be deployed independently of the location of the users. The services are created and stored on a server, meaning that it becomes very easy to change them according to the needs of the customers. By using off-the-shelf authoring-tools services are created with minimum effort which, combined with the fast and convenient way of launching them, enables an extremely short time-to-market. The reduced service development time does of course also imply reduced costs compared to conventional service development in wireless networks.

As users become more and more dependent of services offered on the Internet, one shortcoming becomes increasingly evident - the need for a wire to connect to the Internet. This shortcoming makes itself especially remembered to the millions of users that spend a substantial amount of their time on the move.

The last few years' attempts to make this shortcoming disappear has not made the Internet crossing the chasm as a wireless service platform, only the early adopters has accepted the technologies provided so far. One of the main reasons for this is the lack of a widely accepted standard, a problem being addressed by the WAP Forum. The wide support for WAP will most probably enable the Internet as a means to provide services to wireless devices within a foreseeable future. This includes both services that we recognise from the WWW, and services like telephony services. The two next sections show how the Internet and WAP relate to each other.

The Internet Model

The Internet model makes it possible for a client to reach services on a large number of origin servers; each addressed by a unique Uniform Resource Locator (URL). The content stored on the servers is of various formats, but HTML is the predominant. HTML provides the content developer with a means to describe the appearance of a service in a flat document structure; i.e. the entire content of a page is shown simultaneously. If more advanced features like procedural logic are needed, scripting languages such as JavaScript or VB Script may be utilised.

The figure next page shows how a WWW client requests a resource stored on a web server. As mentioned above, a resource on the Internet is identified by a unique URL, that is, a text string constituting an address to that resource. In the example below, the resource is an HTML document.

Figure 1: The Internet model

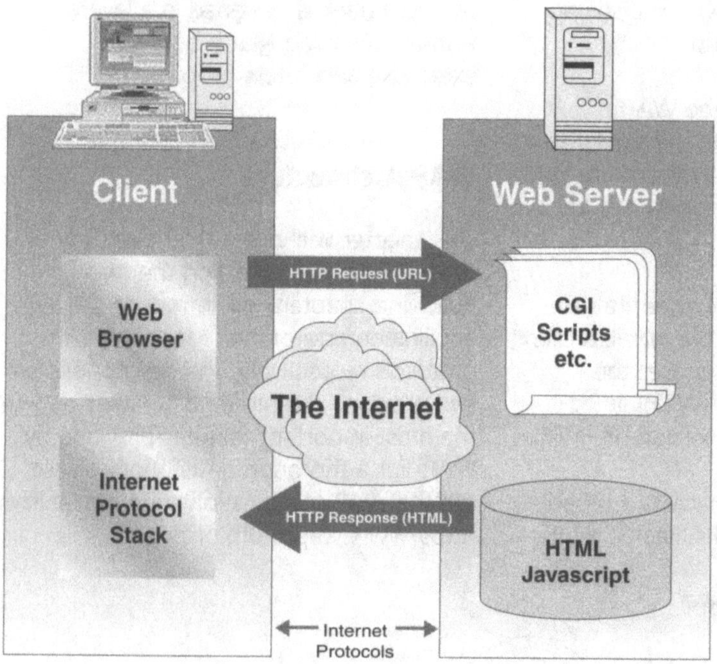

On the Internet, standard communication protocols, like HTTP and Transmission Control Protocol/Internet Protocol (TCP/IP) are used. The content may be static or dynamic. Static content is produced once and not changed or updated very often, for example a company presentation. Dynamic content is needed when the information provided by the service changes more often, for example timetables, news, stock quotes and account information. Technologies such as Active Server Pages (ASP), Common Gateway Interface (CGI), and Servlets allow content to be generated dynamically.

The WAP Model

WAP does also make use of the Internet paradigm to provide a flexible service platform. In order to accommodate wireless access to the information space offered by the WWW, WAP is based on well-known Internet technology that has been optimised to meet the constraints of a wireless environment. Services created using HTML would not fit very well on small handheld devices since they are intended for use on desktop computers with big screens. Low bandwidth wireless bearers would neither be suitable for delivering the rather extensive information that HTML pages often consist of. Therefore a markup language adapted to these constraints has been developed - the Wireless Markup Language (WML).

WML offers a navigation model designed for devices with small displays and limited input facilities (no mouse and limited keyboard). In order to save valuable bandwidth in the wireless network, WML can be encoded into a compact binary format. Encoding WML is one of the tasks performed by the WAP Gateway/Proxy, which is the entity that connects the wireless domain with the Internet.

WAP does also provide a means for supporting more advanced tasks, comparable to those solved by using for example JavaScript in HTML. The solution in WAP is called WML Script.

The figure below shows the WAP programming model. Note the similarities with the Internet model. Without the WAP Gateway/Proxy the two models would have been practically identical.

The observant reader may note that the request that is sent from the wireless client to the WAP Gateway/Proxy uses the Wireless Session Protocol (WSP). In its essence, WSP is a binary version of HTTP.

WAP is designed to scale across a broad range of wireless networks, like GSM, IS-95, IS-136 and PDC. Finally, the WAP protocol stack is designed in a layered fashion, meaning that it becomes extensible and future proof.

WAP Architecture

This chapter will give a brief overview of the WAP architecture, and the two following chapters will introduce the WAP application environment and the WAP protocols respectively. The descriptions are by no means complete; they rather outline the most important features provided by WAP. For a thorough description, please see the WAP specification suite available at http://www.wapforum.org.

Figure 2: The WAP model

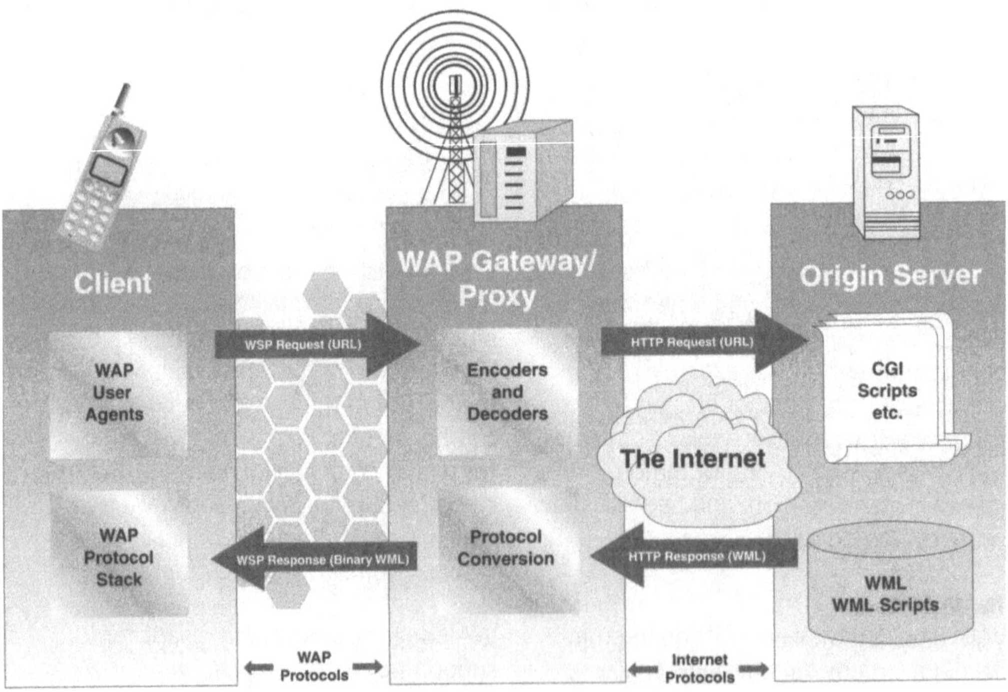

WAP is designed in a layered fashion in order to be extensible, flexible, and scalable. With the Open System Interconnection model (OSI model) in mind, the WAP-stack basically is divided into five layers. They are:

- ***Application Layer***
  Wireless Application Environment (WAE)
- ***Session Layer***
  Wireless Session Protocol (WSP)
- ***Transaction Layer***
  Wireless Transaction Protocol (WTP)
- ***Security Layer***
  Wireless Transport Layer Security (WTLS)
- ***Transport Layer***
  Wireless Datagram Protocol (WDP)

Each layer of the WAP protocol stack specifies a well-defined interface to the layer above, meaning that a certain layer makes lower layers invisible to the layer above. The layered architecture allows other applications and services to utilise the features provided by the WAP-stack as well. This makes it possible to use the WAP-stack for services and applications that currently are not specified by WAP.

Figure 3 shows the WAP protocol stack and how it relates to the protocols on the Internet.

Note that the mobile network bearers in the lower part of the figure above are not part of the WAP protocol stack.

WAP Application Environment

The uppermost layer in the WAP stack, the Wireless Application Environment (WAE) provides an environment that enables a wide range of applications to be used on wireless devices. In the chapter "WAP - the wireless service enabler" the WAP WAE programming model was introduced. This chapter will focus on the various components of WAE:

- ***Addressing model***
  A syntax suitable for naming resources stored on servers
- ***Wireless Markup Language (WML)***
  A lightweight markup language designed to meet the constraints of a wireless environment with low

Figure 3: The Wap architecture

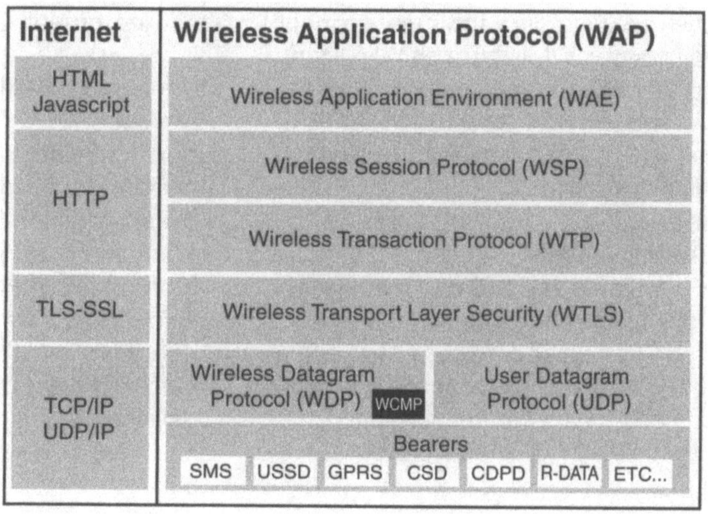

bandwidth and small handheld devices

- ***WMLScript***
  A lightweight scripting language
- ***Wireless Telephony Application (WTA, WTAI)***
  A framework and programming interface for telephony services

In order to make use of the features mentioned above, WAP assumes that two user-agents will be available in the wireless device.

A user-agent is within this context typically an in-device application that interprets content in a well-defined manner and handles user-interactions when necessary. In the wired world, a user-agent is typically referred to as a browser; e.g. Microsoft Internet Explorer and Netscape Navigator used on desktop computers. As mentioned, WAP assumes two user-agents, the WML user-agent and the WTA user-agent.

The wording "WML user-agent" might be a little confusing since this user-agent not only interprets WML, but also WMLScript etc. The wording should rather reflect that this user-agent, in contrast to the WTA user-agent, is not capable of handling telephony services.

Addressing Model
WAP uses the same addressing model as the one used on the Internet, that is, Uniform Resource Locators (URL). A URL uniquely identifies a resource, e.g. a WML document, on a server that can be retrieved using well-known protocols.

In addition to URLs, WAP also uses Uniform Resource Identifiers (URI). A URI is used for addressing resources that are not necessarily accessed using well-known protocols. An example of using a URI is local access to a wireless device's telephony functions.

Wireless Markup Language (WML)

The Wireless Markup Language is WAP's analogy to HTML used on the WWW. WML is based on the Extensible Markup Language (XML).

WML uses a deck/card metaphor to specify a service. A card is typically a unit of interaction with the user, that is, either presentation of information or request for information from the user. A collection of cards is called a deck, which usually constitutes a service. This approach ensures that a suitable amount of information is displayed to the user simultaneously since inter-page navigation can be avoided to the fullest possible extent.

Key features of WML include:

- Variables
- Text formatting features
- Support for images
- Support for soft-buttons
- Navigation control
- Control of browser history
- Support for event handling (for e.g. telephony services)
- Different types of user interactions, e.g. selection lists and input fields

WML can be binary encoded by the WAP Gateway/Proxy in order to save bandwidth in the wireless domain.

WMLScript
WMLScript is based on ECMAScript, the same scripting language that JavaScript is based on. It can be used for enhancing services written in WML in the way that it to some extent adds intelligence to the services, for example procedural logic, loops, conditional expressions, and computational functions.

WMLScript can be used for e.g. validation of user input. Since WML does not provide any mechanisms for achieving this, a round-trip to the server would be needed in order to determine if user input is valid or not if scripting was not available. Access to local functions in a wireless device is another area where WMLScript is used; for example access to telephony related functions.

WMLScript does also support WMLScript Libraries. These libraries contain functions that extend the basic WMLScript functionality. This provides a means for future expansion of functions without having to change the core of WMLScript.

Just as with WML, WMLScript can be binary encoded by the WAP Gateway/Proxy in order to minimise the amount of data sent over the air.

Wireless Telephony Applications (WTA)

The Wireless Telephony Application (WTA) environment provides a means to create telephony services using WAP. As already mentioned, WTA utilises a user-agent separate from the common WML user-agent, at least logically. The WTA user-agent is based on the WML user-agent, but is extended with functionality that meets the special requirements for telephony services. This functionality include:

- ***Wireless Telephony Application Interface (WTAI)***
  An interface towards a set of telephony-related functions in a mobile phone that can be invoked from WML and/or WMLScript. These functions include for example: call-management, handling of text messages and phonebook control. WTAI is divided into three categories: Network Common Functions, Network Specific Functions, and Public Functions. The common functions are available in all types of networks, while the specific functions specify functions that are unique to a certain network type. In contrast to the other two function libraries, the Public Functions library can be invoked from the WML user-agent as well. Currently, the Public Functions library only contains a function for setting up calls, which, in contrast to the corresponding function in the Network Common Functions library, must be acknowledged by the user before it is carried out.

- ***Repository***
  Many WTA services put requirements on real-time handling, implying that it is not feasible to retrieve content from a server since this involves a certain delay. The repository makes it possible to store WTA services persistently in the device in order to enable access to them without accessing the network.

- ***Event handling***
  Typical events in a mobile network are incoming call, call disconnect, and call answered. In order to create telephony services, it must be possible to handle these events. The event handling within WTA enables WTA services stored in the repository to be started in response to such events. Events can also be bound to a certain action in WML in order to make it possible to handle events within a service.

- ***WTA Service Indication***
  A content type that allows the user to be notified about events of different kinds (e.g. new voice mails) and be given the possibility to start the appropriate service to handle the event. In its most basic form, the WTA Service Indication makes it possible to send a URL and a message to a wireless device. The message is displayed to the user, and she is asked whether she wants to start the service indicated by the URL immediately or if she wants to postpone the Service Indication for later handling. The WTA Service Indication should be delivered to a device using push, an area where WAP Forum currently is working actively.

WTAI enables access to functions that are not suitable for allowing common access to them (except for the Public Function Library). For instance, setting up calls and manipulating the phonebook without use acknowledgement can both imply undesired costs and violate user-integrity if the corresponding functions are used inappropriately. The other functions provided by the WTA framework can be considered in the same way.

The WTA framework relies on a dedicated WTA user-agent capable of carrying out these functions, a functionality not provided by the common WML user-agent Only "trusted" content providers should b able to make content available to the WT user-agent, i.e. the operator or content providers trusted by the operator. Thus it must be possible to distinguish between servers that are allowed to supply the use agent with services containing these functions, and those who are not. To accomplish this the WTA user-agent retrieves its services from the WTA domair which, in contrast to the Internet, is controlled by the network operator. Figur 4 shows how WTA services and other services are separated from each other using WTA access control based on port numbers.
The WTA server may be an ordinary web server used for housing content. It may also be able to communicate with other entities, such as IN-nodes or voice mail systems, to provide extended telephony-related functionality. The communication with such entities can be controlled by applications on the WTA server that can referenced within a WTA service using URLs.

Figure 4: WTA access control

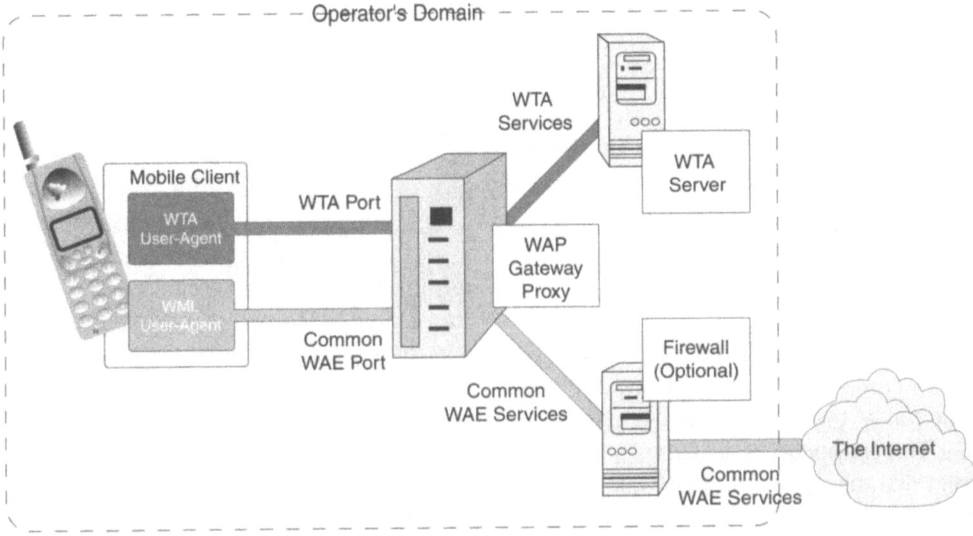

WAP Protocols

The WAP protocol suite contains four protocols for handling the communication between clients and the WAP Gateway/Proxy. These protocols are modelled after protocols used on the Internet, and can be used in four different configurations (each of the protocols mentioned below are explained in the next sections):

- ***Connectionless mode***
  This configuration utilises only WSP on top of WDP. It offers a datagram service, meaning that sent messages are not acknowledged, and hence no guarantee of delivery is offered. The functionality can be seen as a simple "send-and-forget" model.

- ***Connectionless mode with security***
  In addition to what is mentioned above, WTLS is used in order to provide authentication, encryption, etc.

- ***Connection mode***
  The connection mode uses WTP in addition to WSP and WDP. WTP offers reliable transmissions, meaning that sent messages must be acknowledged and may be retransmitted if lost. It also uses a mode of WSP that handles long-lived sessions

- ***Connection mode with security***
  In addition to what is mentioned above, WTLS is used in order to provide authentication, encryption, etc.

Wireless Session Protocol (WSP)

WSP is the interface between WAE and the rest of the protocol stack. WSP is a binary version of HTTP 1.1 with additions such as:

- Capability negotiation.
- Header caching
- Long-lived sessions
- Push

The two main stack configurations (connection and connectionless mode) are named after the session services offered by WSP. The connectionless session service is basically a thin layer that WAE can use when there is no need for reliable delivery of messages.

The main functionality of the connection mode of WSP is to set up a session between a client and the WAP Gateway/Proxy. This session handles capability negotiation at session establishment and also communication interrupts such as change of bearer. It is assumed to be long-lived and can be suspended, and later resumed, instead of disconnected if no communication will be needed for some time. This means that no new capability negotiation is needed when the session is resumed, which lessens the traffic load. Support for asynchronous handling of requests implies that, if several data requests have been sent, the answers may be delivered in a different order.

WSP also supports header caching in order to minimise bearer utilisation. In HTTP, which does not support header caching, about 90% of the requests contain static headers that need to be sent over and ove again.

Wireless Transaction Protocol (WTP)
WTP is responsible for control of transmitted and received messages. It provides a reliable communication path where messages are uniquely identified so as not to be accepted twice and may be retransmitted to the peer if lost in transmission. There is no coupling betweer communications as every communication sequence is only alive during the exchang of an individual message set. WTP works with three different message classes.

- ***Unreliable "send" with no result message***
  No retransmission if the sent message i lost.

- ***Reliable "send" with no result message***
  The recipient acknowledges the sent message. Otherwise the message is resent.

- ***Reliable "send" with reliable result message (three-way communicatio***
  A data request is sent and a result is received which finally is acknowledged by the initiating part.

WTP is also adapted to the constraints of wireless bearers in that it minimises the protocol overhead by introducing functionality to minimise the number of (re-) transmissions, for example, message concatenation and acknowledgement of received data requests. WTP can be extended with functionality for segmenti and reassembling messages. This includes selective retransmission of lost segments.

WSP and WTP are modelled together in such a way that the different functions in WSP have a defined and consistent usage of the message classes in WTP.

Wireless Transport Layer Security (WTLS)
As its name implies, the purpose of WTLS is to provide transport layer security between a WAP client and the WAP Gateway/Proxy. WTLS is based on Transport Layer Security (TLS) 1.0 but optimised for narrowband communication channels. Key features include:

- Integrity through the use of Message Authentication Codes (MAC)
- Confidentiality through the use of encryption
- Authentication and nonrepudiation of server and client, using digital certificates

These features make it possible to certify that the sent data have not been manipulated by a third party, that privacy is guaranteed, that an author of a message can be identified, and that both parties can not falsely deny having sent their messages. A secure connection is set up with an establishment phase where negotiation such as parameter settings, key exchange and authentication is performed. Both parties can abort the secure connection during establishment or at any time later.

WTLS is optional and can be used with both the connectionless and the connection mode WAP stack configuration. If used, it is always placed on top of WDP.

Wireless Datagram Protocol (WDP)
The base of the WAP protocol stack is a datagram layer, WDP, offering a consistent interface to the upper layers of the stack.

If WAP is used over a bearer supporting User Datagram Protocol (UDP), the WDP layer is not needed. On other bearers, such as GSM SMS, the datagram functionality is provided by WDP. This means that whether WAP uses UDP or WDP, it is given a datagram service, which hides the characteristics of different bearers and provides port number functionality. If necessary WDP can also be extended with functionality for segmenting and reassembling datagrams that are to big for the underlying bearer.

It is also possible to extend WDP with an optional protocol for error reporting called Wireless Control Message Protocol (WCMP). This protocol can be used when WAP is not used on an IP bearer (IP has its own control message protocol).WCMP can also be used for informational and diagnostic purposes.

Motivations for WAP

During recent years, both the Internet and wireless voice communication have undergone wide and rapid acceptance. The unification of these two technologies, the wireless Internet, has however not enjoyed the same development even though the Internet provides a means for rapid service development, short time-to-market, ease-of-use, and convenient manageability.

Why a new Standard?
A legitimate question is why the use of wireless data capabilities, in this context wireless Internet access, has not followed the trends of neither the wireless voice communication nor the Internet.

Just as with many other things in life, expectations matter very much. Anyone who have tried to access the Internet by using a laptop and a cellular phone, knows that the expectations we have created by using the Internet at the office or at home are not fulfilled; as a matter of fact it is usually a quite tiresome experience.

WAP addresses this issue by being designed to meet the constraints of a wireless environment. Both limitations in the network and in the client are taken into consideration.

The following sections outline some of the motivations behind WAP, often with parallels to existing Internet technology.

Adapting to the Bounds of the Wireless Network
WAP scales across a broad range of wireless networks and bearers. Hence it is designed to allow access to services via the Internet using simple SMS as well as fast packet-data networks such as General Packet Radio Service (GPRS). The most important issues in the network addressed by WAP are summarised below:

- ***Low Bandwidth:*** The problem with poor performance when using wireless bearers with low bandwidth becomes especially valid if the user is not well aware of what services to access. This since the service must not consume much bandwidth if it should be suitable for wireless access. The larger portion of the mass market, as well as many advanced users, are not aware of this. And besides, the users should simply no have to care about how the services they access are designed in order to have their expectations fulfilled.

WAP addresses this issue by minimising th traffic over the air-interface. WML and WMLScript are binary encoded into a compact form when sent over the air in order to minimise the number of bits and bytes.

WSP, WAP's equivalent to HTTP on the Internet, is also binary for the same reason Moreover, it supports long-lived sessions, that can be suspended and resumed, and header caching; saving valuable bandwidt since session establishment can then be done rather seldom.

The Wireless Transaction Protocol WTP, th analogy to the Internet's TCP, is not only designed to minimise the amount of data in each transaction, but also the number transactions.

- ***High Latency:*** Wireless networks have high latency compared to wired networks. This constraint is relevant in all of today's wireless networks, even f those providing high bandwidth. This is addressed in WAP by minimising the roundtrips between the wireless device and the wireless network. An asynchronous request/response model is also used.

In wired networks the low latency implies that requests and responses can be handled synchronously since the time between them most often do not affect the user experience. In wireless networks with high latency this is not a feasible approach, especially when using high latency bearers like SMS. This issue is among other features addressed by WSP by allowing requests and responses to be handled asynchronously, that is, a new request can be sent before the response to an earlier request has been received.

The application environment in WAP uses the concept of scripting, meaning that roundtrips between a client and a server can be avoided when it comes to e.g. validation of user input. The Wireless Telephony Application environment addresses latency (and low bandwidth) by introducing the repository, which is a persistent storage container used for housing services that should be started in response to an event in the mobile network (e.g. an incoming call). Since these services are available immediately, no roundtrips to the WTA server are needed and thereby real-time handling is made possible.

Compared to TCP, WTP needs a smaller number of transactions for each method invoked, i.e. keeps the number of roundtrips down.

- ***Less Connection Stability/ Unpredictable Bearer Availability:*** Wired network access provides a more or less reliable connection to the network. That is not the case in wireless networks where bearers might be inaccessible for shorter or longer periods of time due to fading, lost radio coverage, or deficient capacity.

As already mentioned, the sessions supported by WSP are assumed to be long-lived. The problem mentioned above is in WSP addressed by allowing lost sessions to be resumed, even when dynamically assigned IP addresses are used. The transaction layer in WAP, WTP, has been kept simple compared to TCP used on the wired Internet. Since no connection is set up the effects of lost bearer and times of inactivity are minimised.

The nature of a wireless connection implies that small segments of a message often are lost. WTP supports selective retransmission of data, meaning that only the lost segments are retransmitted, not the entire message as in TCP.

Adapting to the Bounds of the Wireless Device

WAP is targeted at handheld devices of various kinds. Services should be accessible from a Handheld PC as well as from a small phone. WAP addresses this fact by taking the following issues into consideration:

- ***Small Display:*** When accessing a service from a desktop computer, the size of the screen does not limit the user experience. Wireless devices might also have "big" displays, for example a Personal Digital Assistant (PDA). But

Figure 5: Limitations of a wireless device

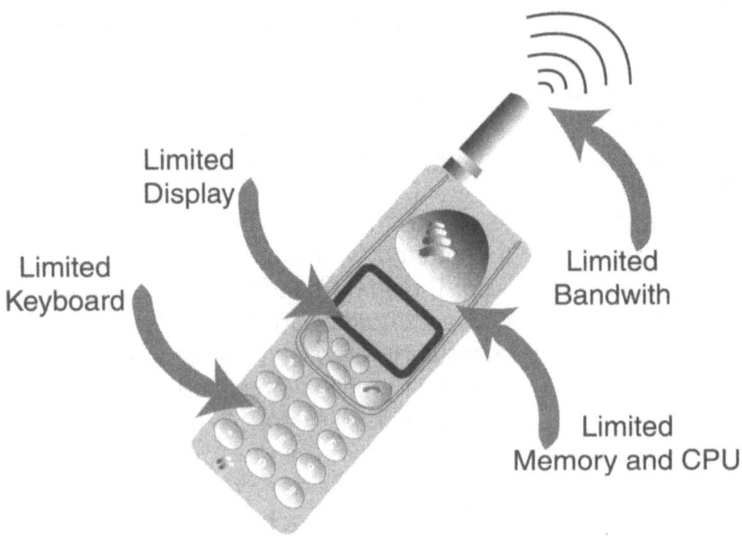

many devices will have smaller displays, for example mobile phones, to provide larger portability. No matter how good these displays will be in the future, the size of the human hand will always limit the size of them. Try to imagine what the experience would be like when accessing a service designed for a device with a big display on a small phone. The result would in most cases be very bad performing services far from what was originally intended. The information the user really wants would probably be drowned in undesired information due to the low perspicuity that the small display of such a device offers.

Instead of using the flat document structure HTML provides, WML structures its document in decks and cards. A card is a single unit of interaction with the end-user, for instance a text-screen, a selection list, an input field, or a combination of them. A card is typically small enough to be displayed even on a small screen. When a service is executed the user navigates through a series of cards. The series of cards used fo making a service is collected in a deck.

- ***Limited Input Facilities:*** Wireless devices do most often not have the same input facilities as their wired equivalents, that is, they lack QWERTY keyboards and have mouse-less interfaces.

WML addresses this issue as well. The elements that are used in WML can easily be implemented so they make very humbl requirements on the keyboard. The use of decks and cards provides a navigation model that call for minimum inter-page navigation since the user is guided trough a series of cards instead of having to scrol up and down on a large page.

Further, soft-buttons are supported by WML in order to provide the service developer with a means to couple desired actions to vendor specific keys.

- ***Limited Memory and CPU:*** Wireless devices are usually not equipped with amounts of memory an computational power (CPU) comparabl to desktop computers. The memory restriction is valid for RAM as well as for ROM. Even though the trend indicates that more memory and more powerful CPUs will be available in a foreseeable future, the relative difference will most probably remain.

WAP addresses these restrictions by defining a lightweight protocol stack adapted to its purpose. The limited set of functionality provided by WML and WMLScript makes it possible to implement browsers that make small claims on computational power and ROM resources. When it comes to RAM, the binary encoding of WML and WMLScript helps keeping the RAM as small as possible.

- ***Limited Battery Power:***
  The stumbling block in wireless communication devices today is the operating time, i.e. the battery power restricts its usage. Even though batteries become better and better, and the radio interfaces are tuned to consume less power, there is still a lot to accomplish in this area.

Access to wireless services will increase the utilisation of bearers (radio interface), and thus will the power consumption also increase. This issue is solved by minimising the bandwidth needed (see above) and thus keeping the bearer utilisation as low as possible.

Roles and Benefits

The operator that runs a mobile network traditionally controls almost the entire value chain for mobile Value Added Services (VAS). Third party alliances are, if existing at all, in most cases restricted to providing the operator with information that it can use as a basis for serving its customers with mobile VAS. Typically, this information can be weather information from the national meteorological office, or stock quote information from a stockbroker. The network operator then conveys this information to the user, either automatically or when the users make an explicit request for it.

This scenario will most probably change when WAP enters the scene. Using the Internet as a service platform opens new possibilities for third party service providers to take part of the value chain at different stages. Third party service providers will be able to create WAP services, put them on the Internet, and thereby making them available to millions of subscribers. They will even be able to create complete suites

Figure 6: Mobile VAS value chain today

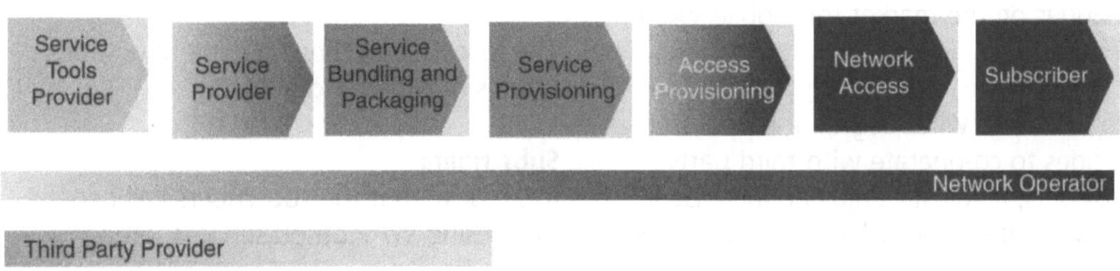

Figure 7: Mobile VAS value chain with WAP

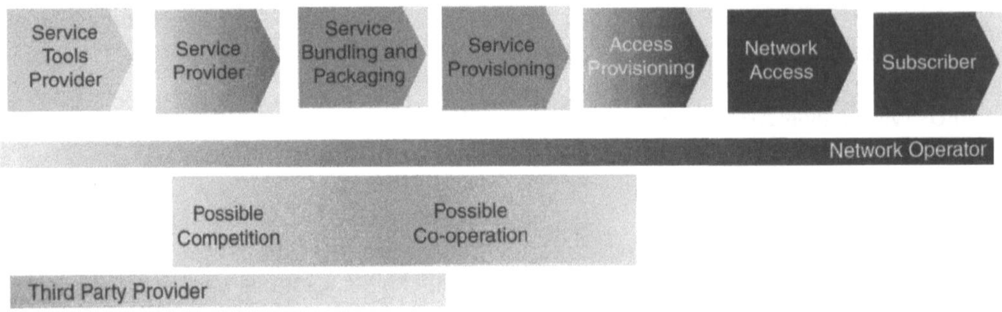

of services and thus also affect the operator's role in bundling services. In co-operation with the operator, third party providers can seize the opportunity to take part of service and access provisioning as well. However, this requires that they make certain investments in network technology. The operators must make important decisions about their role in providing mobile VAS when WAP is rolled out since the role of third party service providers is about to change rather dramatically. With the magnitude of new services that WAP will make available, network operators are unlikely to be able to serve all of its customers with self-made services that attracts each and every one of them. Therefor they have to decide how they should position themselves in the value chain, in order to be able to differentiate themselves from their competitors and have flexibility enough to respond to new preferences among its customers and changes on the market for mobile VAS in general.

No matter to what degree the operator decides to co-operate with third party providers, it will still enjoy an increase in network utilisation, which of course will increase the revenues. Third party co-operation should be considered in order to maximise that utilisation and to provide a well-adapted mix of services thatallows th operator to differentiate itself from its competitors and attract new or underdeveloped market segments. This wi likely reduce churn and improve customer loyalty, and thereby enable increased average revenue per user. As indicated in the figure above, there is a possible risk of competition between the operator and third party providers when it comes to providing services. This since there is no hindrance for either of them to provide similar services.

However, when services like telephony services are considered, i.e. services that require some level of integration with the wireless network, WAP provides the operator with a means to control how these services can be deployed.

The following sections outline how variou groups may gain from WAP.

Subscribers

It is crucial that the subscribers will benefi from using WAP based services, otherwis there will be no incentive neither for WAP as a whole nor for any of the other groups mentioned below. The key-benefits can be summarised as:

- Portability
- Easy to use
- Access to a wide variety of services on a competitive market
- The possibility of having personalised services
- Fast, convenient, and efficient access to services
- To fulfil as many customers needs as possible, WAP devices will be available in various form factors, e.g. pagers, handheld PCs, and phones

Operators

As described in the introduction to this chapter, the network operator can cover the value chain to various degrees. Therefore, many of the advantages mentioned under "Service Providers" are applicable to operators as well. The operator's benefits may include:

- Address new market segments of mobile users by enabling a wider range of mobile VAS
- Deploy telephony services that in contrast to traditional telephony services are easy to create, update, and personalise
- Use the flexibility of WAP as a tool to differentiate from competitors
- Attractive interface to services will increase usage
- Increased revenues per user due to higher network utilisation
- Convenient service creation and maintenance, including short time-to-market
- Replace expensive customer care centres with WAP based services (E-care)
- WAP services are designed to be independent of the network, implying that an operator who runs different types of networks only have to develop its services once
- An open standard means that equipment will be provided by many manufacturers

Service Providers

WAP opens new possibilities for service and content providers since they not necessarily have to come to an agreement with a specific operator about providing services to their customers. The gains are for example:

- Create a service once, make it accessible on a broad range of wireless networks
- Address new market segments by launching innovative mobile VAS.
- Keep old customers by adapting existing Internet services to WAP
- Convenient service creation and maintenance
- Creating a WAP service is no harder than creating an Internet service today since WML and WMLScript are based on well-known Internet technology
- Use standard tools like ASP or CGI to generate content dynamically
- Utilise existing investments in databases etc that are the basis of existing Internet services

Manufacturers

Mobile devices supporting WAP will be available in many different form factors, e.g. cellular phones, pagers, and handheld

PCs. Hardware manufacturers will also need to supply operators etc with equipment, such as WAP Gateway/Proxys and WTA servers. According to WAP Forum, there will be 10's of millions of WAP enabled devices by the end of year 2000. Manufacturer benefits are for example:

- WAP scales across a broad range of mobile networks, meaning that WAP implementations can be used in devices supporting different types of networks
- The expected wide adoption of WAP implies that economies of scales can be achieved, meaning that the huge mass-market can be addressed
- The fact that WAP is designed to consume minimal amount of memory, and that the use of proxy technology relieves the CPU, means that inexpensive components can be used in the handsets
- Reuse the deep knowledge about wireless network infrastructure to develop advanced servers that seamlessly integrates mobile VAS with telephony
- Seize the opportunity to introduce new innovative products

Tools Providers
Today there is a large amount of tools available for creating applications for the web. Content developers have become used to the convenience that tools like FrontPage and DreamWeaver provides. The knowledge of how to develop these tools may be leveraged for developing tools supporting WAP as well. Tools providers will be able to:

- Reuse and modify existing products to support WAP, or even integrate WAP support in existing tools
- Address a new customer base in the wireless community

Services using WAP

When using the Internet from a desktop computer, it is very easy to find new and hopefully interesting services by using search engines, clicking on links and banners, typing in URLs recommended by a friend, etc. This is primarily a fact due to the attractive user-interface (a big screen, a full-sized keyboard, and a mouse). We simply sit down and "surf the net"...

With WAP it is different. While we are on the move, we do not want to go and look for the services we want. We just want the utility they provide, wherever we are. This requires a mindset different to what we are used to from the WWW today. Instead of using advanced search engines and full-fledged portal sites, users will most likely want small portals providing access to the services they really are interested in, no matter if it comes to business or pleasure. This will lead the way to new opportunities for companies that either understand the customers needs very well, or can personalise such portal sites to meet the demands of each and every customer.

So, what kind of services do users want? Of course, most of the services we are used to today can be of interest in the wireless community as well. As indicated above, the key to successfully launching these services is "utility". If the utility is not high enough it is quite unlikely that the services will be widely used. However, one must remember that the utility of, for example, a game might be very high in certain situations. Do also keep in mind that simple-to-use services often are needed to open the door for more advanced ones since the vast majority of the market is not very familiar even with basic mobile value added services today.

Some examples of such services are:

- ***Banking***
  Account statements, paying bills, transfer money between accounts...
- ***Finance***
  Stock quotes, buy and sell stocks, interest rates, exchange rates...
- ***Shopping***
  Buy everyday commodities, books, records...
- ***Gambling***
  Lottery, horse-race betting, poker...
- ***Ticketing***
  Book and/or buy air tickets, cinema tickets, concert tickets...
- ***Weather***
  Weather forecasts, weather on other locations...

In addition, WAP enables a new category of services that we do not find on the Internet today - telephony services. These services will not only bring utility to the user, which she presumably is willing to pay for, they will also increase the operator's revenues due to increased voice traffic if designed correctly. Examples are:

- ***Call management***
  A wide variety of services including:
  - *Incoming call selection*
    Allow the user to choose how an incoming call should be handled. Options could be: answer, reject, forward to assistant, forward to voice mail, etc
  - *Multiparty*
    Provide a comprehensive user interface to multiparty call handling
  - *Call waiting*
    Handle waiting calls with an attractive user interface
  - *Forwarding rules*
    Set and view forwarding rules

- ***Voice mail***
  Provide a menu driven user interface to existing voice mail systems

- ***Unified Messaging***
  Handle e-mail, faxes, voice mails, etc in a unified manner

- ***Enhanced support of legacy SMS services***
  Allows seamless migration from existing text based services into WAP

- ***Attractive interface to DTMF services***
  Increase usage of existing DTMF services by providing a better user interface

- ***Advanced phonebook management***
  Allows the user to update her phonebook, for example, download a corporate phonebook or the personal phonebook managed via a common WWW browser on a desktop computer

In addition, operators will be able to drastically reduce their costs for customer care. Today substantial amounts of money are spent on voice call centres, where people get questions about their bill, features of a service, etc answered. Many operators have today successfully launched WWW based customer care services as well, allowing the users to access support data on-line. These services can also be designed to speed up the process at traditional call centres by, for example, make the user fill out a questionnaire to pin down the problem before the customer care operator is contacted. This approach does however not solve the problem entirely since customers do often not have access to the WWW when they are on the move, and hence they need to call the call centre anyway to find help. With a WAP based customer care service the customers would be able to get help whenever they want, without having to spend a substantial amount of time waiting for their call to be answered.

WAP Forum is today conducting work in several areas that will facilitate mobile VAS, such as persistent storage, use of smartcards, provisioning, external interfaces, billing, data synchronisation, user-agent profiles, etc. Two areas that will have direct impact on the services we will see in the future are push and telematics.

Abbreviations

ASP	Active Server Pages
CGI	Common Gateway Interface
DTMF	Dual Tone Multi Frequency
GPRS	General Packet Radio Service
HDML	Handheld Device Markup Language
HDTP	Handheld Device Transport Protocol
HTML	HyperText Markup Language
HTTP	HyperText Transport Protocol
ITTP	Intelligent Terminal Transfer Protocol
MAC	Message Authentication Code
OSI	Open System Interconnection
PDA	Personal Digital Assistant
PDU	Packet Data Unit
RAM	Random Access Memory
ROM	Read Only Memory
SMS	Short Message Service
TCP/IP	Transmission Control Protocol/Internet Protocol
TLS	Transport Layer Security
TTML	Tagged Text Markup Language
UDP	User Datagram Protocol
URL	Uniform Resource Locator
WAE	Wireless Application Environment
WAP	Wireless Application Protocol
VAS	Value Added Service
WCMP	Wireless Control Message Protocol
WDP	Wireless Datagram Protocol
WML	Wireless Markup Language
WSP	Wireless Session Protocol
WTA	Wireless Telephony Application
WTAI	Wireless Telephony Application Interface
WTLS	Wireless Transport Layer Security
WTP	Wireless Transaction Protocol
WWW	World Wide Web
XML	Extensible Markup Language

Wireless Data Connectivity

Title: *Wireless Data Connectivity*
Author: *Zeus Wireless, Inc. Columbia, Maryland*
Abstract: *This whitepaper will discuss the issues and opportunities of wireless communications in data connectivity applications, with a focus on the benefits of frequency hopping spread spectrum technology.*

To learn more, contact:
Zeus Wireless, Inc.
8325 Guilford Road, Columbia, MD 21046
Phone: (410) 312-9851, Fax: (410) 312-9852
Website: zeuswireless.com, E-mail: info@zeuswireless.com.Wireless Data Connectivity -6- Zeus Wireless, Inc.

Preface: Wireless Technologies

The timely collection and transmission of data will improve the productivity of any business. But it is often too costly or impractical to collect and transmit this data over wires or telephone lines. Therefore, businesses are increasingly turning to wireless alternatives to link dispersed equipment and personnel.

The principal wireless technologies are Infra-red and Radio Frequency (RF). Infra-red cannot operate over distances of more than a few feet. It is limited to applications such as bar code scanning and television remote control. Given this range limitation, most commercial products for wireless applications utilize RF technology instead.

Standard RF is the same technology used in vehicle dispatch, police communication, and citizen band radios. These products are relatively simple and inexpensive to build. However, they typically require Federal Communications Commission (FCC) licenses to operate. Standard RF transmissions also are susceptible to interference from a growing number of sources, including other radios, as well as to interception by readily available eavesdropping equipment. Consequently, this technology is unsuitable for applications where every bit of information transmitted must be accurate, complete, and secure.

Direct Sequence Spread Spectrum (DSSS) is one alternative to standard RF. Direct sequence radios "slice" transmissions into small bits, thus spreading the energy of these bits simultaneously across a wide range, or spectrum, of radio frequencies. But direct sequence is a relatively unreliable transmission medium, because spreading the message greatly reduces the strength of the radio signal carrying it. To find this weakened signal, a receiver must "listen" simultaneously to the entire allotted spectrum and risk severe interference by any high-energy RF source that appears within it. Direct sequence performance also degrades quickly in shared-service environments in which multiple radios operate.

Frequency Hopping Spread Spectrum (FHSS) technology was developed by the U.S. military to prevent interference or interception of radio transmissions on the battlefield. It is employed by the military's best tactical units in situations where reliability and speed are critical. Standard RF and direct sequence cannot match the reliability and security of frequency hopping. Instead of spreading (and diluting) the signal carrying each bit across the allotted spectrum, as in direct sequence, frequency hopping radios concentrate their full power into a very narrow signal and randomly hop from one frequency to another within that spectrum up to several hundred times per second.

Frequency hopping transmitters and receivers coordinate this hopping sequence by means of an algorithm exchanged and updated by both radios on every hop. If they encounter interference on a particular frequency, the radios retain the affected data, randomly hop to another point on the spectrum, and then continue the transmission. There are always spaces without interference somewhere in the spectrum. A hopper will find those spaces and complete the transmission. This ability to avoid interference enables frequency hopping radios to perform more reliably over longer ranges than standard RF or direct sequence products.

Wireless Data Connectivity

Data connectivity involves the transmission of short packets of information from equipment or sensors to a recorder or central control unit. These data packets are transferred as electronic signals via wire, infra-red or RF technologies. Data is received at a central control unit, typically a computer with software that automatically polls and controls the remote devices. This control unit analyzes, aggregates, archives and/or distributes the collected data to other locations via a local area network (LAN) and/or a wide area network (WAN).

Wireless data connectivity offers several advantages over wire. First, wireless systems are easier to install. Second, installation and maintenance costs are lower than with wire. Third, operations can be reconfigured or relocated without re-wiring. Fourth, wireless data connectivity offers improved mobility. Overall, wireless connectivity solutions are more practical and affordable than wire.

Today's Fortune 1000 corporations, government agencies, and other large organizations have overwhelmingly adopted wireless technology. However, the use of wireless technology is rarely uniform. For example, many organizations routinely use it to make telephone calls and move large volumes of information around the world. Yet within these same organizations, data exchange between factory or office equipment is likely to be performed by hand or wire. This disparity in the utilization of wireless technology results in substantially reduced productivity and increased costs.

Opportunities for License-Free Wireless Operations

The Federal Communications Commission (FCC) has designated three, license-free, bandwidth segments of the RF Spectrum and made them available for Industrial, Scientific and Medical (ISM) use in the United States. These three segments are 900MHz, 2.4GHz and 5.8GHz. No one "owns" these license-free frequency bands. Anyone may operate a wireless network in a license-free band without site licenses or carrier fees. Operators are subject only to power restrictions (1 Watt or less) and the type of radio signals transmitted (spread spectrum). Other national and international telecommunications bodies have also agreed to recognize a common, license-free ISM frequency.

2.4GHz: The ISM Band of Choice

Worldwide, 2.4GHz has become the de facto standard for license-free ISM communications. This is not surprising. The ISM band at 2.4GHz has more than twice the bandwidth capacity and is subject to far less congestion and interference than the ISM bandwidth at 900MHz. Several industrial nations do not even have an ISM band at 900MHz. Only a few have such a band at 5.8GHz. But the United States, Europe, Latin America, and many Asian countries have adopted ISM bands at 2.4GHz. It is the only band so many nations offer for license-free operations.

Frequency Hopping Technology

Frequency hopping solutions are ideal for license-free bands because hoppers continue to operate under conditions that shut down other technologies. Frequency hopping wireless data connectivity systems put full power into a narrow signal which randomly hops around interference. Transmitters and receivers coordinate the hopping sequence to update each radio. This allows the hopper to find space within the spectrum where there is no interference and to reliably deliver real-time data. Currently, no other technology compares to frequency hopping for reliability and security.

The Right Speed for Wireless Data Connectivity

There is a common misconception regarding speed and data connectivity: specifically, that faster is better. In computer and telecommunications networks, higher speeds are unquestionably superior. However, this is not the case with wireless data connectivity.

Many industrial and commercial applications deal with lower-speed measurement and control functions. Paying for more speed than what is necessary to implement a specific application can be counter-productive. This is because the broader RF bandwidth needed to transmit higher data rates dramatically reduces range (see figure 1).

In contrast, narrow bandwidth radios operate at lower data rates, such as 9.6Kbps (9600 Baud). They are ideal for intermittent, repetitive data transfer, where the premium is on accuracy and long-range transmission (up to 1,500 feet indoors and line of sight to the horizon outdoors). The following chart clearly shows the superiority of narrow bandwidth operation at 9.6Kbps for many wireless data connectivity applications.

Figure 1

Assumptions	Bandwidth	Range
Outdoor line-of-sight transmission Omni-directional antennas (0dBi) 500W transmit power 6dB noise figure for reception +20 dB detected signal strength to noise (+25dB for 1,6 Mphs only) Link frequency equals 2.442GHz	1MHz for 1.6Mbps	3,081 meters
	1MHz for 800Kbps	5,479 meters
	230KHz for 115Kbps	11,425 meters
	25KHZ for 9.6Kbps	34,655 meters

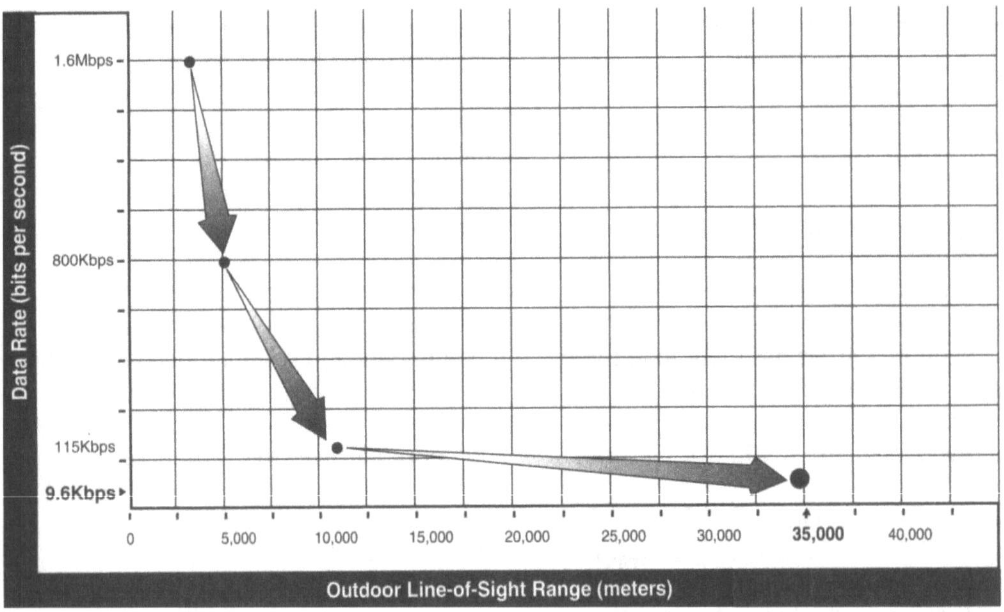

Typical Data Rates

There are hundreds of factory automation, distributed control, remote monitoring, security and alarm, building access, financial transaction, inventory management and health care applications that are ideal for wireless data connectivity. Examples of these applications and their typical data rates follow:

Table 1

Application	Data Size	Data Rate
Access Control	100 Bytes	1.2Kbps - 9.6Kbps
EFT/Point-of-Sale	100 Bytes	1.2Kbps - 4.8Kbps
Inventory Management (Bar Code)	150 Bytes	9.6Kbps
Remote Sensors	100 Bytes	1.2Kbps - 9.6Kbps
Perimeter Security	100 Bytes	1.2Kbps - 9.6Kbps
Vending	500 Bytes	1.2Kbps - 9.6Kbps

Access Control
Wireless transceivers enable remote devices to be connected to a central PC or controller, without the expense or difficulty of wiring.

EFT/Point-of-Sale
Wireless transceivers enable POS terminals to be connected to a central PC or controller without wire. This lets merchants easily add or move POS terminals or printers.

Inventory Management
Wireless transceivers let portable terminals communicate to a central database. Terminals log in shipments of materials from vendors and track those materials in inventory as they are needed.

Remote Sensors and Distributed Control Systems
Wireless transceivers let organizations economically aggregate and report data from remote sensors and replace wire in distributed control systems, indoors and outdoors.

Perimeter Security
Wireless transceivers, communicating to a central PC or controller from various locations, allow businesses to reliably secure indoor or outdoor facilities.

Vending Equipment
Wireless transceivers enable vending equipment operators to minimize product restocking time and maximize cash collection from multiple sites, such as within a large office complex.

Potential applications include any product that transmits or receives data, and any data connectivity network that requires easy deployment or frequent reconfiguration.

Long-Range Solutions from Zeus Wireless
Zeus Wireless, Inc. provides wireless data connectivity solutions for OEMs, VARs, and integrators in need of reliable, secure, long-range, and low-cost alternatives to wire for commercial and industrial applications. Zeus wireless data connectivity products provide the first practical and affordable alternatives to wire for industrial and commercial applications where data must be transmitted reliably and securely in a wide range of indoor and outdoor environments. Zeus products can be linked with existing computer, telephone, and wireless networks—including Ethernet and the Internet—to any device with an IP address.

Zeus Transceivers utilize frequency hopping technology developed by the military and refined by intelligence agencies to avoid interception and interference, making them more reliable and secure than wire. They operate in the license-free Industrial, Scientific and Medical (ISM) band at 2.4GHz and use no sole-source components, making them significantly less expensive than other narrowband wireless options. The design of Zeus Transceivers enables them to transmit data packets 500 to 1,500 feet indoors and line-of-sight to the horizon outdoors, the longest range of any 2.4GHz wireless device.

Pairs of transceivers operate point to point without a host. Zeus Network Software enables transceivers to communicate with a host PC from more than 200 remote locations. Transceivers also can relay data to and from locations that are too far away to communicate directly with a host. Zeus Configuration Manager software enables users to support various wireless data network configurations and accommodate different application requirements.

Conclusion

Zeus is building close working relationships with leading firms in large data connectivity markets. OEMs, VARs, and integrators can profit from the growing demand for secure and reliable wireless connectivity solutions that can be incorporated into customer applications easily and inexpensively. Zeus' business partners benefit from our years of experience developing and deploying frequency hopping and other RF technologies for clients who require the most reliable and secure performance possible.

Appendix: How Zeus Addresses Indoor RF Signal Propagation

In developing the first wireless data connectivity products to provide practical and affordable alternatives to wire, Zeus engineers addressed and resolved a number of important technical challenges. One of those challenges was designing Zeus Transceivers to overcome the problems of indoor RF signal propagation. Propagation is the technical term for the transmission of a radio frequency (RF) signal from the transmitter to the receiver. Indoors, it is impossible to predict RF signal propagation reliably. However, by understanding how RF signals travel inside a building it is possible to estimate the usable range of wireless data systems and minimize the chance of link failure.

To develop such an estimate, it is common engineering practice first to determine how much of the transmitter's output will not reach the receiver—this is referred to as path loss. Within a building, the path loss between a transmitter and receiver depends on a variety of factors. These factors include RF power, the type and placement of antennas, the type of building materials used, and the architecture of the building itself. Let's look at these factors before discussing how the quantification of path loss can help estimate indoor RF signal propagation.

RF Power and Antenna Factors

The range of a wireless data link depends fundamentally on the amount of RF power the transmitter sends toward the receiver. That is one reason why, for example, a Zeus Transceiver can automatically adjust its transmitter RF output power. Another fundamental factor is the type of antenna used. Antennas are essential components of the wireless data link, serving as the physical interface between a transceiver and "free space" (its surroundings).

In most applications, a transceiver is supplied with an antenna that transmits and receives equally well in all directions—an omni-directional antenna. Using omni-directional antennas eliminates the need for the antenna at each end of a wireless data link to be pointed at the other antenna. Omni-directional antennas do not send most of the transmitted RF energy in the direction of the receiver, but this is more than compensated for by the convenience to the end user of being able to relocate the transceiver without re-pointing the antenna. For applications where an extended range is critical, Zeus can supply directional antennas that can increase useful range by several-fold.

Building Factors

Radio transmissions often must pass through various parts of a building before reaching the antenna of the intended receiver. The interaction of radio signals with building materials usually results in phenomena that serve to reduce the effective range of the wireless data link. RF energy leaving the transmitter antenna will be partially absorbed and partially reflected by the building materialsencountered. Zeus Transceivers are designed to minimize the impact of these effects and permit continuing robust data connectivity.

Absorption of RF energy by building materials turns part of the radio signal into

Figure 2: Schematic respresentation of RF signals attenuated and reflected by building materials

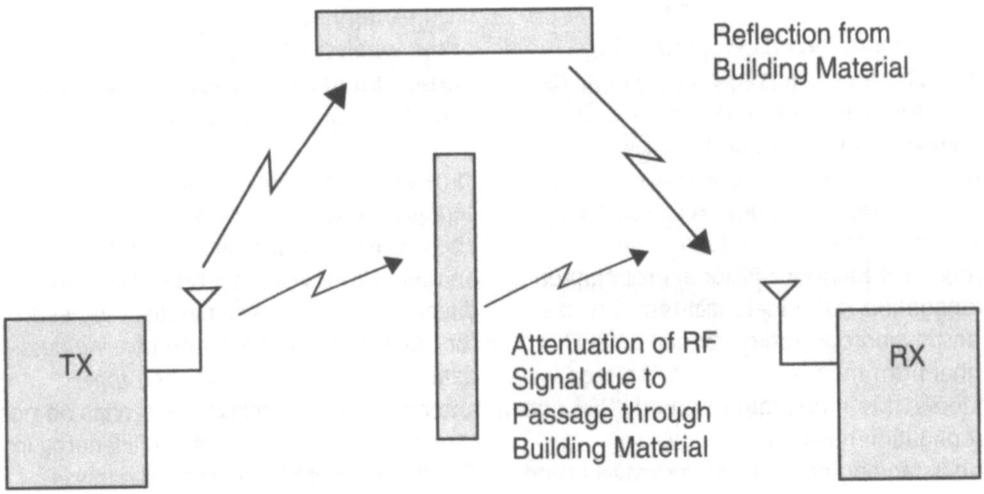

heat and allows the remaining part to continue outward, away from the transmitter. Reflection of RF energy "scatters" the incoming RF energy in directions dependent on the relative geometry of the transmitter and reflector. When struck, building materials both absorb and reflect portions of the RF energy. For example, a 2.4GHz wireless data signal hitting a concrete wall will have part of its energy reflected, part of it absorbed—the remainder will continue through the wall.

The reduced usable range of an indoor wireless link can be analyzed as the result of two distinct effects: increased path loss and multi-path fading, which we address next.

Increased Path Loss

In studying a radio link it is common engineering practice to first determine how much of the transmitter's RF output power will not reach the receiver—this is referred to as path loss. Within a building, the path loss between a transmitter and receiver depends on a variety of factors. These factors include the radio frequency used, the type of building materials used, and the architecture of the building itself.

As path loss increases, RF power into the receiver decreases. If path loss is constant, an increase in transmitter power will cause a corresponding increase in the power at the receiver input. Path loss within buildings has been shown empirically to increase much more rapidly as distance increases (compared to outdoor propagation). This additional indoor path loss means reduced range. For example, a wireless data link that works over a clear line-of-sight distance of ten miles outdoors might only work over a range of 1,000 feet inside an office building.

It is possible to mitigate indoor path loss by increasing the transmitter power or by reducing the received power needed by the receiver. Increased transmitter power is only available until practical and regulatory limits are reached. Of much greater benefit is reducing the data rate of the link, thus reducing the power needed by the receiver to recover the transmitted data. In general, for the same RF power transmitted, a reduced data rate will increase useful range. For example, the Zeus Transceiver, when providing a wireless data link of 9,600 bits per second, achieves ranges several times greater than competing products with equivalent transmitter output power but higher data rates.

Wireless Data Link Range Estimation
The following "rule of thumb" can be applied to yield an approximate idea of the path loss, and thus the useful range of an indoor wireless data link. The quantitative values of RF power ratios are expressed in decibels (dB).

Indoor path loss in the 2.4 GHz ISM band can be estimated to be:
50 dB for the first 20 feet, and 30 dB for each additional 100 feet.

Example:

QUESTION: Can we expect to establish an indoor wireless data link with the following characteristics?

Table 2

Range	320 feet (Using the above "rule of thumb" this implies a path loss of about 140dB)
Receiver sensitivity	-120 dBm (120 dB less than one milliwatt)
Transmitter output power	500 mW (27 dBm, i.e., 27 dB more than one milliwatt)
Antennas	omni-directional at each end of the link (0 dBi, i.e. essentially isotropic antennas)

TX power + antenna gains – path loss = power at RX input

27 dBm + 0 dBi + 0 dBi –140 dB = -113 dBm

-113 dBm > -120 dBm (receiver sensitivity)

ANSWER: Since the RX power is greater than the receiver sensitivity, we expect the wireless data link to work over this range.

Multi-Path Fading

Multi-path fading can also cause performance problems in a wireless data link. When RF power from a transmitter takes two or more separate paths to the receiving antenna, the power, when summed at the antenna output, can be less than the power in each of the paths individually. The result is a much weaker signal for the receiver to process, an event referred to as a "fade." Moreover, a wireless data link may suffer from severe multi-path fading at one frequency yet function unaffected at another frequency.

It is virtually impossible to predict the multi-path fading environment that a wireless data link will encounter in any particular indoor installation. Therefore, it is vital to have enough frequency diversity to overcome multi-path fading. The Zeus Transceiver uses an intelligent frequency hopping technique to automatically minimize the effects of multi-path fading.

Summary

By making predictions of RF propagation based on experience, the maximum indoor range of a wireless data link can be estimated. Implementation of features such as automatic transmitter power adjustment, intelligent frequency hopping, and lower data rates can help overcome the primary impediments to indoors RF propagation. Zeus Transceivers are designed to address these real-world issues effectively and affordably.

Note: Using Decibels to Quantify RF Power Values

In assigning quantitative values to RF

Table 3

Ratio of two power levels	Power ratio expressed in decibels (dB)
2	3 dB
10	10 dB
20	13 dB
100	20 dB
10,000	40 dB
100,000	50 dB
100,000,000,000,000	140 dB

power ratios the decibel is commonly used. The decibel simplifies otherwise numerically unwieldy expressions of the ratios between two power levels.* The tables 3 and 4 show the clear benefit of using decibels to represent the numbers in a typical RF path calculation:

Table 4

	In milliwatts (mW)	In Decibel Notation
Transmitter Power	500 mW	27 dBm**
Receiver Sensitivity	0.000000000001 mW	-120 dBm
Estimated Path Loss	1/100,000,000,000,000	140 dB

Oracle Portal-to-Go, Any service to any device

Title:	*Oracle Portal-to-Go, Any service to any device*
Author:	*Oracle Corporation*
Abstract:	*The basic problem of Internet today is that all its content is designed to be accessed from a PC with the latest release of Internet Explorer or Netscape Communicator. Everybody who has tried to access the web from a wireless connected PDA (personal digital assistant like the 3Com Palm or Psion Revo) knows that either it doesn't work at all because the PDA browser cannot handle the complex page, or that after a painful long loading time, the page cannot be understood or properly rendered because of the small screen of the PDA. The mobile phones represent another problem. Even if WAP solves the problem of connecting the Mobile Phone to the Internet and to overcome the problem with limited bandwith, it creates a new problem by assuming that Internet content be in a new format specific to WAP. There's another problem: PDA's have no keyboard. The solution? Maybe Oracle's Portal-to-Go adresses all of these problems.*

Introduction

Oracle Portal-to-Go is new server technology that revolutionizes the use of mobile devices, such as Palm OS, Windows CE and EPOC handheld computers, personal computers, and wireless phones connected to the Internet. By transforming existing Internet content to any native format for a specific mobile device, Web services that previously could only be accessed from PCs are now accessible from literally any device.

In addition to sophisticated, personalized content transformation, Portal-to-Go also provides a portal interface for the user. The portal allows users to define, organize, and personalize the services that they access from the mobile device.

Portal-to-Go was formerly know by the Oracle codename "Panama".

Background

The expanding cellular phone market, with market penetration in some countries approaching 70%, has created the need to drive data, as well as voice, through cellular networks. Ideally the phone itself should be a terminal for accessing Internet information. During the past few years companies such as Nokia and Phone.com have created proprietary solutions to connect mobile phones to the Internet, and to enable Internet content to be accessed from the phone. Now the world of telecommunications has come together in the new WAP (Wireless Application Protocol) standard.

The market for small handheld devices such as Palm OS, Windows CE and EPOC handheld computers has exploded. It makes sense for these devices to communicate with the Internet over a wireless network as well.

Oracle Portal-to-Go, Any service to any device

Problem

The basic problem is that the Internet today is designed to be accessed from a PC with the latest release of Netscape or Internet Explorer. Most content assumes a high-resolution screen with sophisticated graphical capabilities.

Anyone who has tried to access the Web from a wireless connected PDA knows that either it does not work well because the PDA browser cannot handle a complex page, or that after a painfully long loading time, the page cannot be properly rendered because of the PDA's small screen.

Mobile phones represent another problem. Even if WAP solves the problem of connecting mobile phones to the Internet and overcomes the problem of limited bandwidth, it creates a new problem by assuming that Internet content must be in a new format specific to WAP. This creates a chicken-and-egg situation. Who will buy a WAP phone if there is no content and who will create content if there are no phones on the market?

Another problem common to many PDAs is that they lack a keyboard. This makes it difficult to enter information. The most frequent typing required when browsing the Internet is the entering of URLs; doing this on a mobile phone or with handwriting recognition systems is challenging to say the least.

Existing Solutions

The solution today is to create special versions of a Web site for each type of device. Since very few Web sites provide device-specific content, this only allows access to a small fraction of the Internet. In the WAP world it is assumed that the WAP service providers should convince content providers to build WAP content. There are no real solutions today to the problem of entering information.

Oracle's Solution

Oracle Portal-to-Go addresses all the problems described above and makes the Internet-connected mobile device a reality.

Oracle Portal-to-Go is capable of dynamically transforming existing Internet content to a generic XML format, and then generating any device-specific output desired.. Examples of markup languages supported are HTML, WML, HDML, TTML, Pager text and VoxML. This approach makes the creation of services very simple. When you create a Web service, it is immediately available on all mobile devices. In addition to the service creation environment, Oracle Portal-to-Go has a personalized portal that is the natural starting page of any mobile Internet-connected device.

Figure 1: Portal-to-go architecture

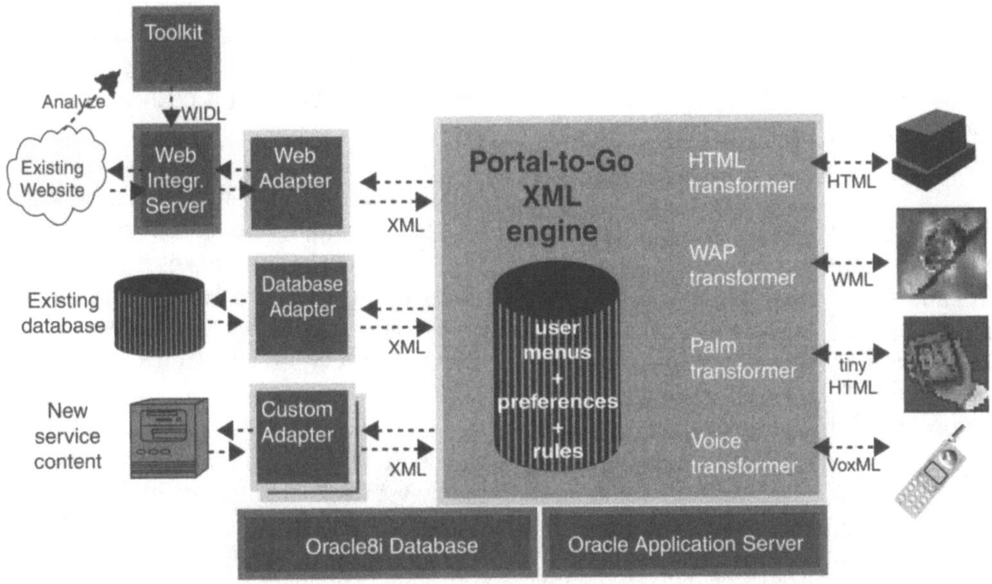

This approach of a generic Web service description in XML also makes the service available in any markup language spanning from WML for mobile phones, to HTML for traditional PC browsers, to VoxML for IVR (Interactive Voice Response) systems. This means that once a service is created, the same information may be accessed through a mobile phone, PDA, Pager or even a PC. The creation of a service is a very simple task that includes analyzing the Web site to be used. The analysis can be done either by a system administrator for more complex transaction-based Web services or by the end user for simpler types of information services.

The architecture of Portal-to-Go allows the creation of extremely advanced services. With the service creation tools of Portal-to-Go, setting up literally any Web-based service, including secure (SSL) services, for mobile access becomes a painless, fast, and cost efficient process. For example, making an online banking Web site a mobile service for all the above mentioned mobile technologies is a couple of hours work for one person.

Portal-to-Go contains portal functionality that allows end users to personalize their start page and the services that they can access from their mobile device(s). Using a browser, users can log on to the portal and add their own favorite links to be used from the mobile device. Users can organize the services or the links as they prefer, effectively creating their own personalized menustructure. Much of the input needed when accessing Internet content includes entering personal information such as email address, frequent flyer numbers, pin codes, etc. The user can personalize individual services by deciding whether information should be entered from the mobile device or stored securely on the portal as part of a personalized Internet service.

There are also interfaces to non-Internet related information such as any database data and any XML-based data. Portal-to-Go has a simple, well-defined XML-based API that allows anyone to create mobile applications. Since Portal-to-Go has knowledge of the different markup languages and devices for the mobile market, content developers do not need to know anything about WML and the mobile devices; Portal-to-Go takes care of that. When new phones and markup languages are developed it is just a matter of downloading the new device transformer and loading it into Portal-to-Go to support the new device without rewriting or changing anything in the existing content or portals

In addition to the described request /response model, Portal-to-Go has an asynchronous push mechanism to allow push of any Portal-to-Go service to the mobile device. Portal-to-Go includes push gateways to SMS and email. Since Portal-to-Go handles all data in XML, event-based push can easily be configured from the end user portal by setting conditions on the data returned from a service before it is pushed to the user.

Portal-to-Go has XML interfaces allowing for provisioning, billing, and monitoring. The billing mechanism makes it possible to create billing records differentiating user groups and information value. The monitoring is based on industry standard SNMP.

Oracle Portal-to-Go is implemented using highly scaleable architecture based on Oracle8i, Java and XML. This architecture allows it to be deployed at small corporations wanting to create their own

Figure 2: The Nokia 7110, a new generation WAP-compliant featurephone. On the screen is a section of th demo shown at Telecom '99 in Geneva, Switserland

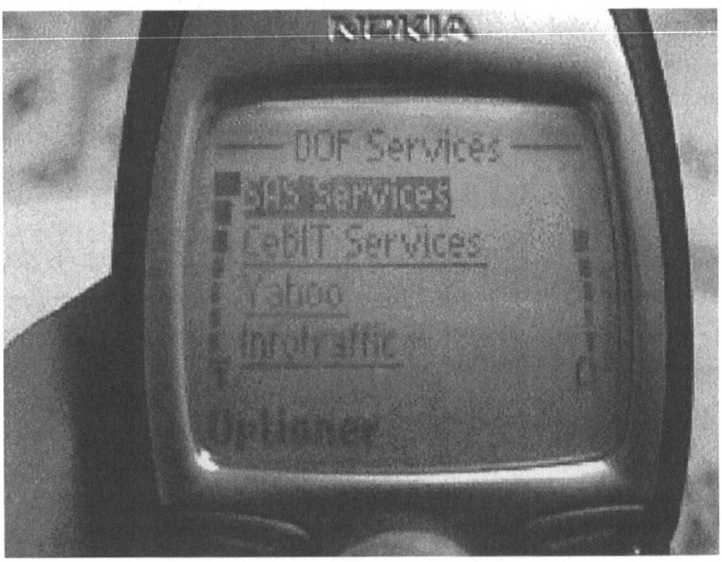

mobile content as well as at very large mobile operators with millions of users.

References

Oracle developed the original prototype of Portal-to-Go over a period of two years together with Telia Mobile, Sweden's largest mobile operator. Telia Mobile offers service to their customers using the Nokia TTML technology, and is a now a trial user for WAP phones using the Portal-to-Go beta. Telia is offering services such as online banking, travel services including airline reservations, news services from CNN, and real-time traffic information services. Portal-to-Go Version 1 is currently tested by leading carriers and content providers in Europe, Asia, and the US.

Availability

Oracle Portal-to-Go Version 1 is now generally available.

Wireless IP – A Case Study

Title: *Wireless IP – A Case Study*
Author: *Peter Rysavy, Rysavy Research*
Abstract: *What if field workers of a public utility had online access to inventory databases, work orders, maps and other essential information from anywhere? What if crew chiefs had access to e-mail and schedules without having to return to their offices? This is the vision that the City of Seattle Public Utilities is making a reality in a project spearheaded by its Information Technology Division. This case study shows not only the issues the utility has faced and the solutions it has found; but, more importantly, how the lessons learned can be applied by almost any organization today to make wireless data an effective and successful tool.*

Biography: Peter Rysavy is the president of Rysavy Research, a consulting firm that works with companies developing new communications technologies and those adopting them.

Figure 1: The utility plans to make applications and information on its intranet available to field workers

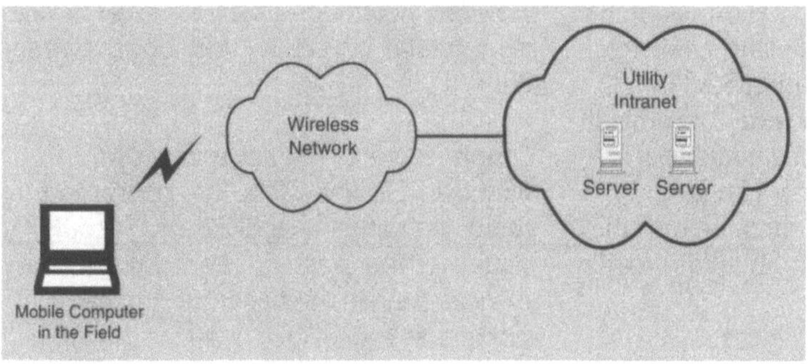

Introduction

The applications the utility is extending to its mobile workers include both work management and office applications, a combination common to many organizations. Developing wireless networking solutions requires special considerations. The utility has identified effective approaches, is about to proceed with a pilot program, and has a plan that accommodates the inevitable changes in applications, platforms and wireless technologies.

Goals

The utility's work and inventory management application is MAXIMO*, a system that uses an Oracle database, developed by PSDI (Bedford, MA, http://www.psdi.com/). Since MAXIMO is developed for specific types of job functions, it can be considered a vertical market-type application. From the point-of-view of providing wireless access to an enterprise database, however, it is similar to any number of other applications based on SQL (Structured Query Language), one of the most common protocols used today for client/server applications.

The office-based applications that the utility intends to extend to the field include Novell's GroupWise* and Web-hosted applications on the utility's intranet. These are horizontal market applications in that their use is not restricted to any particular job function or type of industry. In addition, the utility plans to make a GIS (graphical information system) available eventually, though it realizes that current wireless networks are not well-suited for such intensive graphical content. The goal for all these applications is to provide reliable remote operation with preferably the same user interface as a direct LAN connection. Though slower response times are acceptable and somewhat inevitable, applications must operate in a reliable and effective manner.

The utility has two types of remote workers who will use the wireless system: leads that will use MAXIMO primarily and crew chiefs that will use the office applications in addition to MAXIMO. Because it is difficult to predict what applications may be needed in the future, a key goal is to provide a flexible wireless architecture that allows new applications to be added easily.

Another goal is security. Wireless connections should be no less secure than existing remote access methods based on dial-up connections. Finally, while the utility is willing to commit to a particular wireless technology in its initial deployment, it wants an approach that allows it to migrate easily to other wireless technologies in the future.

Choosing the Computing Platform

The utility recently adopted the Microsoft Windows* 95 platform across multiple departments. For the wireless IP project, it needed to decide whether to use Windows 95 notebook computers or to consider a somewhat more specialized platform such as Windows CE.

Though MAXIMO client software is not available for Windows CE, Windows CE was an option because Syclo Corporation (Barrington, Illinois) supplies middleware that enables Windows CE computers to access MAXIMO databases through a gateway. Using Windows CE would have provided advantages such as lower device cost, greater portability and longer battery life.

Despite some of the advantages of Windows CE, the utility was concerned about the range of applications it could deploy on the platform. Because the utility expects the requirements for mobile workers to evolve over time, and for the types of work performed in the field to expand, it needs the greatest degree of flexibility possible for the types of applications it can deploy. For this overwhelming reason, the utility chose Windows 95 over Windows CE. In addition, because the computers are mounted in the vehicles and not used outside the vehicles, the extra portability Windows CE was not a factor. Finally, ther are a number of ruggedized laptops available that can address the demanding field conditions that utility workers encounter.

Choosing the Wireless Network

The utility faced a bewildering situation when it began evaluating the wireless networks available. There was the analog cellular network, new digital cellular and PCS technologies, and four wireless packet networks with service in the Seattle area.

The utility decided to base their applications on TCP/IP communications, so this quickly disqualified the BellSouth Wireless Data and ARDIS networks which do not directly support TCP/IP. Moreover, the utility believed that a packet-based approach would better support the frequent communications that workers in the field require. This requirement eliminated circuit-switched cellular connections. Since packet-based services are not yet available for digital PCS networks, the remaining choices were CDPD and the Metricom Ricochet* network. CDPD and Metricom Ricochet* are both IP-based packet networks. However, data services for GSM and CDMA digital PCS networks are expected to be deployed in the 1999 time frame and so may be candidates in the future.

Since wireless data services are evolving rapidly, the utility decided to implement an architecture that insulates its applications from the actual network used to the maximum extent possible. Using an IP-based approach, where applications make no assumptions about the nature of the physical connection, achieves this goal. This is not unlike Internet-based communications, where packets may flow across copper cable, one moment; fiber optic cable, the next; and a satellite. It should be possible to deploy applications using one wireless network; and with minimal effort, migrate the application to another wireless network in the future, should that network become more desirable.

Migrating between network types is indeed possible, though some adjustments may be necessary for each network. For example, CDPD uses fixed IP addresses and Metricom Ricochet uses dynamically assigned addresses. This difference could affect how firewalls are configured. The effective throughput rates of Ricochet and CDPD also differ, with Ricochet operating at 20 to 30 Kbps and CDPD at about 10 Kbps.

Figure 2: The utility chose a wireless network that is IP-packet based for greatest flexibility

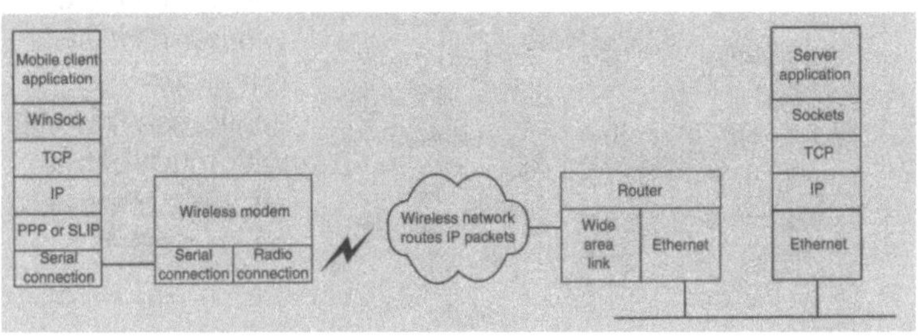

Software Approaches

In an ideal world, a computer connected over a wireless network would work just like a computer on a LAN. But wireless networks operate at lower speeds with higher latency, and connections can be lost at any moment, especially when mobile. The utility has considered a number of software approaches, seeking to strike a balance among these factors: ease of use, performance, reliability, and cost of deployment. To complicate matters, it discovered that the best approach for supporting one application is not always the best approach for another.

The first approach is to use all applications in their native form, with client software installed on the mobile computer and communicating using TCP/IP protocols. Because some workers will be working with the same applications both in an office environment and in the field, the advantage of this approach is that the user interface stays the same in both environments. Also, IT managers can set up mobile computers in the same way as desktop computers. A disadvantage is that this approach does not address some of the connectivity issues associated with wireless, such as throughput and latency. Another disadvantage is the requirement for software installation on field computers, which can add to maintenance and support.

Another approach is to use Citrix MetaFrame* (combined with Microsoft Terminal Server), where applications run on an application server at a central location, and mobile nodes operate as terminals (thin clients). The utility has already deployed Citrix MetaFrame to support dial-up users. The advantage of this approach is that installing and maintaining mobile computers is simplified because they only need the Citrix client software to access multiple applications. The disadvantage is that Citrix MetaFrame has some significant limitations when operating over wide area wireless connections. We learn about these limitations in the next section when we look at test results.

The third approach is to use wireless middleware (specialized software installed on a mobile computer and on a centralized server that acts as an intermediary between client applications and server processes) to optimize communications. The utility has looked at wireless middleware designed specifically for MAXIMO, as well as general purpose middleware that optimizes IP communications over wireless links. The advantage of wireless middleware is it allows applications to run with much better response times and much greater

Figure 3: Wireless middleware adds a mobile server that handles transactions on behalf of the mobile client

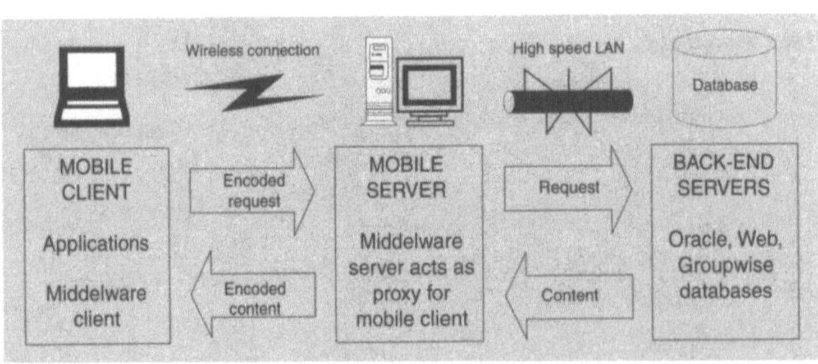

reliability, however, it increases complexity and adds cost.

Because wireless coverage is not always available everywhere the workers spend time, an approach also considered was Oracle Lite where workers can download a subset of the database they need, operate on it locally during the day, and then synchronize at the end of the day. This approach reduces the demand for wireless connectivity, but it is not as flexible as the other approaches where field workers remain in constant communications during the day and can respond quickly to changing circumstances.

By using a flexible computing platform such as Windows 95 on a laptop, the utility realized it could also consider a mix of approaches. Perhaps one application would work best in its native form, and another would work best using a thin-client approach. This indeed was the case as we see next.

Test Results

Whereas architecting different wireless approaches on paper may be an entertaining diversion, it is difficult to predict how the approaches will actually perform until tested in the real world. The utility has tested the architectures discussed earlier with results that did not always match expectations. For instance, the utility expected that an IP-based client would perform reasonably well over a wireless IP connection. This was the case for certain applications but not for others, which performed better using a different approach. Testing emphasized three scenarios:

1. Given the slower speed of wireless connections, what are the issues when starting (and restarting) sessions and applications?
2. How do applications perform once started, under normal operating conditions?
3. What happens when a connection is lost due to driving outside a coverage area or to strong interference?

Interestingly, every application and every software approach performed somewhat differently under the three different test scenarios.

Here are the various configurations and how they performed.

Remote IP-based Clients

The first configuration tested was with remote clients, specifically MAXIMO and Web browser clients installed on laptops using TCP/IP communications. The version of GroupWise used at the time did not provide a TCP/IP client, so GroupWise could not be tested using this configuration.

Because both Metricom Ricochet and CDPD are based on IP, the applications operate in the same fashion as if installed on LAN-based workstations using TCP/IP protocol stacks. What is different, of course, is the slower speed of wireless connections. Also, the mobile nodes are not necessarily always in wireless coverage. The first comprehensive series of tests used the Metricom Ricochet network. Compared to CDPD, Ricochet has higher average throughput but it does not support seamless hand-offs between base stations. This means that active applications may lose their connections when the vehicle drives out of range of the original base station.

In looking at the first test scenario (how applications started), Web applications experienced no problems. But MAXIMO would sometimes require more than five minutes during the logon process. Subsequent research revealed that because MAXIMO is an Oracle database application, large data dictionaries are downloaded at startup. This is clearly not acceptable in a field environment. Fortunately, it is possible to cache local versions of these dictionaries on a local hard disk. Such up-front synchronization is common to many applications and is often a performance issue for wireless communications.

Once connected (the second test scenario), the Web client performed acceptably as long as the content was more textual than graphical. MAXIMO, in contrast, ran extremely slowly. Opening new modules (e.g., the inventory module or the work-order module) within MAXIMO would take 60 to 90 seconds. Once a module opens, a screen update (such as looking at a new order) would take about 30 seconds. It is easy to understand why operations were so slow. Oracle transactions, based on SQL, involve a considerable amount of back-and-forth traffic. The slow screen updates make a remote MAXIMO client practically unusable. However, users entering text in either application posed no problems.

The last operating scenario examined the effect of lost connections. The Web client was highly tolerant of intermittent connections, which was expected since HTML applications are stateless; each page entails a new TCP connection. With MAXIMO running, a dropped connection would generate an error message for transactions in process and result in the module closing; but the overall session is maintained. If no transactions were in progress, MAXIMO readily tolerated the underlying connection being lost and regained.

Citrix MetaFrame

The second software scenario tested was the thin-client approach using Citrix MetaFrame. Starting a remote MetaFrame session over a wireless connection took about 60 seconds. Once the session was started, application startup was not an issue at all, which was expected since the applications run on an application server that has a high speed LAN connection to back-end services. Screen updates for all applications tested (MAXIMO, Web client and GroupWise) ranged from 10 to 15 seconds. This was about the same speed a a remote Web client but significantly faste than a remote MAXIMO client.

The biggest problem with using MetaFrame, however, is that it is not tolerant of intermittent connections. Even in the absence of any application processing, driving out of range of the Metricom base station would terminate th MetaFrame session as well as all the application sessions. With this architectur text input proved very slow for all applications—not surprising since every character typed by the user would have to be echoed over the wireless link by the application server.

Wireless Middleware

The last architecture tested was wireless middleware. The particular middleware chosen for testing was Smart IP* from Nettech Systems, Inc. Smart IP has a number of different capabilities, but the one of greatest interest is its ability to make IP communications more efficient over wireless networks. It achieves its efficiency through a number of mechanisms, including compression as well as replacing TCP with its own wireless-optimized transport protocols. These transport protocols are used over the wireless connection between the middleware client software that is installed on the mobile computer and the mobile server as Figure 3 shows. The net result is transmission of fewer and smaller packets.

Actual test results with Smart IP showed noticeable data transfer gains. Using a browser application, Smart IP reduced the time required to download pages by an average of about 25%. For example, a page that took 20 seconds to download without Smart IP would on average take 15 seconds with Smart IP. The utility tried to configure Smart IP to operate directly with MAXIMO, but tests were postponed due to configuration difficulties.

The utility found that different applications worked better using different approaches. Only through testing could the utility determine how their applications would function in a wireless environment. Though applications generally ran slower than over dial-up modem connections, with the right approach, applications run well enough to be deployed in the field. The utility also found that the number of software approaches available increased during the course of its project.

The Changing Application Landscape

Computer technology continues to evolve rapidly, as software vendors keep revising and improving their applications. The utility experienced a number of changes that had implications on their wireless strategy. In particular, the number of software approaches available to support wireless networking expanded.

The utility upgraded from GroupWise version 4.1 to version 5.2. With the older 4.1 version there was no easy way to provide remote access other than by using Citrix MetaFrame. But version 5.2 includes TCP/IP support as well as a Web browser client thus adding two new paths for providing remote wireless access. The most attractive approach appears to be the TCP/IP client; testing is under way to confirm this.

Another change involves PSDI, the maker of MAXIMO. Realizing the importance of wireless communications for field service workers, PSDI began to architect their next generation of software to better support wireless networking. In the new version, MAXIMO offers a Web-based interface to mobiles using HTML protocols. HTML is a far more efficient approach than extending the SQL database protocols all the way to the mobile computer. This new "wireless friendly" version of MAXIMO is release 4 and the utility plans to upgrade from release 3. Once it does, the Web interface to MAXIMO will probably be the preferred approach for mobile field workers.

As the utility proceeds with its deployment, and as it expands the number of field workers using wireless networking, it will need to rely on a hybrid set of approaches to address their application needs. In some cases, the utility will run applications in their normal LAN-based or modem-based modes. In other cases, the utility will take advantage of wireless middleware products. And for other applications, a thin-client software approach may be the most effective.

By using a strategy that includes an IP-based communications infrastructure and a flexible computing platform, combining the various software approaches is completely feasible. Such a strategy provides the utility, and any organization for that matter, maximum flexibility when supporting both field workers with specific job functions and office workers with more generalized computing needs.

Chapter 3: Business Solutions

WAP – The wireless application protocol

Title: *WAP – The wireless application protocol*

Author: *Christer Erlandson and Per Ocklind*

Abstract: *Today, the wireless network is mainly used for voice communication, where voice mail is the most popular value-added service. However, a new buzz word is increasingly being mentioned in the marketplace: WAP. With the WAP, a new dimension will be added to the use of mobile phones, through the introduction of new data-oriented mass-market services. The authors describe the WAP concept, it's historical and technological background, and the wide range of applications offered by this new approach to wireless data communication.*

Biography: see end article

The wireless application protocol (WAP) is a completely new concept. It provides data-oriented (non-voice) services to the mass market and is capable of benefiting—anywhere and at any time—far more end-users than the personal computer.

The wireless application protocol is a global standard for all wireless systems (Box B). The number of WAP subscribers is expected to skyrocket in the next five years: from its debut in 1999 to about 800 million in the year 2003 (Figure 1).

Box A Abbreviations

CDMA	Code division multiple access
CDPD	Cellular digital packet data
D-AMPS	Digital advanced mobile phone service
GSM G	lobal system for mobile communication
HDML	Handheld device markup language
ITTP	Intelligent terminal transfer protocol
PDC	Personal digital cellular
TTML	Tagged text markup language
URL	Universal resource locator
WAP	Wireless application protocol
WML	Wireless markup language (WAP Forum script)
WTA	Wireless telephony application
WTAI	WTA interface

The WAP stack

One of the objectives of specifying the wireless application protocol was to make the mobile phone a "first-class citizen" of the Internet. Therefore, it was only natural that an Internet-oriented

Figure 1: Estimated increase in the number of wireless subscribers in the next five-year period

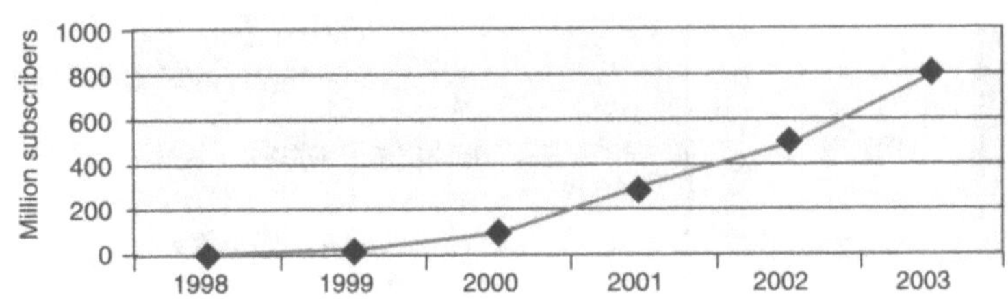

approach be adopted. As Figure 2 shows, the WAP stack is similar to the layers used in the Internet.

The following entities are defined in the wireless application protocol:

- Micro-browser; Can be compared to a standard Internet browser; for example, Netscape Navigator or Microsoft Internet Explorer.
- WML Script (wireless markup language, specified by the WAP Forum), similar to JavaScript. The script provides a means of reducing airtime by enhancing the capability of the handset; that is, it enables the handset to process more information locally before sending it to the server.
- Wireless telephony application/WTA interface. The telephony part of the wireless application protocol. Makes it possible to create call-control and call-handling applications; for example, the definition of call chains and various options when a call is received.
- Content formats. Includes business cards, calendar events and so on.
- A layered telecommunication stack. Includes transport, security and session layers.

Figure 2: A comparison between the layered communication stack in the wireless application protocol and the corresponding Internet hierarchy.

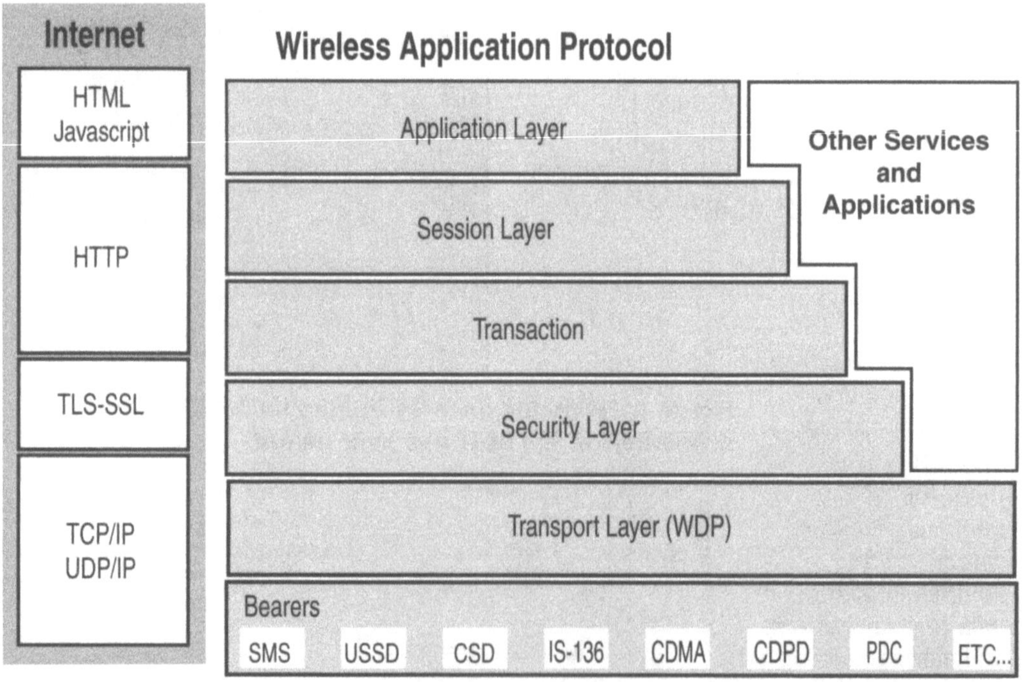

WAP applications

The wireless application protocol encompasses the entire wireless community.

Examples of applications include:

- Information retrieval on the Internet
  The wireless application protocol can be employed to reach information on the Internet. However, the WAP browser cannot be used in the same way as an ordinary "surfing tool," since the mobile phone sets some limits on input and output capability, memory size, and so forth.

- The "serviceman application"
  With a WAP-enabled mobile phone, servicemen on duty can access their company inventory to check whether or not a spare part is available and directly inform customers about the situation. Of course, they can use the same application to order spare parts, and will immediately receive a confirmed delivery date.

- Notification applications
  By means of agents residing in servers, users can be notified of e-mail and

Figure 3: Via her terminal, a user chooses a particular service, which is the same as instructing her terminal to download the "deck of cards" associated with that service. She then follows the directions for the service, making selections and entering relevant information, as necessary. To finish, she enters a command and the requested action is carried out.

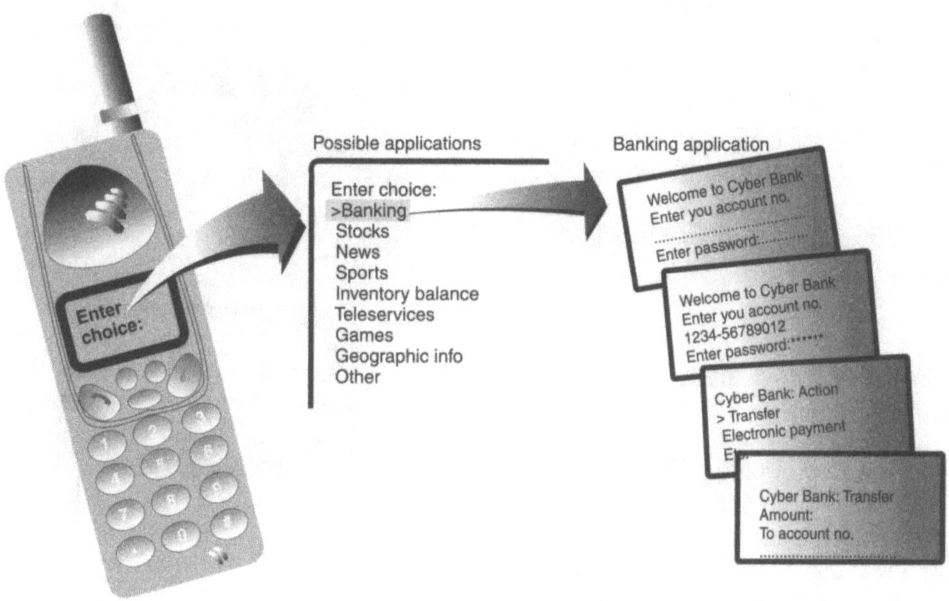

voice-mail messages that have been sent to them. They can interactively request that more information be sent to their phones, or order a printout on a fax machine of their choice.
Users interested in buying or selling shares can define a buying/selling profile that shows, for example, what stocks they are interested in, and at what quotation they might be willing to buy or sell. When a specific quotation has passed a defined trigger value, the agent notifies them and asks if they want to make a transaction.

- Mobile electronic commerce
  Users can have access to payment services for bank transactions (Figure 3), and to ticket offices and wagering systems.

- Telephony applications
  A user can have access to services that handle call setup, in combination with other services provided by a wireless operator. A typical example involves a menu (Figure 4), defined by the user, that is displayed for each incoming call. This menu allows the user to decide whether to answer or reject the call, or to forward it to another extension or to a voice-mail service.

Optimized for wireless communication

In addition to providing end-users with new services, the wireless application protocol has been designed for the economical use of the resources available in the telecommunications network. Over the air interface, communication is binary-coded so as to use the bearer services as efficiently as possible. Headers and parts of messages that are frequently sent are represented by figures. The original content is then restored in the receiver.

Figure 4: User-defined menu for handling incoming calls

Service management

Deck of cards

The handling of services is based on a "deck-of- cards" metaphor. A "deck" is sent from the network to a user's terminal when he or she enters a command to invoke a service. The user can then navigate through the complete deck to make a choice. If the desired choice is not included in the deck, another deck can be requested by command. When the user has made his or her choice and entered the relevant command, the requested action is performed or information is retrieved.

Depending on the capacity of the mobile phone, the decks and cards can be cached in the phone (terminal) for future use.

Benefits and market opportunities

Benefits to the content provider

The content provider writes or creates content that be read from practically anywhere.
The content, which is available on a server, can be accessed from any wireless device almost anywhere on the globe.

Benefits to the network operator

Network operators can offer a new category of services to end-users. They can create new, unique services, and provide access to services that are available on the Internet.

Operators can reduce their cost of customer services and Help Desks by providing access to information residing on their networks. With the introduction of the wireless application protocol, they will also remotely be able to tailor the menus and the interface of customers' telephones to further differentiate their services.

End-user benefits

More and more people are using their PCs to retrieve information from global sources. Many users have found increased possibilities of executing services on their fixed phones, around the clock using an interactive voice-response system. Now, even greater possibilities are facilitated through services based on WAP technology (Figure 5). To make it practical to use a

Figure 5: WAP technology will bring any number of services directly to your fingertips.

Box B Background

About one year ago, on Ericsson's own initiative, a consortium consisting of Motorola, Nokia and Unwired Planet started to define a new protocol for mobile devices. The objective was to offer new wireless data communication services to end-users, both in the form of telecom-related and Internet-oriented applications.

Over the years, several protocols have been defined by various players in the market, for various types of application. Unwired Planet, a California-based company, has specified an Internet access protocol—called the handheld device markup language (HDML)—for Internet access, to be used over cellular digital packet data (CDPD) networks. At CeBIT 1997, Nokia launched another protocol, called the tagged text markup language (TTML), in some respects a similar protocol but incompatible with that of Unwired Planet. Ericsson, in turn, has been defining a protocol that mainly focuses on call handling and call control: the intelligent terminal transfer protocol (ITTP). These three protocols represent only a fraction of the different protocols defined by different organizations and available in the marketplace.

This fragmentation has prevented, and still prevents, the market for wireless applications from "taking off." Early on, Ericsson realized that the company most certainly would be the only manufacturer to offer ITTP-supporting products. Consequently, the market would take up an unresponsive attitude to these products. Ericsson therefore contacted Motorola, Nokia and Unwired Planet and suggested that a joint venture be established, whose aim was to define a protocol capable of use for call control, call handling, messaging, and Internet access. To make it a global, generic protocol, Ericsson stressed that it should be independent of the underlying air-link bearer. Initially designed for use over the GSM network, the application should also operate on D-AMPS, CDMA and PDC, and even on non-voice systems, such as Mobitex and paging systems. Ericsson's suggestion was accepted, which resulted in the formation of a joint venture for the purpose of specifying what was to be called the wireless application protocol (WAP).

In December 1997, the four parties formed a company, WAP Forum Ltd., which controls and manages the continued development of the wireless application protocol. Early in 1998, the WAP Forum opened its doors to new members.
On April 30 this year, the first revision of WAP 1.0 (still uncompleted in some parts) was released. As of November 1998, WAP Forum Ltd. Comprised more than 70 members, who continue to work on finalizing the first complete revision, keeping an open mind on new ideas for future revisions.

phone for accessing services, the user must be able to personalize the services; for example, from a Web page.

Thanks to the wireless application protocol, end-users can access the following services, and many more, by entering the relevant choice via their terminal:

- banking;
- stock exchange quotations;
- news;
- sports;
- weather;
- inventory balance;
- teleservices;
- games;
- geographic information.

Challenges to the network operator

In the world of new services and opportunities, network operators are likely to meet new challenges. They run high-quality networks with good coverage as a basis for their business, but they must also decide on the next step in developing their business:
Should they remain pure network operators only, providing access to a multitude of new services, or should they take an active part in providing the services themselves? With the wireless application protocol, operators can choose to open up access to all sites on the Internet, by allowing all universal resource locators (URL) to be accessed by their subscribers. Or they can opt to exercise full control, solely allowing access to those sites that are included in a service package.

Creating a service package
Network operators can do everything themselves; for instance, developing all sorts of services and bundling them in an attractive fashion. It is doubtful whether anyone will choose to do so, however, since a wide variety of companies already offer sets of services with attractive content.
The other extreme is to outsource all services to a partner that handles the services, including subscriptions.

To keep control and yet allow access to content that is already available on the Internet, operators will negotiate directly with content providers, or with content brokers that provide an assortment of services. In addition, the operators will have to introduce a service-management system that allows end-users to select the services they want.

Mass market

The wireless application protocol enables network operators and content providers to reach a mass market. Although WAP-enabled phones have not yet been manufactured, there are today approximately 200 million subscribers using wireless devices. Looking less than five years ahead, this figure will probably have risen to 700 or 800 million, and we have good reason to believe that the majority of handsets in use will support the wireless application protocol.

The Ericsson view

Ericsson holds firm to the belief that the wireless application protocol will be a major enabling component in the telecom business, and is the obvious technology for providing data communication services to mobile phones. However, this process will not happen by itself. We are therefore doing our part to develop and offer WAP-supporting products to the market, and to make sure that there are applications in need of them, in order to stimulate the use of the wireless application protocol. Recently, Ericsson launched its first two WAP products: a generic service development kit and a WAP gateway for GSM.

Conclusion

Ericsson, in cooperation with Motorola, Nokia and Unwired Planet, has defined a new protocol for wireless data communication. The wireless application protocol provides end-users with new services in a wide range of applications, such as information retrieval on the Internet, mobile electronic commerce, and telephony applications. At the same time it is designed for the economical use of the resources available in today's telecom networks. The new wireless application protocol enables network operators and content providers to reach a mass market. By combining the wireless application protocol and Bluetooth—Ericsson's universal radio interface for ad hoc, wireless connectivity—we will have an even more powerful concept.

Box C The WAP and Bluetooth

When Bluetooth has been implemented in WAP phones and in small electronic devices, the WAP concept will be strengthened further. The combination of the wireless application protocol and Bluetooth technologies enables content providers, such as banks, to distribute small, branded calculators with keys that are dedicated to different banking services. For example, a user will be able to press a certain key on the calculator to retrieve the balance of his or her bank account. By pressing another key and entering the appropriate account number, he or she will be able to transfer money to or from the account—of course, security functions will require the user to enter a personal identification number for access to be allowed.

WAP – The wireless application protocol

Biography:
About Ericsson
Ericsson is the leading provider in the new telecoms world, with communications solutions that combine telecom and datacom technologies with freedom of mobility for the user. With more than 100,000 employees in 140 countries, Ericsson simplifies communications for its customers - networkoperators, service providers, enterprises and consumers - the world over. Additional information WAP, Wireless Application Protocol, is the de facto worldwide standard for providing Intnernet communications and advanced mobile services on digital mobile phones, pagers, personal digital assistants and other wireless terminals. WAP is compatible with GSM 900, 1800 and 1900 MHz, cdmaOne, TDMA wireless standards, as well all of the proposed 3G communication systems. Already in June 1999, Ericsson delivered the world's first end-to-end WAP 1.1system - the first true WAP standard. Since then, Ericsson has already delivered more than 40 WAP 1.1 compliant systems to mobile network operators, worldwide. To promote Internet access using a mobile phone, Ericsson was part of establishing the Wireless Application Protocol (WAP) Forum. The general objective of the forum was to enable the wireless industry and content developers to provide compatible products and services across a wide variety of platforms. WAP-compatible devices are capable of translating the incoming Internet data to match the capabilities of the mobile terminals that are unable to display HTML-based Web pages.

About the authors
Christer Erlandson, who joined Ericsson in 1971, is resonsible for WAP and mobile electronic commerce at Ericsson Wireless Internet AB. He is also Alternate Director of the board of WAP Forum Ltd. During his more than twenty-five years with Ericsson, he has held various positions in product development, product management and business management, covering data transmission, data networks and value-added services for mobile communication. He holds an MSc in electrical engineering from the Royal Institute of Technology in Stockholm and has studied business administration at Stockholm University.
Per Ocklind works as the WAP Coordinator at Ericsson Wireless Internet AB. He joined Ericsson in 1989, working as a Product Manager af radio modems and terminals for Mobitex. In 1996, he transferred to Ericsson Radio Sytems (GSM), where he functioned as a "human interface" to the terminal side. Since 1997, he has been involved in creating and administering the wireless application protocol. He holds an MSc in industrial engineering.

Business goes Mobile - Mobile Business Applications

Title:	*Business goes Mobile - Mobile Business Applications*
Author:	*Brokat Infosystems AG*
Abstract:	*This document is intended for people interested in mobile communication technologies, especially in the field of transactional financial services.*

1. Preface

Mass Market Data Applications
More and more people, no matter were they happen to be at a given moment, want mobile access to information - regardless of where that information resides and in what format it appears. Voice service is no longer sufficient to satisfy mobile customers' business and personal requirements; they also need mobile access to data. Among the factors accelerating the demand for mobile data are the explosive growth of the Internet and the rising popularity of mobile computers.

In addition, that demand is no longer coming simply from the high-end market. Increasing numbers of mass market customers want mobile access to information services. In fact, the overall number of mobile data users is expected to climb dramatically between 1998 and 2005, from three million customers, each trans-mitting about 1.1 Mbps per month, to 77 million customers, each transmitting around 30 Mbps a month. Clearly, mobile data will be a major revenue generator as the wireless industry and GSM technology evolve towards the third-generation UMTS. Yet until recently, customers faced three significant barriers in their efforts to use mobile data: it was too complicated, too expensive and too slow. However, the industry is making significant progress in developing plug-and-play, easy-to-use terminals, based on such standards as the Wireless Application Protocol.

With GPRS, the amount of data sent rather than the connection time determines the cost. In terms of network resources and subscriber costs, this means that GPRS is well suited for bursty, mass market applications typical of Internet-type services.

1.1 Target Audience
This document is intended for people interested in mobile communication technologies in connection with Twister, especially in the field of transactional financial services.

Business goes Mobile - Mobile Business Applications

2. Executive Summary

Demand for Mobility

In the present dynamic business environment, companies are forced to become more competitive in order to survive. Tightening competition, globalization and changes in customer behavior present formidable new challenges to the companies.

One the most recent and significant changes in the business environment has been growing demand for mobility. This means that customers, partners, employees should be able to access the information resources and services of a company wherever they are and whenever they want.

In 1998, 73% of European corporations were using some kind of mobile data solution and 91% of those who were not said, that they would within 1999 (The Yankee Group).
This means that there will be an explosion in the number of services that can be provided to a mobile phone. As a result, markets, brands and customer loyalities are on the move. Constant availability is the key concept for future competitiveness.

Mobile phones are the key to numerous services

The wireless communication market is growing rapidly, reaching new customers and introducing new services. To enable service operators, manufacturers and also content providers to meet the challenges in providing advanced services, service differentiation and fast/flexible service creation there are a lot of different wireless applications necessary.

This will provide a huge market potential for corporations. Besides offering their existing services via a new mobile channel, they have unlimited possibilities to create new services and products for their customers. Users of mobile phones can now be offered relevant personal services that suit their needs. Today people must carry with them at least a wallet, a calendar and a phone. Soon it will be only a phone, which enables you to make calls, pay bills, buy tickets, check e-mail and manage one agenda.

Internet goes Mobile

For several years industry watchers have been forecasting an explosion in wireless Internet applications and users, based upor the success of the Internet and that of mobile communications. The assumption has been one of translating the fixed world of Internet access to that of the mobile. A new report from the ARC Group presents a reversal of this theory; the wireless Internet industry will provide the direction for the future development of fixed Internet access.

According to the ARC Group the number of mobile subscribers in 1999 (428 million) already greatly exceeds that of Internet users (241 million), and is expected to grow to more than one billion by 2003. To most of this group wireless access to the Internet has proven unattractive because of the low bandwidths and high costs of access such that mobile data users only represent 7% of mobile subscribers in 1999.

The situation is starting to change with the introduction of IP based value added services. This does not include Internet browsing as we know it today but the provision of a bundle of information services which source their data from the Internet. Through the adoption of these the ARC Group forecasts the number of mobile data users in Western Europe to exceed those of the fixed Internet in 2003. This discrepancy in growth levels will continue such that in 2004 the num-ber of mobile data users in Western Europe will exceed that of the fixed Internet by over 40%.

In the fixed domain the Internet is commonly accessed through one of two browsers, Microsoft's Internet Explorer and Netscape's Navigator. These, and much of the content on the Internet, have been designed by IT oriented developers with a bias towards the use of jargon, technical terminology, and total flexibility for the user. They are not user friendly and there is a significant learning curve to be overcome in operating both the computer and the interface to the Internet that is not conducive to mass consumer acceptance.

It is well recognised in marketing circles that consumers do not buy technology, they buy benefits. They want the benefits of the Internet but not the Internet for its own sake. These benefits include:

- Easy communication through email
- Ready access to information
- Entertainment, such as games
- Improved lifestyle through e-commerce and home banking

These can all be provided without any knowledge of the technology that is delivering them. Imagine a terminal which is ready as soon as it is switched on, providing a simple hierarchical menu structure into these categories, with services filtered to the personal requirements of the user and his local area. This model is currently being rolled out by

Figure 1: Internet and Mobile Subscriber Penetration -2004

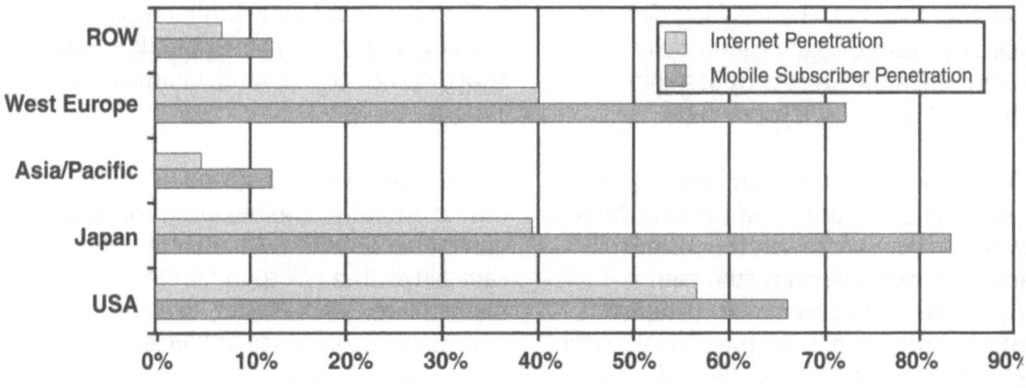

Source: The Arc Group

many mobile network operators through the mobile phone handset.

Successful services provided through wireless Internet will necessarily be oriented to the unique characteristics of mobile access. These are the timely provision of information when it is most relevant, the ability to access information from any location, and the relating of information to the current location of the user. These include pull services, where the user consciously connects to obtain information, and push services, where information is provided under predetermined conditions, often using SMS on GSM networks.

3. What is Wireless Communication?

The evolution beyond voice traffic in the new Information Age presents a great challenge to GSM operators and their suppliers because it involves a merging of two industries. New markets for wireless data will expand where information technology and telecommunications meet, bringing the Internet and the GSM network together. The resulting demand for personalized solutions in wireles data will reveal new business opportunities to the GSM operator beyond voice.

The GSM operator is in an intrinsically advantaged position in which to build new wireless data services on top of their existing voice telephony to retain customers, and attract new customer groups such as mobile professionals, who constitute some 40% of the professional workforce. The GSM operator can make the shift from being a pure bandwidth supplier towards being a value-added service operator with minimal investment.

Being first to market with wireless data solutions will make it possible for the operator to sustgain a significant competitive edge over later entrants. The industry trend shows that a GSM operator starting today and following an aggressive leadership position in wireless data can command a share of 20-30% of revenues with wireless data around the year 2000.

GSM is the only network which can support a global wide area solution for wireless data. GSM provides the best platform on which to build personal messaging applications for both business and customer use.

Mobile Banking is only one example of the use of novel applications based on GSM. The mobile phone is increasingly developing into a multifunctional platform for a wide range of value-added services. This is no accident, for in order to gain an advantage over their competitors, network operators and service providers are trying to diversify their range of offerings. In this strategy to gain a competitive edge, value-added services are playing a particularly important role.

Services like wireless banking or stock-buying are targeting the dozens of next-generation phones with Internet capabilities that will soon hit the marketplace (Mark Plakias, director of voic and wireless commerce at Princeton, N.J.-based Kelsy-Group).

Information and Transaction Services
There are two service offers which are to be differentiated: information services and transactions.

Information services are sending unchangeable contents from a database only to inform the mobile phone user. These additional services typically start with mainstream content such as news, travel, weather and sports and over time, new information providers are sourced that offer lifestyle services such as horoscopes and jokes.

Transaction services allow an access into the processing center of a content or service provider (e.g. a bank). Transaction services represent often a better conveniance than information services.

3.1 What is SMS?
The Short Message Service, as defined within the GSM digital mobile phone standard, has several unique features: With SMS the mobile phone user is able to create short messages via mobile phones. A single short message can be up to 160 characters. The SMS is a store and forward service, short messages are not sent directly from sender to recipient. Short messages are always sent via an SMS Center, supported by mobile operators.

Attractive additional Services
Via the SMS, information can be transmitted to the mobile phone, such as stock exchange prices or football results. Other possible functions are the sending or receiving of faxes, database searches, and e-mail. Applications in the telematic field are also conceivable.

Difficult Usage
With standard mobile phones mobile applications are not menu-guided. They are not as convenient as SIM Toolkit-based mobile phones. Furthermore no local storage of user data (e.g. account numbers, bank number) is possible. The end user enters direct his commands (e.g. "BRJ" for BROKAT stock rate).

User-friendly menu with SIM Toolkit
Another technology that will become increasingly important in the successful introduction of wireless Internet services on GSM networks is SIM Application Toolkit. It adds intelligent support to services on the SIM card itself. Menus displayed on the phone, and the behavior on selection of menu items, are controlled by instructions stored on the card accessible through a SIM toolkit enabled phone. Applications are more convenient to use and the local storage of user data on the phone becomes possible.

3.1.1 Approaches to Implementing SMS-based Applications
There are two approaches used in implementing SMS-based applications: push and pull. The push approach involves the information provider sending an SMS message to the customer's GSM telephone without the customer requesting that specific message. For example, a bank sends their subscribing customers their account and account balance on a daily basis without being prompted to do so. The pull approach requires the customer to send an SMS message to the bank before a response is sent. For example, a customer could request a latest account balance or a foreign exchange rate.

3.1.2 SMS Support on different networks
Since its inclusion in the GSM standard, SMS has also been incorporated into many other mobile phone network standards, including Nordic Mobile Telephone (NMT), Code Division Multiple Access (CDMA) and Personal Digital Cellular (PDC) in Japan. Each of these standards implements SMS in slightly different ways and message lengths do vary.

The past few years have failed to see the realization of the full potential value-added non-voice mobile network services such as Data or Short Message Service.

Over the next several years, we expect that the number of banks offering SMS-based services will increase markedly. The bulk of this growth will be centered on the European market, where greater reliance on the GSM standard will allow banks to reach a large audience.

In North America, a variety of analogue and digital standards exists and many of the standards do not have a messaging component. This will severely limit the number of banks offering GSM networks. The GSM standard has only recently made any sort of impact on the North American market.

At the same time, as the number of banks offering SMS banking increases, the range of services offered will rise.Currently, most of the offerings are limited to very simple reporting capabilities. Very few institutions are offering the ability to actually initiate transactions with GSM phones. This will change. As the number of GSM phone subscribers continues its high rate of growth, more and more banks will start to allow their customers to make payments or buy and sell securities using SMS. Meridien Research expect that by 2002 over 30% of Europe's major retail banks will be offering SMS banking to their clients. Progress in both the USA and Japan, with their more limited use of GSM systems, will be far more limited.

3.1.3 Technology Requirements
SIM Tool Kit enabled mobile phones
Access to these services is facilitated by a novel technology which is ready to be applied on an international basis. The user needs a GSM mobile phone (that supports SMS) with an integrated SIM Toolkit and a SIM card which supports the new value-added functions.

Using an active SIM card, banks and servic providers, for example, can offer numerou additional services with a high degree of user-friendless. The SIM Application Toolkit Standard extends the specifications of the interface between the mobile phone and its SIM card. This facilitates convenient operator guidance based on the SMS.

Over-the-Air-Technology
Over-the-Air Server, an advanced fault-tolerant system for 'over-the-air' customisation, allows operators and/or subscribers to remotely download new Jav application programs or applets' to SIMs via GSM's SMS (short message service) channel.

Two further server options substantially broaden the ability to reach customers an are expected to accelerate the development and usage of mobile value-added services.

POS Server supports applet downloading and SIM card configuration management using certified point-of-sale terminals such as those used by banks, retailers, petrol stations, etc. Similarly, Web Server allows applet downloading and SIM card configuration via the Internet, to enable end users to customise their mobile services directly from their home or work PC.

The major distinction is the ability of the platform to be configured by both the network operator and the subscriber. Unlike targeted loyalty schemes, users can now configure their SIM cards themselves to create their own highly personalised tools. The back-office database allows operators to track and manage a subscriber's application profile, and can use and mine this data to market other services, sending targeted marketing or advertising messages, or information updates about related new services.

3.2 What is WAP?

The third generation of wireless technology is just around the corner. It has to provide wireless mobile, high quality multimedia communication services to the mass market with global roaming and boundary-free coverage. The wireless terminal will be the personal gateway into the world of voice, data, video, mobile Internet and interactive multimedia communications.

Mobil phone communications is presently limited mainly to voice services. Now that the leading mobile phone manufacturers have reached a worldwide agreement on Internet access for mobile phones, there are no further obstacles to state-of-the-art applications. One of the most promising services for the financial sector and ist customers is banking by mobile phone.

The Wireless Application Protocol (WAP) is a result of continuous work to define an industry-wide specification for developing applications that operate over wireless communication networks. WAP is an attempt to define the standard for how content from the Internet is filtered for mobile communications. WAP was developed because content is now readily available on the Internet, and there needs to be a way of making it easily available to mobile terminals. One of the reasons why

Figure 2: Percentage of major banks offering SMS Banking

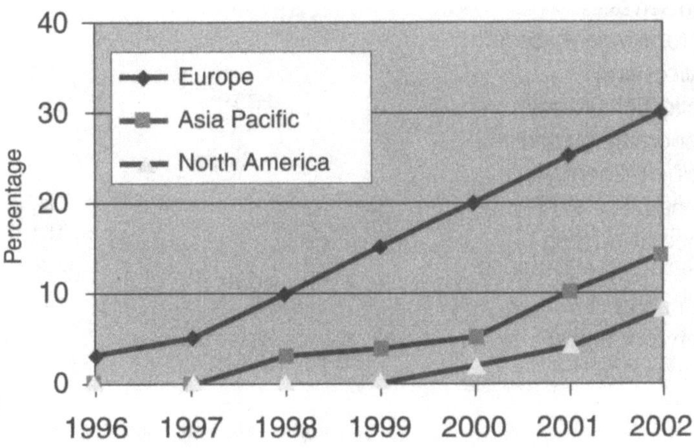

the mobile industry became so excited about WAP is because it combines two of the fastest growing industries: wireless and the Internet .

3.2.1 Technology Requirements
WAP is a WWW oriented method of deploying applications that is independent of the network technology. Applications and content are kept centrally on web servers and downloaded as required. WAP is a particularly good choice when deploying applications that also have an HTML version for desktop use.

Mobile Phones with Micro-browser
The Wireless Application Protocol incorporates a relatively simple micro-browser into the mobile phone. As such, WAP's requirement for only limited resources on the mobile phone makes it suitable for thin clients and early smart phones. WAP is designed to add value-added services by putting the intelligence in the WAP servers whilst adding just a micro-browser to the mobile phones themselves. Microbrowser-based services and applications reside temporarily on servers, not permanently in phones. The WAP is aimed at turning a mass-market mobile phone into a "network-based smartphone".

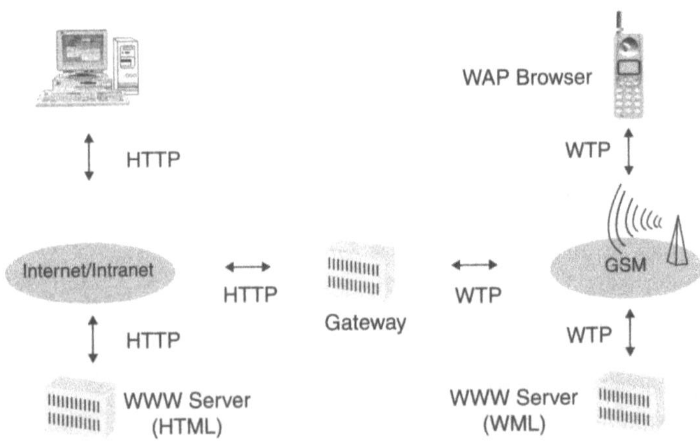

3.2.2 WAP Forum
The WAP Forum is a nonprofit industry association open to all industry. It has published a global wireless protocol specification based on existing Internet standards such as XML and IP for all wireless networks. Founders of the WAP effort: Nokia, Ericsson, Motorola and phone.com/Unwired Planet chartered the Forum and opened it to general membership in January 1998. At the moment there are over 130 mebers and more are joining constantly.

WAP Forum members represent over 90 percent of the global handset market, carriers with more than 100 million subscribers, leading infrstructure providers, software developers and other organisations providing solutions to the wireless industry.

4. Market Potential

The mobile communication market will undoubtedly be the No. 1 of the telecommunication branch. In four years already up to 50 % of the turnover will be achieved by mobile telephone networks.

In 1998 there were 300 million mobile phone subscribers worldwide, about 100 million more than just at the end of 1997.

Analysts at UK-based ARC Group say in a new report that the wireless industry is showing us the future of the Internet. Pointing out that the number of mobile subscribers in 1999 (428 million) already greatly exceeds the number of Internet users (241 million), the report's authors suggest that wireless will beat the "fixed Internet" by becoming more user-friendly.

Market predictions foresee a global market of 800 million users by the end of 2003 - a market that is likely to exceed the number of installed wired phone. The problem is not so much the market size, but the change of application orientation from purely voice type usage towards more bandwidth consuming high speed data traffic. As the age of the mobile phone as a mere voice tool is expected to pass, the most critical challenge facing mobile telecommunication technology in the next decade is to develop the ability to handle the burgeoning growth of information traffic. Early examples of value-added services, such as voice-mail and SMS available in the second generation networks, have established themselves as the important features creating an extra revenue stream for network operators and an added-value service for end users.

Cellular phone giant Nokia said it has raised its forecast for the world's mobile phone subscribers to one billion users in 2005.

Value-added Services via Mobile Phones

Although only about three million people will subscribe to a wireless data service this year, more than 36 million are expected to subscribe by 2003, making the market worth about $ 3 billion, according to a Dataquest study.

Furthermore, Dataquest's market researchers calculate that by 2001, 18 percent of all mobile phone users will be taking advantage of value-added services as well. These so called "power users" will then account for 80 percent of all connection time.

Internet-capable wireless communications devices like the next-generation cell phones are expected to grow from an estimated 12.2 million in 2000 to 56 million by 2002, according to International Data Corp., based in Framingham, Mass. Another research company, Jupiter Communications, based in New York, predicts, that more than 10 million mobile phones will be capable of accessing Internet-based data by 2002.

Industry projections indicate 1.2 billion cell phones to be use by 2004, with 700.000 to 800.000 of these providing Internet access. Nigel Deighton, analyst with the Gartner Group, is optimistic that by 2004,

30 to 50 per cent of business-to-customer (B2C) transactions will occur via mobile devices.

5. Wireless Solutions

Most of the recent attention on electronic banking has been focused on the Internet. However, electronic banking via GSM mobile telephones has quietly been gaining ground. Unlike their analogue counterparts, GSM telephones rely on digital communication and can, therefore, offer a range of additional services, including the transmission of short e-mail messages using the SMS. While the number of financial institutions offering SMS-based access to accounts is lower than the number offering Internet banking, the medium term prospects for SMS banking are excellent.

5.1 Projects at BROKAT

The majority of banks offering SMS banking offer read-only services, including a customer's latest account balance or the past three transactions. An innovative improvement and extension in functionality is realized by using transactions. In doing so it will become necessary to integrate the SMS-based banking in the existing Internet-based banking infrastructure at the bank.

5.1.1 Mobile Banking with DVG Karlsruhe

In summer 1999 the DVG Karlsruhe has realised a mobile banking application. The banking customer uses his standard mobil phone to get the account overview, foreig exchange rates, turn over and other banking information via SMS.

5.1.2 Mobile Banking with 1822direkt

A consortium of the direct bank 1822direkt, smart card expert Schlumberger, mobile operator VIAG Interkom and BROKAT have shown a new mobile banking pilot project at CeBIT 99. The pilot is the result of five months' work on applying the GSM SMS protocol to

Figure 4: Growth of European mobile telecommunication market per anno

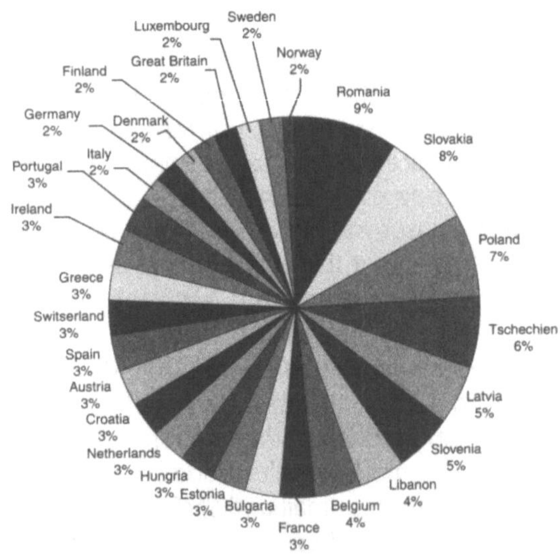

Source: Mobilcom 10/98

bank transactions, and the first implementation in Germany of continuous (end-to-end) encoding. A high level of transaction security is ensured by the additional use of PIN/TAN data.

Mobile banking builds on telephone banking and on-line banking to open up a new, easy-to-use service channel, requiring only a mobile phone and a personal bank account. Bank transactions ca be carried out at any time, quickly and easily, from anywhere at home or abroad using a mobile phone over VIAG Interkom's mobile phone network E2. Using the simple menue-driven handset display, users can for example initiate transfers and request detail such as an account balance or stock exchange prices.

5.1.3 Internet Administration of SMS-based brokerage information with Zürcher Kantonalbank

The Zürcher Kantonalbank in Switzerland offers a new brokerage service to administer your watch list via Internet. The end user who owns a mobile phone is now able to set up different agents and functionalities for the push service: time-, value- and event-related agents can be define by the user to receive the actual share price.

5.1.4 Mobile Insurance Application with a big German Insurance Company

A big German Insurance Company has realised a pilot project regarding Online Automobile Rating Application: The Sales Person of the Insurance Company uses a NOKIA communicator to connect into the intranet of the Company via SMS, and to administer an insurance application.

5.2 Projects Worldwide

5.2.1 Mobile Banking

Beginning in 2000 Bank of America will allow its 30 million customers to do their banking over the Web via wireless devices such as cell phones and 3Com's Palm computing device.

Judging from the experience at Bank of Montreal, customers are attracted to the freedom of banking from anywhere and not just from a PC. When the bank asked for volunteers for the trial, more than 100,000 customers applied (Rick Kuwayti, spokesman for Bank of Montreal).

T-Mobil together with T-Online will offer a mobile banking service. In September 1999 D1 customers will be able to request the account balance via SMS.

Gemplus, Citibank Singapore and the mobile operator MobileOne have announced a mobile banking service: Besides access to the latest account balances and payment information, customers are able to conduct interbank and intrabank fund transfers, pay bills and request for a temporary credit line increase.

Barclaycard and its partner Cellnet have announced the world's first consumer trial of 'mobile electronic cash machines'. The trial will enable 1,000 triallists to download e-cash over Cellnet's GSM network onto their Barclaycard by inserting it into a specially designed Motorola StarTAC mobile phone. Triallists can also use the card as part of the on-going Visa Cash UK trial in Leeds (a computer chip embedded in the card that is able to store up to £50 worth of cash electronically).

5.2.2 Mobile Brokerage

While trading over the Internet has exploded in the past year, with more than 450,000 online trades a day in the first quarter of 1999, online investors are tethered to their desktop computers.

Still, wireless trading appears to be the next wave of innovation for brokerages, letting investors make trades from the sidewalk, the gym, or the back of a cab.

Mydiscountbroker.com, InvestN.com Securities, and U.S. Discount Brokerage join 1800Daytrade.com, R.J. Forbes, InvesTrade, and The Wall Street Discount in offering a wireless-trading system.

5.2.3 Mobile Payment

At Finland's Helsinki Airport customers can even pay to quench their thirst from a soft-drink machine by mobile telephone without using cash. Anyone wanting a softdrink just needs to dial a number on the machine, and a can containing the thirst-quenching refreshment drops down.

France Telecom is going to launch an open mobile commerce pilot in August 1999, with a view to full rollout by the end of the year. The pilot, to be operated in conjunction with the Groupement des Cartes Bancaires (CB), will enable consumers in possession of a Motorola dual-slot StarTAC D handset to use chip-based credit cards for secure transactions. Following an initial trial during the second quarter of 1999, France Telecom Mobiles customers will be able to make secure transactions for goods and services such as gifts and travel scheme tickets 'over the air to merchants.

During payment, the merchant connects t the mobile France portal to confirm orders via SMS, whereupon the consumer pays with a PIN-protected card. The SIM card within the handset subsequently writes th amount to the chip embedded in the CB card, before this information is relayed back to the portal.

In 1999 in Finland, Visa, Nokia and MeritaNordbanken launched a mobile pilo involving the SET protocol.

Merchants signed up with Westamerica Bank can avail of high-speed wireless cred card authorization services following a two-year transaction processing deal with US Wireless Data (USWD). Transactions processed by the Wireless Express Paymen Service from USWD. The service also provides encryption, diagnostic and reporting capabilities.

Cellphone giant Nokia has also invested ir a device intended for use as a substitute for credit cards. From this fall, PalmPilot users will be able to "beam" funds between handheld devices, effectively conductiong person-to-person transaction

Scotiabank has announced the first mobil secure debit and credit transactions via portable POS terminals this year. Mobile point-of-sale terminals will be used at the Pan Am Games in Canada - over the Mobitex Data Network. Spectators will be able to make purchases with their credit

Table 1

Bank	Country	Service Description	Charges
Bank Austria	Austria	Current balance and last three transactions	No charges by bank.
Postsparkasse	Austria	Inter-bank transfers up to ATS 50,000. Current balance and available overdraft limit. All transactions since last statement.	No charges by bank.
Expandia Bank	Czech Republic	Balance accounts information, accounts credits and debits, access to actual exchange rates, information about interest earned on deposits with the bank; payments from subscriber's private account	CZK 1,50-3,00 per SMS
OKO Bank	Finland	Information provision and transacting	-/-
Banques Populaires	France	Daily account and card balance updates	FF 20 per month
Dredsner Bank	Germany	Current balance and last three transactions	DM 0,17 per message
1822direkt	Germany	Current account balance and money transfer	-/-
Inter-Europa Bank	Hungary	Current balance and previous transactions	-/-
Banco Ambrosiano Veneto	Italy	Current balance and previous transactions	ITL 198 per message
Credito Italiano	Italy	Current balance and previous transactions	-/-
(six large banks)	Japan	Information access and transfers between accounts	-/-

Banco Santander	Spain	Current balance and previous transactions. Automatic alerts when movements above a set limit occur. Automatic account updates at dates and times specified by customer	-/-
Post Girot Bank	Sweden	Pay domestic bills, transfer funds between accounts.	-/-
Co-operative Bank	UK	Current balance and previous transactions.	-/-
Barclays Bank	UK	Current balance and previous transactions.	-/-
National Westminster Bank	UK	Mobile banking services.	-/-

Source: Meridien Research and BROKAT Infosystems AG

debit cards at concession stands utilizing the mobile terminals deployed by Scotiabank.

Proton World and Gemplus announced that they had successfully loaded value onto a Proton card using a mobile phone. The Proton card was loaded using dual-slot mobile phone handsets over the Belgian operator Mobistar's network.

5.2.4 Mobile Auction

The Internet auction company versteigern.de has launched a new service for its customers: Using the auction manager button the participants will be informed about the current precept via mobile phone. After receiving the current SMS, the customer is able to increase his offer.

5.2.5 Mobile Betting

The biggest comsumer application in Hon Kong is placing bets on horses via the Internet. Now it seems to be possible to bet via mobile phones in up to few seconds before the race starts.

5.2.6 Mobile Ticketing

Since mid 1999 Mobilkom Austria and th Österreichische Bundesbahn ÖBB offer a new service: The mobile phone will be a mobile payment device to buy the ÖBB ticket via SMS. The amount will be charge to the telephone bill.

5.2.7 Mobile Information

CNN Interactive has teamed up with BellSouth to launch its CNN Mobile servic - 24 hour news service for digital cellular phones - in the U.S. CNN launched the

service in February this year and the service has been launched by several European and Asian operators, but BellSouth is the first to offer the service in North America. The service delivers targeted regional content to handsets using a SMS initially, to be followed by delivery using the WAP once compatible handsets are available. News headlines and market information are updated every 30 minutes, and sports scores and weather forecasts are updated regularly throughout the day.
CNN Mobile's nine operators in Europe are EuroTel Praha (Czech Republic), Orange Israel, Sonera (Finland), Sonofon (Denmark) and Telia (Sweden); and in Asia, M1 (Singapore), Hongkong Telecom, New World Mobility and SmarTone (Hong Kong).

News and information company Reuters Group and Ericsson will begin trials this year to deliver news and data to mobile phones using wireless Internet technology.

Reuters and mobile telecommunications company Vodafone UK also said they had started trials to deliver personalized news and information to mobile phone subscribers. Stories corresponding to chosen categories or keywords would be broadcast to users' phones at pre-selected times or whenever the stories were published.

MobileOne Asia (M1) is now able to offer its customers Info-on-Demand-Services with support of De La Rue. This latest innovation offered by M1 allows its customers access to timely information on stocks, currencies, banking services, movies, restaurants, travel timetables and weather reports quickly and easily through a user-friendly menu-driven interface.

AirFlash.com will offer a new type of service for mobile phone users this year that provides business, entertainment, and travel information specific to your location.

Telecel customers in Portugal will be able to access a range of information services, such as: direct access to voicemail service, management of bank accounts –using a defined PIN number, current bank balances, order statements, authorisation for payment of utility invoices and transfer funds between accounts, direct access to weather, travel, sports, lottery games, TV programmes, cinema listings, stock exchange and currency information, e-mail and fax communications on the move. access to standard Telecel services, such as Customer Service, Telephone Directory, Anti-Thief, Express Assistance and the new and originally featured Wake Up Call service.

Lufthansa Systems offers a special tender for its customers to coordinate their travel management by themself - by using a Nokia 9110. It will be possible to call the current flight schedules, to book flights, or to request the miles & more account information.

5.2.8 Mobile Internet/Portal

Yahoo has joined forces with German mobile telephone operator Mannesmann Mobilfunk to deliver Internet services to telephone handsets. Subscribers with a correctly enabled mobile phone would have quick and easy access to Yahoo Germany's Internet programming.

America Online has announced similar efforts to capitalize on the proliferation of wireless devices and smart consumer appliances that one day rival personal computers as a way to plug in to the Internet.

Nextel is going to offer a wireless data package with Netscape Communications, that combines voice, data, and the Internet. The new service will enable customers to use their digital phones to access a personalized Internet gateway, send and receive e-mails, and browse the Web.

Microsoft has launched a wireless portal called MSN Mobile, which delivers wireless content to cell phones, handheld devices and pagers. Microsoft will be offering customizable wireless information services such as news, stock quotes, sports reports, and other information.

The Company is working on a voice-activation system to enable services that otherwise would be difficult or impossible using the limited keypad and screensize of small devices such as cell phones.

Lucent Technologies has announced Zingo, a new portal designed as an Internet start page for traveling professionals.

Meanwhile, handheld leader 3Com offers a Web clipping service with ist new wireless Palm VII. Users access Internet information by sending requests to pro-vider sites via the Palm.net Network.

Ericsson has partnered with several application providers to bring a range of new services to the company's wireless application protocol-based Mobile Internet portal. New features include restaurants and travel tips, news, weather forecasts and stock updates, as well as limited online shopping.

5.2.9 Mobile Advertising

Imagine a trip through a shopping mall. While passing Macy's department store, your cell phone beeps, and on the screen flashes a message: "Ten percent off at Macy's."

A handful of service providers and Internet portal companies aiming at the wireless phone market are toying with the idea of subsidizing portal-like cell phone services with highly targeted advertising. The idea is to deliver messages that pertain to what the customer is interested in, at a time when they are most valuable to the client.

Direct Marketing

As on the Internet, the content and advertising would be targeted at a user's demographic group and personal interests. When signing up for a service, a customer would likely give up enough information to allow customized targeting of advertisements and personalized content.

Since cell phones can be used to pinpoint a user's location, stores will be able to push their advertisments when a customer comes within shopping range. This location targeting facility is currently limited to an average of about a quarter mile, but may shorten as technology improves.

5.2.10 Mobile Enterprise Applications and Corporate Database

3Com, the No. 2 maker of network equipment, said it will form a new company to supply wireless Internet service for cell phones, 3Com's PalmPilot, and other handheld devices.
The service will let customers download e-mail and information from corporate databases over the Web.

Enterprises with mobile users can extend their business critical applications to handhelds with AvantGo. Leveraging Web technology and mainstream database access, AvantGo's enterprise products enable the rapid deployment of new and existing applications across both PalmOS and Windows CE handhelds to mobile corporate users. For example, sales representatives can use AvantGo products to stay up-to-date with their most important customer service logs and orders, as well as current product pricing and availability. Field service technicians with a wireless connection can instantly review their schedules and customer support histories, as well as maps and driving directions for service calls. Busy executives can also have current business metrics in their pocket at all times.

Wireless Knowledge, Microsoft and Qualcomm announced a service offering called "Revolv". Revolv allows access to e-mail, Microsoft Exchange based calendars and contact databases, and corporate networks for Windows CE based devices, smart phones and pagers, and other PDAs.

Nokia and Yellow Computing have developed a software to record order data via the Nokia 9110. The order data can be sent via ftp, WWW, Fax, e-mail, or SMS to company's headquarter.

5.2.11 Mobile SAP

With the Nokia 9110 it is possible to create a link between mobile phone and SAP R/3. In this case the sales representative will be kept informed of all offers by the headquarter via SMS.

Microsoft and German software giant SAP are teaming up to connect mobile devices with enterprise software. The two companies announced plans to develop technology for connecting Microsoft Windows CE-based and other mobile devices to SAP's business applications using the Internet and the Microsoft's BizTalk Framework.

5.2.12 Mobile Multimedia

NEC Corporation has developed the world's first MPEG-4 coder/decoder (CODEC) which enables the transmission and reception of both high-quality audio and video over the Internet and also next-generation mobile telephones. The MPEG-4 CODEC will enable mobile video-phones, video e-mail and cordless video cameras to become a realistic possibility.

The growth of the Internet and mobile communications allow for the incorporation all forms of audio, graphic and video data. From 2001, next generation mobile communication services will get underway and working mobile multimedia communications are expected to become available.

5.2.13 Mobile Digital Signature
The development of the digital signature as the legal equivalent of the traditional signature as a guarantee of authenticity, non-repudiation and integrity of electronic messages highlighted the fact that the use of Smart Card-based signatures created problems in the end device. Current Smart Card reading devices are still too expensive for widespread distribution to be possible. In addition, Smart Card reading devices also have the problem of requiring a fixed location, since a PC usually is required to view the document to be signed. Increasing demands being made on mobility and the fact that transaction services are on offer outside a client's own home - in Kiosk Terminals for instance – provided the impetus for the BROKAT concept of using mobile phone end devices as wireless Smart Card reading devices.

In the second half of 2000, the Asia Mobile Electronic Services Alliance, is to combine smart card functionality with cell phones to provide customers with mobile electronic services. Under the auspices of Singapore bank, Standard Chartered, in partnership with Singapore Telecom, Hong Kong telecom firm Smar-Tone, Nokia, Ericsson, Motorola and smart card vendor, Gemplus, the project will be tested in Hong Kong and Singapore.

In Hong Kong, the multi-application smart cards are to include a Visa Cash ePurse suitable for reloading via GSM channels. All cards issued in the pilot are to contain loyalty applications and digital certificates for use in an intended public key infrastructure. Additional industry sectors such as airlines, ticket firms and retailers are likely to join the consortium once a sufficient infrastructure has evolved.

6. Customer Benefits

Companies can offer end-users new channels and new mobile services. Besides offering their existing services via a new mobile channel, they have unlimited possibilities to create new services and products for their customers. Users of mobile phones can now be offered relevant personal services that suit their needs. Services that make everyday life much more convenient and are easily accessed via a mobile phone.

This means that there will be an explosion in the number of services that can be provided to a mobile phone. As a result, markets, brands and customer loyalities are on the move. Constant availability is the key concept for future competitiveness.

Better customer relationship
It seems just as certain that the mobile phone will become another cornerstone of multimedia communications. If implemented systematically, mobile phone banking ensures a user-friendly, permanently available customer service. Banks are not expecting a windfall from potential mobile phone services, but are hoping to lessen the burden on branches, while at the same time improving their customer relations. After all, better service has always paid off in the past.

Mobile Commerce Solutions via Over-the-Air-Platforms provide a step-function improvement in flexibility compared with other current efforts at personalised marketing. The one-to-one approach is probably at its most developed in the food industry for instance, where companies are producing databases and systems that track buying behaviour and allow an individual's loyalty to be rewarded with points and targeted offers related to their buying patterns. Cyberflex Mobile So-lution supports this type of one-way marketing facility and more.

7. Summary

One of the changes in the business environment has been the growing demand for mobility. This means that customers and employees should be able to access the information resources and services of a company wherever they are and whenever they want.

The size of the mobile community will soon exceed that of the fixed Internet. The characteristics of the handset and the bearer require that the service providers change the way in which services are supplied. Interfaces have to be simplified and a package of services offered which take advantage of the personal nature of the mobile phone. With the reduced opportunities for advertising and the branding of the service package by information provider or operator, new business models need to be employed.

The wireless community has, by necessity, redesigned the nature of Internet access to a more user friendly format, with instant access to a set of services personalised to the profile and location of the user.

Benefits are finally exceeding technology in the importance placed on the marketing of the Internet. The fixed Internet community would do well to integrate some of the ideas being developed in this sector.

Reaching mobile phone users with your services

By 2003, the number of mobile phones is estimated to be over one billion. So the market is really big. And fast-moving. The brands, partnerships and customer loyalities are on the move as mobility increasingly permeate everyday activities and lifecycles. To get their share of this totally new opportunity, the corporates need to add mobile phone into their service channels immediately.

Industry Trends

The future growth of the telecommunications industry will be driven by the growth of various value-added services. Users will appreciate the possibility to access valuable information and services regardless of time and place. These services will soar as a result of the digitalization of content and because the Internet Protocol provides a uniform way of delivering services to the customer.

Abbreviations

ACS	Access Control List
API	Application Programming Interface
BAS	Bestellabwicklungssystem
BDC	BROKAT Data Capsule
BLZ	Bankleitzahl
BTX	Bildschirmtext
CA	Certification Authority
CDMA	Code Division Multiple Access
CD-ROM	Compact Disc-Read Only Memory
CGI	Common Gateway Interface
COM	Component Object Model
CORBA	Common Object Request Broker Architecture
CPU	Central Processing Unit
CRL	Certificate Revocation List
CSP	Commerce Service Provider
ESDP	Electronic Services Delivery Platform
FTP	File Transfer Protocol
GB	Gigabyte
GDP	Generic Device Presentation
GPRS	General Packet Radio Service
GSM	Global System for Mobile Communications
GUI	Graphical User Interface
HBCI	Home Banking Computer Interface
HD	Hard Disk
HTML	Hyper Text Markup Language
HTTP	Hyper Text Transfer Protocol
ICR	Internet Relay Chat
IDEA	International Data Encryption Algorithm.V
IDL	Interface Definition Language
IIOP	Internet InterORB Protocol
IMAP	Interactive Mail Access Protocol
IOR	Interoperable Object Reference
IP	Internet Protocol
IT	Informationstechnologie
ITSEC	Information Technology Security Evaluation Criteria
JECF	Java Electronic Commerce Framework
JVM	Java Virtual Machine
MAC	Message Authentication Code
MB	Megabyte
MD5	Message Digest Vers. 5
NMT	Nordic Mobile Telephone
OMG	Object Management Grou
ODBC	Open Database Connectivity
ORB	Object Request Broker
OTS	Object Transaction Service
PCI	Peripheral Component Interconnect
PDA	Personal Digital Assistant
PDC	Personal Digital Cellular
PHS	Personal Handyphone System
PIN	Personal Identifcation Number
PKCS	Public-Key Cryptography Standard
PLZ	Postleitzahl
POP	Point of Presence
RAM	Random Access Memory
RSA	Rivest, Shamir, Adleman/Data Security Inc
RDO	Repository Defined Object

SCSI	Small Computer System Interface
SET	Secure Electronic Transaction
SHA	Secure Hash Algorithm
SMS (C)	Short Message Service (Center)
SMTP	Simple Mail Transfer Protocol.VI
SNMP	Simple Network Management Protocol
SQL	Structured Query Languag
SRT	Secure Request Technology
SSL	Secure Socket Layer
TAN	Transaction Number
TCL	Tool Command Language
TCP	Transmission Control Protocol
T & T	Track & Trace
UMTS	Universal Mobile Telecommunications Systems
URL	Uniform Recource Locator
WAE	Wireless Application Environment
WAP	Wireless Application Protocol
WDP	Wireless Datagram Protocol
WML	Wireless Markup Language
WSL	Wireless Session Layer
WSP	Wireless Session Protocol
WTA	Wireless Telephony Application
WTLS	Wireless Transport Layer Security
WTP	Wireless Transport Protocol
WTS	Wireless Transaction Protocol
WWW	Worldwide Web
ZVT	Zahlungsverkehrsterminal.

Let's WAP! With Ericsson Business Consulting

Title:	*Let's WAP! With Ericsson Business Consulting*
Author:	*Ericsson Business Consulting*
Abstract:	*Consumers are adapting to technology increasingly faster as they become more and more experienced and the services become easier and easier to use. Industries are responding to this development by introducing new technologies. Ericsson Business Consulting delivers the business solutions.*

Biography: see end article

Ericsson Business Consulting knows WAP

Technology and globalization are transforming business environments and trends into both personal and professional mobility. We're all becoming increasingly mobile and our efficiency has notably increased with the introduction of mobile business solutions. Adopting new technology has become a viable alternative to traditional service delivery chains.

Consumers are adapting to technology increasingly faster as they become more and more experienced and the services become easier and easier to use. This increased awareness is extending the reach of mobile information and services.

Industries are responding to this development by introducing new technologies that lower their costs, provide better information access and further control for their customers. Ericsson Business Consulting works with combinations of WAP, e-Commerce, mobile location services and other elements to deliver business solutions that suit the needs of your particular industry.

Examples of mobile network benefits:

- We're already benefiting from instantaneous, up-to-date, personalized and geographically based industry information.
- As consumers increase their use of mobile networks to execute services on their own, industry costs will decrease.
- Our customers benefit from convenient "in your pocket" information.
- Our customers can align service requests with their geographic location by implementing mobile location services.

Are my competitors using WAP?

Lots of WAP applications are being developed around the world at this very moment. Ericsson Business Consulting is constantly developing and perfecting business solutions where WAP plays an important role in enhancing business processes, customer relationships and collaboration techniques. Chances are that your competitors already are fit-for-fight for WAP. You have to make sure that you're just as fit and trim.

The flexibility of the WAP platform allows for a variety of potential applications, including advertising, e-Commerce and banking & financial services. The openness of the platform makes it possible for third parties and carriers to develop many additional applications.

How will this effect my business?

Mobile networking can help your business become more efficient while increasing productivity. WAP allows instant access to e-mail and allows your employees to stay in touch while they're out of the office. With WAP, the focus is on delivering simple, useful information to WAP Phones or PDA's. Users don' t need to unpack devices, boot-up or logon to get quick access to services like e-mail, headlines or customer information.

Ericsson Business Consulting's WAP solutions help you generate additional revenue now, while protecting your future investments.

Here are just some of a broad spectrum of possible applications:

- e-mail
- Information search and retrieval
- e-Commerce
- Calendar and office applications
- Maps, position-based services
- Voice, messaging, communication
- Call control
- Intranet/Internet access

In addition, by working together with well-known application providers, Ericsson will have a broad range of services available fo all kinds of consumers.

Banking & Finance Services

Work is currently being done to make banking services available for WAP. Some examples of these services are:

- Keeping track of your payments and account information
- Transferring money between accounts
- Paying your bills
- Applying for loans
- Use your WAP Phone to pay in any sho or restaurant just like you would with a credit card
- Subscribe to the latest financial news
- Stock trading
- Monitoring specific stocks

Ericsson collaborates with Östgöta Enskilda Bank

Östgöta Enskilda Bank, affiliated with Den Danske Bank, has signed a collaboration agreement with Ericsson Business Consulting on a unique solution enabling secure transactions over the Internet. The solution is built around wireless e-Commerce products from Ericsson that us a SET-based standard to connect to the bank's systems.

Travel & Ticketing

Work is currently being done to make Travel & Ticketing services available for WAP. Some examples of these services are:

- Ticket purchasing and sales
- Information gathering on special offers
- Booking hotel rooms and rental cars
- Viewing travel schedules
- Monitor travel schedules and receiving messages when a travel schedule is altered
- Access to guide books and tourist information
- Finding travel companions
- Locate nearby entertainment no matter where you are
- Receiving weather information on any location in the world

Ericsson cooperates with CitiKey

CitiKey, the leading mobile platform for city information and services, is enhancing its services by using WAP to deliver information and services to people on the move. CitiKey users will be able to reserve and pay for their movie or theatre tickets directly on their WAP Phone. Afterwards, they can locate nearby restaurants and have directions presented on digital maps. Users can access guides to hotels, restaurants and shopping, receive information on public transit and review a calendar of events.

"Mobile technology will revolutionize city information and services and will change the way people visit, experience and work in cities",
– Ziad Ismail, CEO of CitiKey.

Media & Entertainment

Work is currently being done to make media services available for WAP. Some examples of these services are:

- Subscriptions to news and monitoring specific news events
- Event ticket Purchasing and Sales
- Purchasing and sales of music
- Receiving information about, for example, the theatre; scheduled show times, actors, reviews, etc.
- Purchasing or subscribing to specific magazines, newspapers, book clubs, etc.
- Receiving programs, and schedules of upcoming events, etc.

That's entertainment and Ericsson Business Consulting

Ericsson and application providers have announced a range of new services that will be available on Ericsson's Mobile Internet portal, which is built for WAP. Consumers will be able to shop for CD's, get tips on good restaurants, the local sights and also get international news, weather forecasts and stock information all via their WAP Phones

Ericsson cooperates with Pactive

Pactive is a news broker, offering custom-filtered information from approximately 8,000 of the leading news providers in the world such as Dow Jones, CBS and AFP. It's now possible to get the latest news with a service that Pactive has developed as a part of Ericsson's mobile Internet Website. The news service allows visitors to read the latest international news regarding sports, finance, telecommunication and IT, weather forecasts from 40 cities around the world and stock quotes from the NASDAQ list.

Ericsson cooperates with Boxman

Boxman is Europe's largest Internet-based entertainment retailer and will soon be launching its first WAP application – in close cooperation with Ericsson on this project.

Biography: Ericsson is the leading provider in the new telecoms world, with communications solutions that combine telecom and datacom technologies with freedom of mobility for the user. With more than 100,000 employees in 140 countries, Ericsson simplifies communications for its customers - networkoperators, service providers, enterprises and consumers - the world over. Additional information WAP, Wireless Application Protocol, is the de facto worldwide standard for providing Intnernet communications and advanced mobile services on digital mobile phones, pagers, personal digital assistants and other wireless terminals. WAP is compatible with GSM 900, 1800 and 1900 MHz, cdmaOne, TDMA wireless standards, as well all of the proposed 3G communication systems. Already in June 1999, Ericsson delivered the world's first end-to-end WAP 1.1system - the first true WAP standard. Since then, Ericsson has already delivered more than 40 WAP 1.1 compliant systems to mobile network operators, worldwide. To promote Internet access using a mobile phone, Ericsson was part of establishing the Wireless Application Protocol (WAP) Forum. The general objective of the forum was to enable the wireless industry and content developers to provide compatible products and services across a wide variety of platforms. WAP-compatible devices are capable of translating the incoming Internet data to match the capabilities of the mobile terminals that are unable to display HTML-based Web pages.

Mobile stock trading via WAP

Title: *Mobile stock trading via WAP*
Author: *Ericsson*
Abstract: *Ericsson, Postbank and Libertel are cooperating to realize mobile stock trading via WAP. In the long term this will provide Postbank customers with mobile access to all financial services the bank offers.*

Biography: see end article

One of the leading banks in the Netherlands, Postbank has signed a contract with Ericsson Business Consulting and Libertel to start mobile banking. Mobile Stock Trading will be the first application available. Over time, clients of the bank will be able to make mobile payments.

"We feel proud that Ericsson Business Consulting has been selected as a partner to the Postbank to bring these innovative Mobile Internet banking solutions to the Dutch market," says Haijo Pietersma, Executive Vice President and head of the business segment Enterprise Solutions at Ericsson. "We strongly believe that Mobile Internet will create new ways for companies to develop new businesses and interact with their customers. Mobile Internet will bundle communication and information to and from the mobile phone user and drive the development of a whole new range of personalized services."

Postbank will select their most active stock traders to take part in a pilot that will run for 6 months, starting in April. With the Ericsson R320 WAP mobile phone they will have access to real-time stock information and be able to place their orders live on the Amsterdam Stock Exchange, whenever they want, wherever they are. This way they can manage their stock portfolio anytime, anywhere through the Ericsson mobile WAP phone.

In this project, Ericsson Business Consulting is responsible for the design and implementation of the WAP application. Postbank selected Ericsson because of the solutions the company offers today and its complete product and service range. Dutch mobile operator Libertel (part of Vodafone/AirTouch) has been chosen as the launching operator.

This WAP stock trading pilot is the beginning of a cutting-edge development which, in the long term, will provide Postbank customers with mobile access to all financial services the bank offers.

"Postbank has always been a leader in remote banking. Our aim is to offer no-nonsense financial services in a multi-channel environment," says Kees van Rossum, Executive Vice President of Postbank. "Keeping the success of former innovations in remote banking in mind, we are determined to keep pace with new technologies, such as Internet and mobile telephony. We believe that the WAP-technology enables us to develop even more convenient banking services. Postbank will always be within reach for everyone."

For more information on Ericsson Business Consulting's WAP campaign, please visit http://www.ericsson.se/letswap Ericsson is the leading provider in the new telecoms world, with communications solutions that combine telecom and datacom technologies with freedom of mobility for the user. With more than 100,000 employees in 140 countries, Ericsson simplifies communications for its customers - network operators, service providers, enterprises and consumers - the world over.

About WAP

Wireless Application Protocol (WAP) is the main worldwide industry standard for providing Internet communications and advanced services on digital mobile phones, pagers, Personal Digital Assistants and other wireless terminals. WAP is compatible with all mobile standards, including cdmaOne, GSM, PDC and TDMA, as well as 3G systems. Ericsson, together with Nokia and Motorola, has introduced a new initiative to simplify recognition of mobile Internet applications. The Mobile Media Mode (WWW:MMM) will enable users, content providers and operators to recognize services, Internet sites and devices such as smart phones, which provide access to these services.

More information about WAP is available at http://www.wapforum.org/. Ericsson's Mobile Internet portal is: http://mobile.ericsson.se/mobileinternet

About Postbank

With some 7 million account holders, Postbank - an autonomous company within the ING Group - is one of the largest financial service providers in the Netherlands, serving a substantial number of Dutch households. The bank also maintains business relations with some 75% of all businesses in the Netherlands. ING/Postbank's payment services are an important aspect of the bank's overall services. At Postbank more than 50% of all payment traffic in the Netherlands is handled. Postbank is a modern and independent financial service provider, which is leading the way in the supply of a complete range of services.

About Libertel

Libertel N.V., domiciled at Maastricht (NL), is the second biggest provider of mobile telephone services in the Netherlands. At the end of December 1999 its number of customers totaled 2,100,000. Libertel has been listed on the official market of the

AEX stock exchange since 15 June 1999 and 70% of its shares are in the hands of the world-wide operating Vodafone /AirTouch Group.

Biography: Ericsson is the leading provider in the new telecoms world, with communications solutions that combine telecom and datacom technologies with freedom of mobility for the user. With more than 100,000 employees in 140 countries, Ericsson simplifies communications for its customers - networkoperators, service providers, enterprises and consumers - the world over. Additional information WAP, Wireless Application Protocol, is the de facto worldwide standard for providing Intnernet communications and advanced mobile services on digital mobile phones, pagers, personal digital assistants and other wireless terminals. WAP is compatible with GSM 900, 1800 and 1900 MHz, cdmaOne, TDMA wireless standards, as well all of the proposed 3G communication systems. Already in June 1999, Ericsson delivered the world's first end-to-end WAP 1.1system - the first true WAP standard. Since then, Ericsson has already delivered more than 40 WAP 1.1 compliant systems to mobile network operators, worldwide. To promote Internet access using a mobile phone, Ericsson was part of establishing the Wireless Application Protocol (WAP) Forum. The general objective of the forum was to enable the wireless industry and content developers to provide compatible products and services across a wide variety of platforms. WAP-compatible devices are capable of translating the incoming Internet data to match the capabilities of the mobile terminals that are unable to display HTML-based Web pages.

WAP Banking and Broking - Software System White Paper

Title: *WAP Banking and Broking - Software System White Paper*

Author: *Macalla Software*

Abstract: *According to surveys, there will be over 530 million wireless subscribers by the year 2001. By 2005 the number of wireless subscribers will break the one billion mark. Financial services are a key commercial driver for the mobile commerce market in Europe and beyond. Macalla's Mobility software products allow any investment bank, retail bank or brokerage to exploit their existing online banking operation and backbone to deliver new mobile banking applications in a rapid and straightforward way.*

Introduction

Financial Services and E-Commerce

According to the Strategis Group there will be over 530 million wireless subscribers by the year 2001. By 2005 the number of wireless subscribers will break the one billion mark, and a "substantial portion of the phones sold that year will have multimedia capabilities."

These multimedia capabilities include the ability to retrieve e-mail, and push and pull information from the Internet. In order to guide the development of these exciting new applications, the leaders of the wireless telecommunications industry formed the Wireless Application Protocol (WAP) Forum (www.wapforum.org).

The Wireless Application Protocol is the de-facto world standard for wireless information and telephony services on digital mobile phones and other wireless terminals.

Handset manufacturers representing over 75% of the world market across all technologies have committed to shipping WAP-enabled devices.

Carriers representing nearly 100 million subscribers worldwide have joined WAP. This commitment will provide tens of millions of WAP browser-enabled products to consumers by the end of 2000. WAP allows carriers to strengthen their service offerings by providing subscribers with the information they want and need while on the go. The WAP Forum has published a global wireless protocol specification based on existing Internet standards such as XML and IP for all wireless networks.

The WAP Forum has the following goals:

- To bring Internet content and advanced data services to wireless phones and other wireless terminals
- To create a global wireless protocol specification that works across all wireless network technologies
- To enable the creation of content and applications that scale across a wide range of wireless bearer networks and

device types
- To embrace and extend existing standards and technology wherever possible and appropriate.

The demand for WAP based services is destined to be a significant growth area. The potential for services delivered through mobile Internet technology requires new technology to facilitate the development and deployment of these new services.

Mobile Financial Services
Financial services are a key commercial driver for the mobile commerce market in Europe and beyond. Delivering their online financial services to mobile devices creates a powerful new service channel for the financial institutions. In a recent survey by Nokia, looking at which application types the market might demand, mobile banking was the top application –demanded by more than 85%.

Macalla's Mobility software product allows any investment bank, retail bank or brokerage to exploit their existing online banking operation and backbone to deliver new mobile banking applications in a rapid and straightforward way.

Mobile Banking
Mobile banking is a subset of online banking, a service that is being currently offered by most banks in Europe. The motivator for mobile banking from the bank's perspective is to have an additional distribution channel and to further cut costs, as every transaction on the internet, fixed or mobile, is saving money especially in the traditional retail operations.

The typical services to be provided through mobile banking include:

- check account and credit card balances
- check transaction history
- check exchange rates/interest rates
- transfer funds
- pay bills
- order currency

Mobile Broking
Access to stockbroking services through a mobile device has high value for both private and professional traders alike. The ability to receive real-time price or news information 24 hours a day, and then act on this information – irrespective of their current location is a new facility widely available only through the use of WAP and mobile devices. Mobile broking services need typically to provide the following key functionality:

- buy and sell stocks, options, mutual funds, other financial instruments
- receive alerts about price-movements
- receive message when order is executed
- check quotes
- manage portfolio
- browse and delete existing orders.

Mobility - Macalla Software's WAP Banking and Broking Product
Mobility represents a breakthrough in applying WAP and XML technology in financial services.

This comprehensive end-to-end software product provides the means for investment banks, retail banks and brokerages to rapidly deploy applications that securely deliver personalised content, transactions and payment services across multiple channels including WAP enabled handsets, PDA's, Pagers and GSM mobile phones and the Internet.

Mobility utilizes core software infrastructure and functional components developed by Macalla and leverages market leading encryption, PKI and telephony products.

Incorporating XML, WAP, Java (JSP, EJB's) and CORBA technologies Mobility facilitates application interoperability and enhanced data representation. It provides a robust and scalable platform for delivering m-enabled banking and broking services.

Mobility can be deployed on any of the predominant Internet server and messaging environments and provides a Host Level Interface facilitating seamless integration with Host applications.

Mobility is the culmination of Macalla's real-world experience having already designed, developed and deployed integrated WAP and web solutions for leading retail banks, investment banks and stockbrokers. The base level functionality available via Mobility include following (see table 1).

With growth potential that far outstrips the Internet, mobile devices offer ubiquitous connectivity between businesses and their customers. The need to architect for this channel is immediate and Macalla's Mobility addresses that need.

About Macalla Software
The Company
Macalla Software (www.macalla.com) is a leading provider of software products and e/m-commerce solutions for the global financial services sector. Our XML/WML and WAP expertise, software and products (Mobility, ECHO and DynamiX) provide the

Table 1

Investment Banking	Retail Banking	Stockbroking/Securities
Research	Bill Payments	Order placement and review
Indications of Interest	Account Review	Real-time Market Information
Real-time Market Information	Statement Details	Portfolio Valuation/ Statements
Positioning/P&L Monitoring	Funds Transfer	Stock Watch Lists
Order Entry	Transaction Search	Alerts
Portfolio Management	Alerts	Analyst Reports

essential infrastructure for banks, brokers and other financial institutions to rapidly extend their service reach to include new Internet and wireless (WAP, pagers, PDA's, GSM mobile phones, etc) channels and markets. Headquartered in Dublin, Ireland Macalla has offices in London and New York.

Core Competence

Macalla's expertise is in the design, development and delivery of applied products and e/m-commerce solutions for the global financial markets. Our offerings exploit the distributed object and componentware paradigm allowing sector participants better respond to the rapidly changing demands of their existing business.

Macalla has a deep understanding of both the technical and market issues that are unique to the wireless environment when enabling content and transactions access from handheld devices. We have proven our ability to successfully work within the innate constraints of WAP devices (restricted CPU, memory and battery life, and a very small/simple user interface), wireless networks (constrained by low bandwidth, high latency, and unpredictable availability and stability). And importantly, we understand and know how to address the uniquely different demands of wireless rather than desktop or even laptop Internet users.

Macalla's raison d'être is to enable financial service organisations gain access to a whole new market of content and transactions hungry users while protecting and leveraging their current investments in web technologies. By specifically engineering our products, solutions and delivery capabilities to integrate wireless media such as WAP, Macalla is assisting these organisations give their customers real anytime, anywhere information and service access on a variety of networks and devices such as mobile telephones, pagers, personal digital assistants (PDAs) and other wireless terminals.

Alliances

Macalla is a member of the WAP Forum (www.wapforum.org) and a number of industry XML initiatives, and is actively working with leading global financial organisations in specifying the standards for next generation trading and investment banking systems. Our technology relationships include Baltimore Technologies, CMG, IONA Technologies, Phone.com, Nokia, Oracle, Phone.com, Reuters, SilverStream, Sun Microsystems and Sybase.

Corporate

Macalla is privately owned and in early '99 accepted funding from 3i Group plc., Europe's leading investor and the Guinness Ulster Bank Equity Fund. Further investment followed from the Irish Government, via its company development initiative – Enterprise Ireland.

Mobility – Software Architecture

Macalla's Mobility consists of a number of components that communicate securely with each other to provide a complete solution when developing mobile banking and broking solutions.

The core components of Mobility are as follows:

- XML based Communication Channel
  Extensible Mark-up Language (XML) is a mark-up language for documents containing structured information. A mark-up language is a mechanism to identify structures within a document. The XML specification defines a standard way to add mark-up to documents. In the financial markets, XML has rapidly become the universal language for structuring business critical information and is defining the rate at which application interaction can be achieved.

  XML has industry wide support – IBM, Sun, Microsoft and Oracle all have endorsed it as a technology, and have implemented support for XML in many of their products.

Figure 1

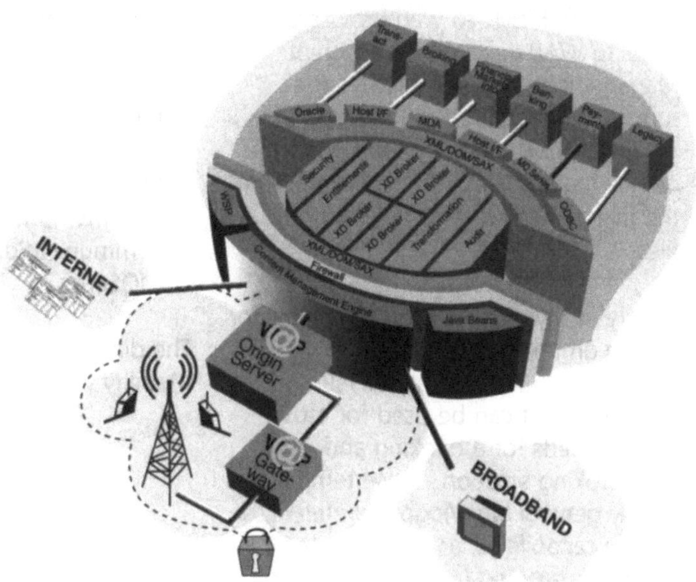

Because Macalla's Mobility uses XML as the representation of information, developers can utilise the many available products specifically geared to solving problems associated with developing and deploying XML based solutions, which means a faster time to market.

The communications channel utilises Macalla Software's XDBroker and provides a scalable, secure and reliable method of communication of XML documents between different applications.

- Java Beans for encapsulating requests and responses
  All XML documents are communicated through Java Beans.

Java Bean encapsulation encourages more readable and maintainable JSP code, as the Java Beans use the vocabulary of the application for class and method names.

Java Beans decouple the Java code from the precise structure of the XML documents, increasing maintainability of the JSP code.

- Content Management Engine
  Mobility's Content Management Engine is an extensible Java component that is deployed on the web server. It acts as the main controller of the XML requests that get sent to the XML communications channel, and also controls what is output to the connecting client. The Mobility Content Management Engine can be configured to handle requests from different client devices and respond to the devices based on the requirements of that device, e.g. HTML to web browsers, WML to WAP devices. The Content Management Engine separates the functionality required to interface with the Communications Channel from the display requirements, thereby simplifying the maintainability of the JSP output files.

- Library of Java Server Page templates for display on the mobile devices
  Mobility includes a library of JSP templates that can be used for most display needs for a banking and stockbroking solution. These templates supply general login/logout/preferences display capabilities, as well as specific pages for banking (account details, transaction history etc) and stock portfolio (list of stocks, graphing history data etc).

- Tools for automatically creating and visually editing Java Server Pages
  Mobility provides tools that integrate with existing JSP tools such as IBM's WebSphere and WebSphereStudio, to facilitate the construction of JSP files, and custom XML documents.

  The tools facilitate the rapid development and extension of Mobility application content with the use of XML based applications for the construction of new JSP templates.

  The tools also facilitate the creation of new XML documents and their associated Java Beans.

- Host Interface Application, for integration within an existing banking environment. This Mobility application module interfaces between the XML Communications Channel and the banking/broking host systems. It is extensible and can interface to a number of different backend host systems via the available communications protocol, e.g. IBM's MQSeries.

The diagram overleaf (figure 2) shows ho Mobility works into a WAP environment.

Figure 2

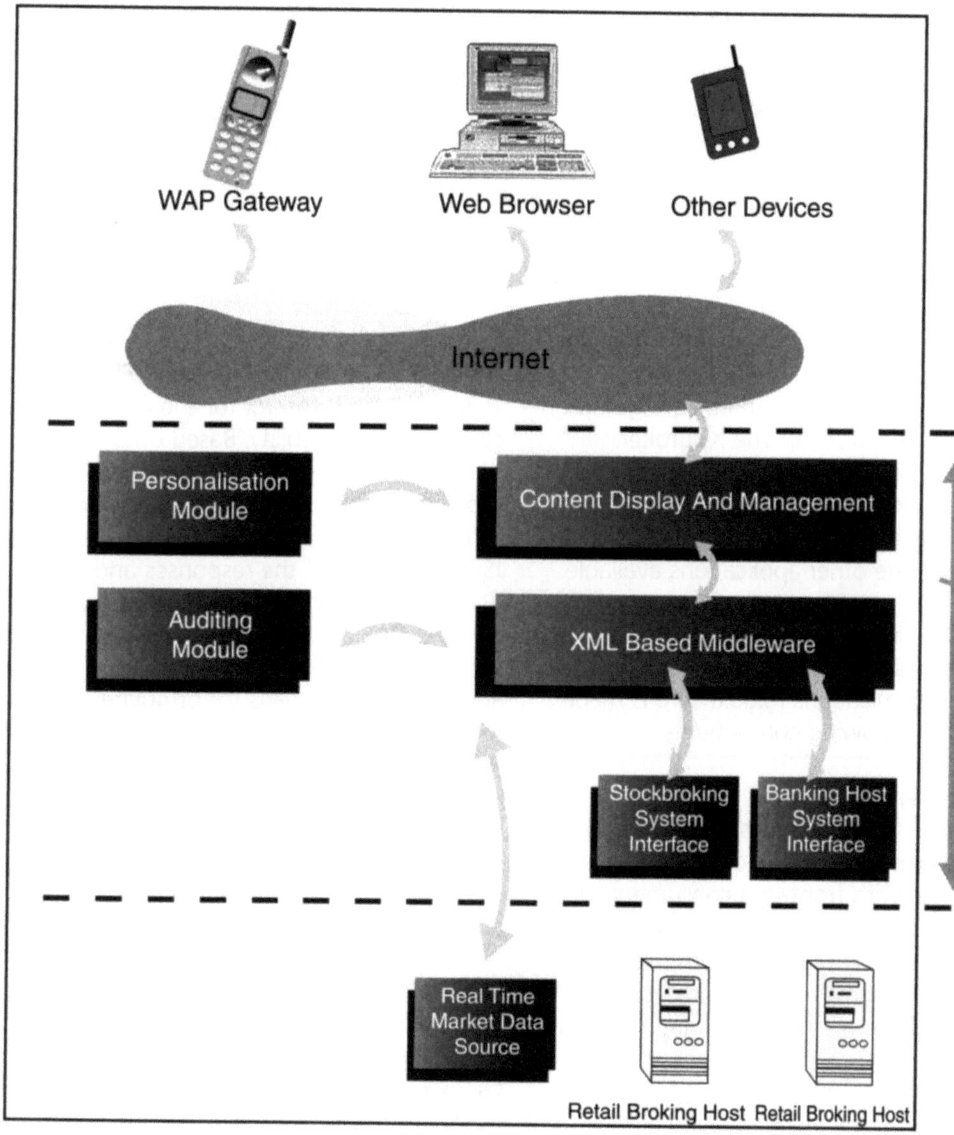
WAP Gateway
Web Browser
Other Devices
Internet
Personalisation Module
Content Display And Management
Auditing Module
XML Based Middleware
Mobility
Stockbroking System Interface
Banking Host System Interface
Real Time Market Data Source
Retail Broking Host
Retail Broking Host

A complete sample mobile banking/broking implementation is included with Mobility. It can be tailored to suit a particular bank/ broker's needs. This sample implementation consists of a sample host integration module and an extensive library of Java Server Pages (JSP) that are used to display the banking solution on the WAP phone.

Scalability and fault tolerance are provided and Mobility specifically allows multiple alternates for each component within the system. For instance, multiple Host Interface applications can be configured on a particular XDBroker. The XDBroker distributes requests automatically between these applications. If one application goes down requests are automatically forwarded one of to the other applications available.

Mobility – XML Communications Middleware
The communications middleware is made up of the following components:

- Macalla's XDBroker
  The XDBroker provides the capability to link back-end sources of content and services with front-end consumers. The XDBroker allows for content and services to be retrieved from multiple sources. It provides a scalable, secure and reliable method of communication between the web server and sources of content and services. By linking a number of XDBroker's together, a large-scale deployment is possible that would link multiple sources of information to many consumers of that information.

- A set of Java Beans that allow for the simple communication between the front-end and the back-end.
  These beans allow an XML Request to be easily created and passed through the XDBroker to the back end system and receive the XML Reply from the host system. These beans also act as a local cache of data from the host system. For instance, a request for a transaction history may return more data than is displayable at a time on the mobile device.

Mobility – Content Management Engine
This is the main control for the content display side of Mobility. Based on a Model/View/Controller pattern, the Mobility Content Management Engine acts as the controller within this pattern, the JSP as the view of the responses and the XML requests and responses as the model. It takes WAP requests for content and services from the phone, calls the appropriate Java Beans to formulate XML requests passes this request to the XML Channel and waits for responses. When a response arrives it then decides based on the contents of the response and the capabilities of the phone which JSP file to use to format the appropriate response to be sent back to the phone.

It also interfaces with the Mobility Personalisation module to decide what content should be displayed at what time. For instance, a user may have a preference to always see his current balance after he has logged in. These details are stored in the Personalisation module.

All errors returned from the host system (generated internally) are returned as XML responses. These responses are then handled specially by the Engine. These responses are forwarded to special JSPs used for displaying errors. In this way, error handling is separated from normal content display.

Figure 3 below displays how this Engine controls the client side of the Mobility system.

Mobility – Library of JSP Templates and Tools
A library of JSP templates is supplied that can be used for most display needs for banking/stockbroking solutions. These templates supply general login/logout/preferences capabilities, as well as specific pages for banking (account details, transaction history etc) and stock portfolio (list of stocks, graphing history data etc).

JSP tools allow the developer to take the JSP templates and automatically generate JSP based on the specific XML information stored within the responses. A JSP editor allows for WYSIWYG editing of these JSP to customize these pages for a particular banking needs.
Mobility delivers a truly dynamic data-driven application maximising WAP's limited bandwidth /resource capabilities.

Figure 3

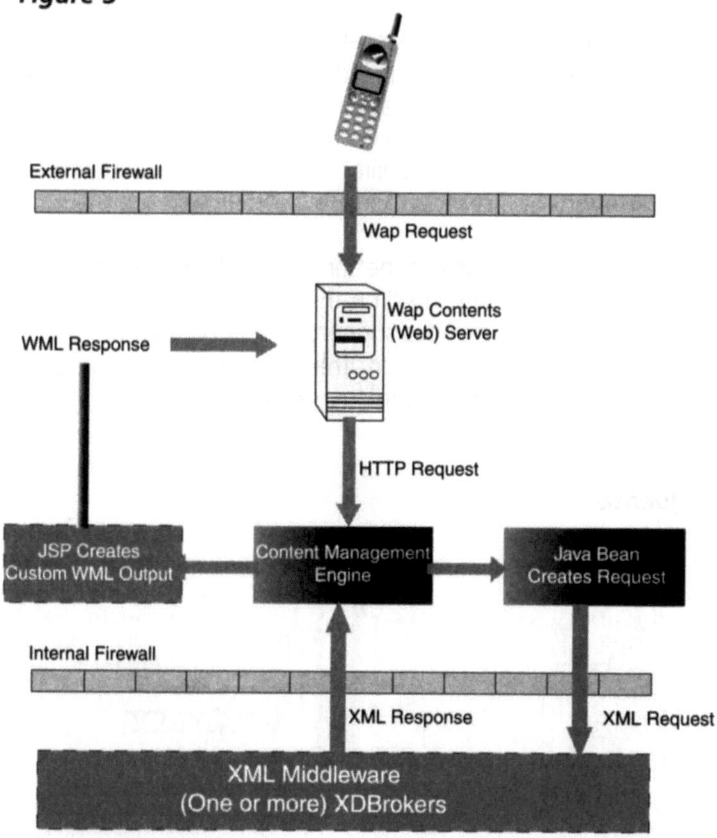

Mobility - Personalisation Module
The Mobility Personalisation module allows each user a customized and configurable interface, essential creating a "Personal Financial Portal". In conjunction with the Content Management Engine this module can be used to store personal preferences of what is display at particular times. The Personalisation module also has configuration information about

how content can be tailored for particular WAP devices, e.g. Support for tables, colours etc.

Mobility – Auditing Module
A full audit trail is available for every message exchanged. The Content Management Engine logs every request and every JSP that is used to display the replies. The XDBroker can be configured to log every message exchanged on the XML Channel and the Banking Interface logs every request received and every reply returned.

Mobility – Host Interface Application
This module communicates between the XDBroker middleware and the banking / broking host system. The developer can extend this application to allow requests to be passed to the back end host system. Java, C and C++ extensions are supported directly within this extensible application to allow for easy integration within an existing environment.

Mobility – Common banking and broking Services

There are a number of requirements for a banking/broking solution. Investment banking demands ready access to research data, positions, [real-time] market information as well as functionality in the areas of order entry and portfolio management. Typical retail banking requirements are to provide customers with services that allow them to view bank account details, check statements, payment of bills and transfer funds between accounts. Stockbroking requirements include allowing order entry and review, portfolio valuation, stock watch lists, alerts and the like. Other services include access to real time /delaye market data information and access to market news. Macalla's Mobility provides the capability to easily source content and services from varied sources.

Each of the services implemented is described on the following pages. Mobility includes support for the common requirement that cross both m-banking and m-broking services.

Figure 4: A typical login sequence

Login
The Login request must be able to authenticate the user against the backend host system. At the start of a session the user is required to input login information, this may vary depending on the security requirements of the Bank. Mobility supports multiple methods of authentication, such as registration ID/password (pin) model, challenge /response model or a PKI model. Below is a sample implementation that uses the Registration ID/Pin method for authentication. When a customer connects the initial screen that they see depends on what personal information that they have previously configured on the system (see profiling below).

Logoff
The logoff terminates the current session for the client. The client receives a confirmation message telling them that their session has ended.

Figure 5

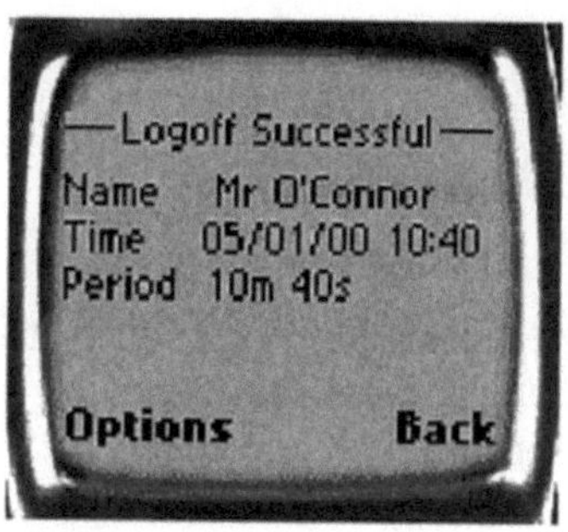

Profiling
Mobility's profiling gives the customer the ability to personalise the content and services that they wish to receive and how they wish it to be formatted based upon a stored profile for the customer. This is an important feature as it allows the customer a degree of individual attention. With profiling the customer can decide what bank accounts that they wish to receive information on, decide what groups of news items they wish read or create a personal watch list of stock indices: creating a true wireless "Personal Financial Portal".

Mobility - Retail Banking Services

The retail banking service allows customers to easily monitor and manage their accounts with a bank.

There are a number of common requirements that a mobile retail banking solution would contain. Customers should be able to retrieve their account balances, view account statements, transfer funds between accounts and schedule payment of bills from their accounts.

Mobility's inherent allows for further facilities to be easily added.

Balance Information
A request for balance information returns a list of accounts and the corresponding balance information for each account. The customer could pick an account and retrieve more detailed information for that account. The customer can configure the level of detail that is displayed. See figure 6.

Account Information
The account information displays information on a selected account. Balance

Figure 6

information, account type, name of account, the facilities that are available for the account and any general related information are some of the fields that are contained in an account information response.

Figure 7

Statement Information

Statement information returns a statement of transactions that have taken place on a particular account within a specified time period. See figure 8.

Current Day Transactions

This contains a statement of transactions that have taken place on the current banking day for a selected account.

Figure 8

Transaction Search

Clients are allowed to search by date or amount for transactions that have taken place on any of their accounts. The result is a statement of transactions that match the search criteria.

Figure 9

Customers use the bill payments facility to schedule, modify or delete payments to payees that appear on an approved list. The payees are usually larger service providers that would have many clients, typically telephone or electricity suppliers. Bill Payments can be immediate or be scheduled. The Scheduled payment should be able to be repeated at specified intervals. So the options are:

Figure 10

- Immediate Bill Payment.
  When selected, this means that the funds will be transferred from the selected account with immediate effect.

- A once off scheduled Bill Payment.
  Schedule the transfer of funds to the approved payees account to take place on a specific date.

- Scheduled Bill Payment, recurring at a specified interval.
  Scheduled bill payments are used when the payment is to be made at some known recurring interval.

There is also the facility to cancel a recurring bill payment if the payment has been scheduled to occur some time in the future.

Funds Transfer

The funds transfer facility allows a client to transfer funds between two accounts that they hold at the same bank. The request specifies the source and destination accounts and the amount that should be transferred.

Like bill payments, the funds transfer can be immediate or scheduled and could also be recurring.

Mobility m – Broking Services

Market Information

This allows for the sourcing and provision of market information such as quotes, news or charts. This information can be sourced from a wide selection of the most popular vendors. This module also allows for manipulation and entitlement of this information so that it can be intelligently distributed and utilized. It is enabled to offer many levels of service from delayed

Figure 11

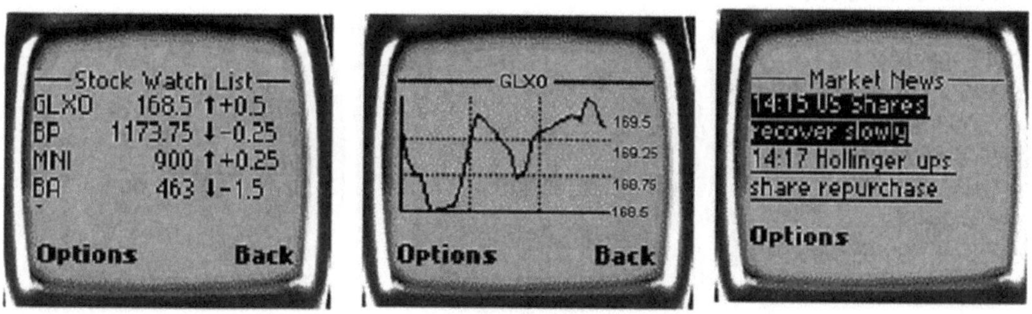

snapshot to real-time information. The market data can also be used to drive portfolio valuation and other areas of functionality such as messaging.

Messaging
Mobility's messaging functionality allows sophisticated notification of movements in stock prices or portfolios. This messaging is driven off market data feeds and can be set to send notifications based upon limit or percentage breeches. Also notification of order fulfillment can be dispatched. The messages can be distributed using SMS, e-mail or WAP.

Figure 12

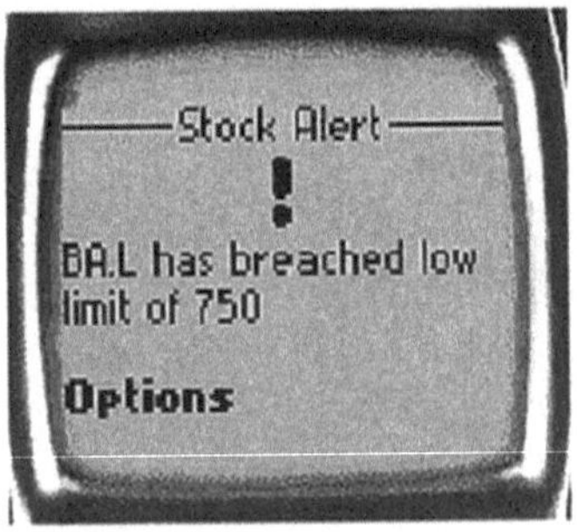

Account views
This facility enables customers to view a number of different reports relating to their holdings. Macalla's component architecture allows the sourcing of information from varied sources and the subsequent construction of reports based on pre-determined logic. The layered approach of having data sourcing into pre-determined JSP templates means that the construction of further reports is made simple: (See figure 13).

Figure 13

- Portfolio Valuation: This report is used to provide valued information across all the various holdings in a customer's account. This is perhaps the most important of all the reports. All calculations are done on the fly as the information is moved through Macalla's XDBroker. See figure 14.

- Transaction Listing: This is a display of a activity on the customer's account over given period of time. It shows all buys

Figure 14

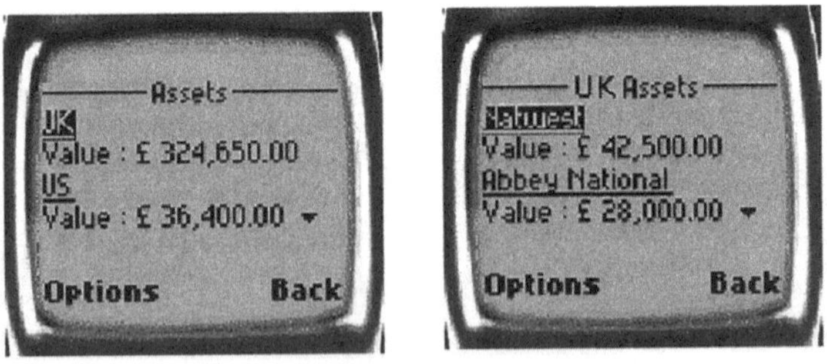

and sells along with interest and dividends accruing.

Figure 15

- Cash Statement: This shows all cash movements across an account both in and out.
- Income Statement: This is all income that is accruing to the account over a given period of time.
- Dividend Statement: All dividend payments on equities held plus all associated tax related outflows.

Order Placement

This Mobility feature allows customers to directly place orders on the order book for filling. They can be provided with indicative prices. Macalla's software architecture allows for direct integration to the back office systems and facilitates the implementation of straight through processing. Orders can be placed, deleted and reviewed.

Example 1 – Statement Information Enquiry

The Statement Information enquiry returns a statement of transactions that have taken place on a particular account within a specified time period. This example shows the sequence of events, and the XML Requests/Responses, that take place when a customer makes a request for a statement of a particular account that they have with the banking institution.

Figure 16: Sequence Diagram for a Statement Enquiry

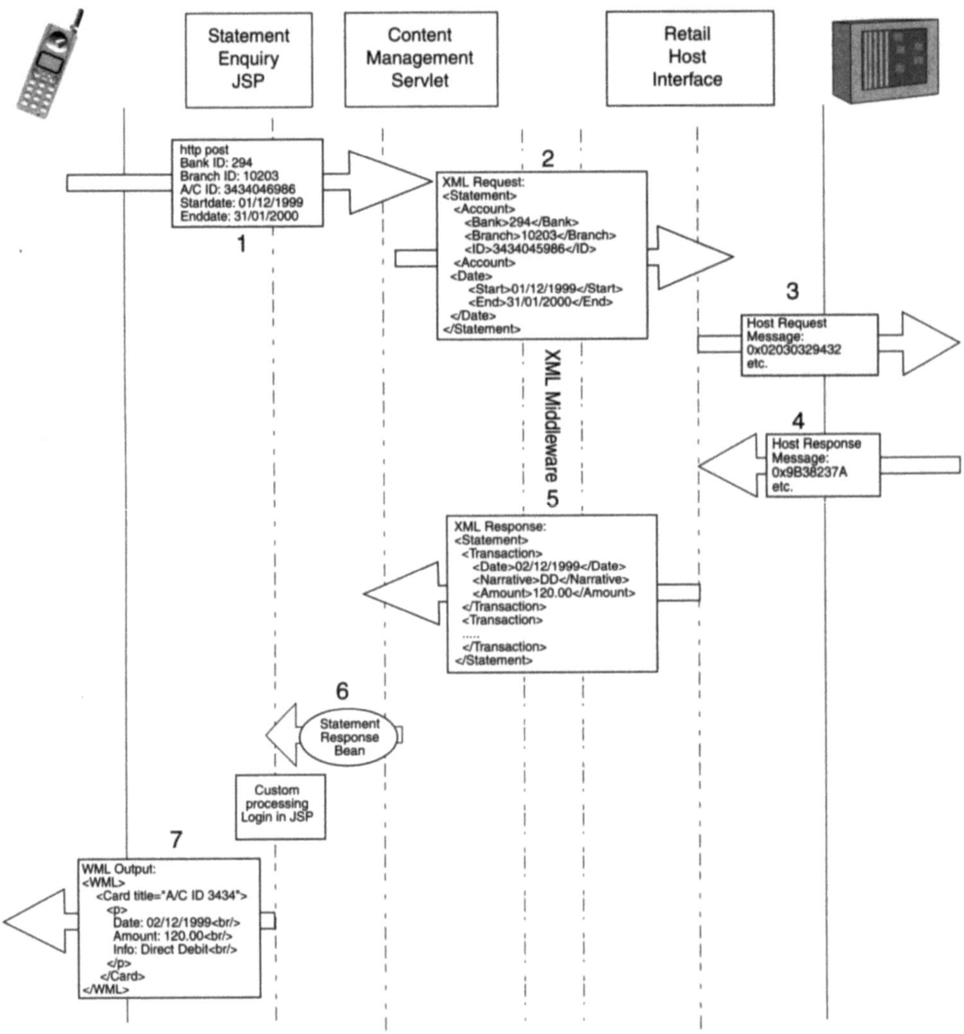

Here the diagram is broken down in order to explain each step more thoroughly.

1. When a customer selects to view a statement of transactions from a selected account, they perform a HTTP POST of the relevant account details into the Mobility Content Management Engine. The Content Management Engine has been configured to call the Statement Enquiry Java Bean for this POST message.
   The Statement Enquiry Java Bean contains methods that reflect the XML Request document that will be submitted to the Communication Channel. The generated XML Request has the following format:

Figure 17

```
<?xml version="1.0" ?>
<RBS>
<Statement>
  <Account>
    <BankID>Macalla Bank</BankID>
    <BranchID>92-22-56</BranchID>
    <ID>29154999</ID>
  </Account>
  <Date>
    <Start>01/09/1999</Start>
    <End>20/01/2000</End>
  </Date>
</Statement>
</RBS>
```

2. The Mobility Host Interface Application receives the request document and calls the onRequest() method on the implementation of the DocumentRequestHandler interface with the above request document.

3. A StatementRequest helper object is created, this provides methods for easy retrieval of the request information.

4. A handler for the statement request is called. This method generates a request to the existing backend back end host system or database and returns the generated response document using the StatementResponse helper class. See figure 18.

Figure 18

```
StatementResponse handleStatement(StatementRequest request)
{
    // Decode the request parameters here and
    // create an appropriate message to send to the host system.
    String strBankID = request.getBankID();
    String strBranchID = request.getBranchID();
    String strAccountID = request.getAccountID();
    ... ... ...
    // Send Request to host (e.g. using MQSeries etc.) and wait for the response
    ... ... ...
    Vector translist = (Vector) m_Host.getStatement(strBankID, strAccountID, ... );

    // Create a response based on the response from the host
    StatementResponse xmlResponse = new StatementResponse();

    for(int index=0; index< translist.size(); index++)
    {

        HostTrans hostTrans = (HostTrans)translist.elementAt(index);

        // Insert the transactions into the XML Response.
        Trans responseTrans = new Trans();.
        responseTrans.setDate(hostTrans.getDate());
        responseTrans.setNarrative(hostTrans.getInfoText());
        responseTrans.setAmount(hostTrans.getAmount());
        xmlResponse.InsertTransaction(responseTrans);
    }
    // Return the
    return xmlResponse;
}
```

5. The XML response that is generated has the following format:

Figure 19

```
<?xml version="1.0" ?>
<RBS>
<Statement>
   <Transaction>
      <Date>1/1/99</Date>
      <Narrative>Direct Debit</Narrative>
      <Amount>300.00</Amount>
      <Currency>GBP</Currency>
      <Type>DR</Type>
   </Transaction>
   <Transaction>
      <Date>2/1/99</Date>
      <Narrative>Lodgement</Narrative>
      <Amount>1500.00</Amount>
      <Currency>GBP</Currency>
      <Type>CR</Type>
   </Transaction>
   <Transaction>
      <Date>2/1/99</Date>
      <Narrative>Bill – British
      Gas</Narrative>
      <Amount>85.98</Amount>
      <Currency>GBP</Currency>
      <Type>DR</Type>
   </Transaction>
</Statement>
</RBS>
```

6. The Mobility Content Management Engine receives the response. From the response it decides which helper Java Bean is appropriate to call, in this case it is the Statement Response Bean. The statement response bean provides a set of methods that allow for the simple retrieval of the response information.

7. This Mobility Statement Response Bean is now available for display purposes. The Content Management Engine decides on an appropriate display module to handle the response to the client. An example of a JSP file below shows the retrieval of the statement information from Statement Response Bean and returning WML code to a WAP device.

Figure 20

```
<%
Response.setContentType("text/vnd.wap.wml");
%>
<BEAN NAME="statementResponse"
TYPE="com.macalla.WAPBank.StatementResponse"
SCOPE="session">
</BEAN>

<?xml version="1.0"?>
<!DOCTYPE wml PUBLIC "-//WAPFORUM//DTD WML 1.1//EN"
                              "http://www.wapforum.org/DTD/wml_1.1.xml">
<wml>
   <%
   int count = statementResponse.getCount();
   for(int index =1; index<count; index++)
   {
   %>
   <card id="transaction<%=index%>"
title="A/C:<%=statementResponse.getNumber(index)%>
                                             <%=index%>/<%=count%>">
   <p>
      Date <%=statementResponse.getDate(index)%><br/>
      Amount <%=statementResponse.getAmount(index)%><br/>
      Info <%=statementResponse.getNarrative(index)%><br/>
      <a href="#transaction<%=index+1%>"><%=index+1%> of <%=count%></a>
   </p>
   <do type="prev"><prev/></do>
   </card>
<% } %>
</wml>
```

8. The result to the WAP device.

Figure 21

Example 2 – XML Content Management using XML Composer

Mobility uses XML as the standard data representation format. In addition to the extensive library of JSPs and XML schemas to address the requirements of banking applications, Mobility also provides an information content management tool, called XML Composer to facilitate the creation of XML documents from existing ones. These new XML documents can then be used to drive dynamic content creation.

This can be useful in the following situations:

- When the required information is a subset of the available information, and it is required to hide information that is not relevant.
- Where the required information is sourced from multiple XML documents, and the preference is not to manually construct the composite XML document when it is required.

This tools works with the JSP files and Java Beans that drive the dynamic content generation. When used, this tool generates a bean that is capable of dynamically composing the required XML document in response to a request. This simplifies the code required in the JSP page, as it is presented only with the information it requires, and does not need to perform searching/filtering of the response data in order to generate the desired content. The composite XML document is constructed in response to a series of XQL queries on existing XML documents.

The example screen below shows how portfolio summary information can be extracted from a full portfolio XML document, and used as the basis of a summary portfolio page. A HTML view of the Portfolio is presented first (Figure 22), highlighting the regional stock groups. In this example a full portfolio XML (Figure 23) document contains a complete list of all stock holdings in multiple regions, and is more detailed than what is required. The desired information is just the total stock holding value for each region (Figure 24).

New XML documents are created via drag and drop operations, from source XML documents. New XML elements and attributes can also be added to the new document as necessary.

When the new XML document has been created, it is available as a JavaBean, which can be used within a JSP page, as illustrated in Figure 24.

Figure 22: HTML view of portfolio

Holding	Description	Cost	Price	Value	Gain or Loss	Div Yield %
Irish	**Equities**	EUR	EUR	EUR	Eur	EUR
	Direct Equities					
5,000	Irish Life and Permanent Plc.	21,700	9.10	45,500	23,800	6.35
12,500	Abbey Plc.	0	3.40	42,500	0	4.32
7,000	Cpl Resources Plc.	14,700	3.20	22,400	7,700	3.10
8,000	Abbey National	13,360	3.50	23,000	14,640	2.76
25,000	Ovoca Resources Plc.	6,200	0.35	3,750	2,550	2.31
	Total in EUR.	195,673		324,650	86,477	
Irish	**Special Sector**	EUR	EUR	EUR	EUR	
	Special Sector					
8,000	Aerospace Industries Plc	25,000	2.40	19,200	-5,800	1.60
	Total in EUR.	25,000		19,200	-5,800	
UK	**Equities**	GBP	GBP	GBP	GBP	
	Direct Equities					
2,750	Glaxo Wellcome Plc.	15,433	4.34	11,935	-3,498	2.56
3,350	Unilever Plc.	9,345	3.56	11,926	2,581	5.67
4,200	Smithkline Beecham	12,456	2.43	13,206	-2,250	4.76
	Total in GBP.	37,234		34,067	-3,167	
	Total in EUR.	57,411		52,527	1,881	
UK	**Equities**	GBP	GBP	GBP	GBP	
	Equities					
10,000	Aea Technology	34,000	1.50	15,000	-19,000	3.50
	Total in GBP.	34,000		15,000	-19,000	
	Total in EUR.	52,424		23,128	29,296	
US	**Equities**	USD	USD	USD	USD	
	Direct Equities					
4,375	Microsoft Corp.	11,346	4.50	19,688	8,342	1.11
2,500	Medtronic Inc.	8,346	2.76	6,900	-1,446	7.45
5,00	Compaq Computer Corp	12,045	2.45	12,250	205	1.57
5,250	Bank of America	18,345	9.45	59,063	40,717	7.43
	Total in USD	50,082		97,901	47,818	
	Total in EUR	46,333		90,571	44,238	

Irish Equities

UK Equities

US Equities

Figure 23: Original portfolio XML document


```
- <Portfolio>
  - <Client>
      <Name>John Smith</Name>
      <Address>Macalla Software Ltd., 172 Merrion Road, Dublin 4, Ireland.</Address>
    </Client>
    <ValuationDate>29/09/1999</ValuationDate>
  - <Stockgroup region='Irish' type='Equities' subtype='Unit Linked'>
    + <AbbeyPlc>
    + <AbbeyNational>
    + <AlliedDomecq>
    + <Greencore>
      <EUR_cost_total>223,297.55</EUR_cost_total>
      <EUR_value_total>324,650.00<EUR_value_total>
      <EUR_gain_total>101,352.45</EUR_gain_total>
      <EUR_income_gross>1,640,247.50</EUR_income_gross>
    </Stockgroup>
  - </Stockgroup> region='US' type='Equities' subtype='Direct Equities'>
    + <Compaq>
    + <BankofAmerica>
    + <Microsoft>
      <USD_cost_total>50,082.45</USD_cost_total>
      <USD_value_total>78,462.50</USD_value_total>
      <USD_gain_total>28,380.07</USD_gain_total>
      <USD_income_gross>181,250.65</USD_income_gross>
      <EUR_cost_total>47,292.21</EUR_cost_total>
      <EUR_value_total>4,091.13</EUR_value_total>
      <EUR_gain_total>26,798.94</EUR_gain_total>
      <EUR_income_gross>171,152.65</EUR_income_gross>
    </Stockgroup>
  - </Stockgroup> region='U.K.' type='Equities' subtype='Unit Equities'>
    + <Unilever>
    + <SmithklineBeecham>
    + <Glaxo>
      <GBP_cost_total>37,234,48</GBP_cost_total>
      <GBP_value_total>34,067.00</GBP_value_total>
      <GBP_gain_total>-3,167.49</GBP_gain_total>
      <GBP_income_gross>146,754.59</GBP_income_gross>
      <EUR_cost_total>58,599.63</EUR_cost_total>
      <EUR_value_total>53,614.65</EUR_value_total>
      <EUR_gain_total>-4,985.00</EUR_gain_total>
      <EUR_income_gross>230,962.38</EUR_income_gross>
    </Stockgroup>
</Portfolio>
```

Figure 24: New XML document

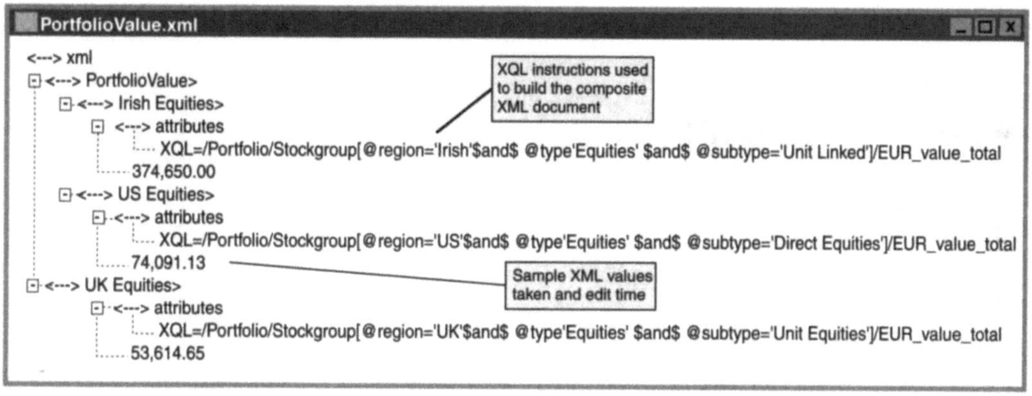

An example JSP file using the Bean created above, is as follows:

Figure 25

```
<%
Response.setContentType("text/vnd.wap.wml");
%>
<BEAN NAME="accountValue" TYPE="com.macalla.WAPBank.AccountValue"
SCOPE="session">
</BEAN>

<?xml version="1.0"?>
<!DOCTYPE wml PUBLIC "-//WAPFORUM//DTD WML 1.1//EN"
                "http://www.wapforum.org/DTD/wml_1.1.xml">
<wml>
   <card id="AccountValue" title="Portfolio Valuation">
      <p>
                  <b>Account: </b><%=accountValue.getName()%><br/>
                  <b>Irish: £</b><%=accountValue.getIrishEquities()%><br/>
                  <b>US: $</b><%=accountValue.getUSEquities()%><br/>
                  <b>UK: £</b><%=accountValue.getUKEquities()%><br/>
      </p>
      <do type="prev"><prev/></do>
   </card>
</wml>
```

When the JSP is executed the following display would be created:

Figure 26

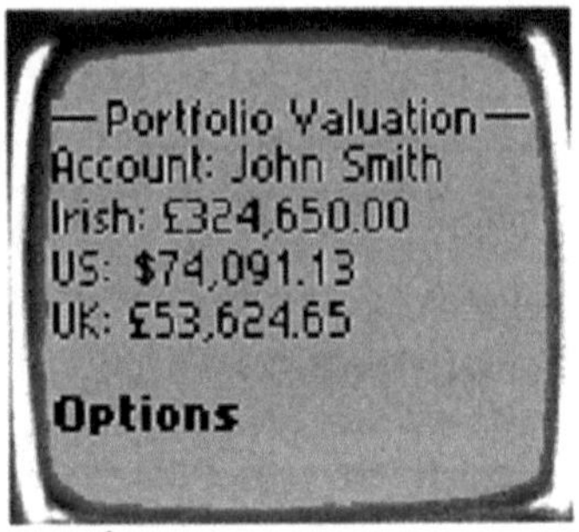

GSM: Worldwide Connection For Mobile Work Force -- Technology Includes Business Features, Easy Installation For Global Connections

Title: *GSM: Worldwide Connection For Mobile Work Force -- Technology Includes Business Features, Easy Installation For Global Connections*

Author: *Alyson Behr, Information Week*

Abstract: *Most multinational enterprises require global communications for a significant portion of their work force. From the CEO to product managers, people need to connect effectively. They need this whether they're in Hong Kong overseeing product manufacturing, in Switzerland attending a conference, or connected to their LAN in their home office.*

Introduction

Global System for Mobile Communications provides a worldwide connection for a mobile work force. It's widely used in Europe and Asia and will be readily available in the United States by year's end.

GSM, introduced four years ago in Europe, is a digital cellular technology that uses radio frequencies to provide wireless global data computing capabilities that include sending and receiving E-mail and faxes, secure LAN and intranet access, as well as Web browsing and Short Messaging Service capabilities.

The good news is that GSM delivers these applications over a single voice channel per user, using a notebook computer with a PC Card connected to a GSM-enabled cellular phone, in addition to standard cellular voice calling. The bad news is that circuit-switched data delivery speeds are only 9,600 Kbps.

There are more than 200 GSM networks with three different frequency ranges in use: The 900-MHz range is most commonly used in Europe; the 1,800-MHz range, also known as the Personal Communication Network, is growing in western Europe, Russia, and the United Kingdom; and the 1,900-MHz range in the United States, where it's known as Personal Communication Services and Data Communication Services. The term PCS is not used strictly for a GSM network; it can also refer to a digital network that operates in the 1,900-MHz range.

GSM cellular-service subscribers equipped with a multiband phone with a Subscriber Identity Module, or SIM card, can use the GSM roaming feature anywhere their service providers have formed a roaming agreement with another network.

Digital networks are still moving toward full-scale U.S. coverage. Meanwhile, several developments will benefit global cellular data services. Manufacturers will achieve data transfer rate increases to 2 Mbps, a bandwidth expansion that will make more enterprise applications in the next five years.

Quicker Transfer

Two technologies that will mature in the next two years are High Speed Circuit-Switched Data and Packet-Switched Data for GSM (GPRS). Early next year, HSCSD will extend single-channel data-transfer rates to 14.4 Kbps and provide the ability to use two channels simultaneously, which will double transfer speeds to 28.8 Kbps.

The introduction of GPRS, probably in 2000, will improve GSM transfer rates. Currently, GSM requires a dedicated radio channel, whether it's transferring data or not. GPRS requires it only when transferring data, so other traffic won't wait for the user to disconnect the phone before it uses the channel. When it does, expect carriers to base fees on data-transfer volume.
Multiband handsets recently were made available in the United States. They operate between either 900-MHz and 1,800-MHz networks or the 900-MHz European band and 1,900-MHz U.S. band. Bosch, Ericsson, Motorola, and Siemens are shipping GSM handsets.

As GSM becomes more widely available in the United States, home banking, E-commerce, and online trading applications will be modified for the mobile environment. Database applications for businesses will increase significantly as the mobile work force accesses the company intranet. Eventually, GSM capacity is expected to support videoconferencing.

Travel Light

To pack for an international trip, you need multiple PC Cards, cables, adapters, and a checklist of connection options stored on your notebook. Omnipoint Communications Services Inc.'s GSM network and associated wireless GSM products decrease your IT packing to one PC Card with a cable, a notebook computer, and mobile phone.

I tested Option International's GSM-Ready 33.6-Kbps PC Card Modem and the GSM-enabled World 718 phone from Bosch Telecom over Omnipoint's GSM digital network with a Micron Electronics Transport XKE notebook.

The testing environments included my Los Angeles lab, where I receive a strong signal, and the mountainous area behind Malibu, Calif., where products are more likely to come in contact with fires, mudslides, and rattlesnakes than an acceptable cellular connection. I also tested in Switzerland. I included an editorial meeting with U.S. business partners from a remote location (a two-hour hike from the nearest road) high atop a mountain. From a cow pasture with a view of Mont Blanc, my reception was exceptional, with only slight intermittent transmission delays.

Overall, these are strong products that require a little work to learn to use them. But the global connection benefits and simplicity that GSM can deliver far outweigh the minimal time it'll take to learn them.

World 718 Multiband Phone

Bosch's World 718 900/1900 Multiband phone comes with a removable SIM card, a battery that gives you 3.2 hours of talk time with 80 hours of standby time, an A/C battery charger adapter, a belt clip, and a user manual.

The phone provides support for new services, including conference calling, call diversion, call barring, call waiting, and caller ID. It also sends and receives SMS, supports data and fax transmissions with a compatible PC Card, and has a calculator. The phone also has a phone book feature and supports speed dialing.

World 718 has some other useful features. I turned the handset microphone off during a conversation so I could have a private conversation with someone else in the room. I transferred a call after I had completed my discussion on a conference call so the other parties could continue without me.

World Phone's backlit display is large enough to read SMS messages easily and see descriptive icons that indicate when you have SMS and voice messages. It weighs 6.8 ounces and is small enough to fit inside a jacket pocket.

Reception in the Los Angeles area using Omnipoint's alliance with the Pacific Bell PCS network registered maximum signal strength. The Malibu area, notoriously brutal on cellular users, varied tremendously, ranging between one bar and four, sometimes dropping the connection. Calls from Switzerland to Clifton, Va., using Omnipoint's alliance with Swiss Telecom were excellent overall with occasional intermittent transmission delay.

Option International's GSM-Ready 33.6-Kbps PC Card modem is compatible with the World 718 and other GSM-enabled cellular phones, such as the Ericsson 788. It ships with BVRP Software's Phone Tools 1.05 telephone management software and a multilanguage user manual. Option makes a 56-Kbps/v.90 GSM-ready PC Card, but compatible cables were unavailable for testing because the World 718 was released so recently. A flash upgrade for the 33.6-Kbps card can be downloaded from Option's Web site. Other than landline speed, there were no significant operational differences. Installation of the PC Card drivers was automatic in the Windows 95 plug-and-play environment. The product is also available for the OS/2 and Mac OS operating systems.

Phone Tools sends and receives data as well as faxes, configures your PC and phone as an answering device, and manages your hardware configuration and voice mail options. After installing the software, I was able to send faxes. I was able to receive faxes sent from another GSM phone with little trouble, but I was unable to receive faxes sent from a standard line. Unfortunately, the manual did not offer troubleshooting suggestions. Option says technical support will be available for Phone Tools this month.

I connected to my analog dialup account from Los Angeles using the RJ11 cable and connected at speeds between 28.8 Kbps and 31.2 Kbps. Using the Bosch, a compatible adapter, and GSM technology, I connected at 9,600 Kbps every time, whether I was in the United States or Switzerland. This speed is satisfactory for E-mail download, but any attempt to download Web pages or files is punishingly slow, especially in Switzerland, where cost is a factor. This is not the PC Card's fault, but a temporary limitation of an otherwise useful first-generation technology.

Option includes a lifetime guarantee on the flash ROM upgradable PC Card.

Wireless e-SecurityTM

Title:	*Wireless e-SécurityTM*
Author:	*Baltimore Technologies*
Abstract:	*The growth in the wireless market is being driven by the immense, universal popularity of mobile phones, personal digital assistants (PDA) and handheld PCs (HPC). Services and applications for these devices are increasing rapidly. In order to deliver these services and applications in a secure, scalable and manageable way, new architectures and protocols are being designed. The Wireless Application Protocol (WAP) is a result of continuous work to define an industry-wide specification for developing applications that operate over wireless communication networks. The WAP specification is developed and supported by the wireless telecommunication community so that the entire industry and most importantly, its subscribers can benefit from a single, open specification.*

Why WAP?

WAP has been designed to work within the constraints that mobile wireless devices have to operate in. These devices have limited display capabilities and simple user interfaces. They have limited processing power, battery life and storage capabilities. Additionally the network provision is inherently more unreliable relative to the wired world of the Internet. Overcoming these hurdles is no easy task, but it has been done. This year has heralded a spate of WAP enabled devices, service and content providers, network operators and infrastructure providers. WAP is here.

The WAP-enabled wireless world represents a huge new market for anyone involved in e-commerce. Essentially the number of users is no longer constrained to PCs connected to the Internet, which itself is a massive market. The e-commerce market is set to explode to new limits fuelled by both the astronomic growths of the Internet and the WAP-enabled wireless world.

The WAP-enabled wireless world represents an opportunity for Enterprises to benefit from new levels of communications and remote access. Now employees will be able to access applications and information from anywhere. This leads to a streamlining of business processes and makes the company more competitive e.g. sales people can have access to vital data such as the latest pricing, competitive data and product availability from customer sites, which in return reduces the sales cycle and leads to faster revenue streams.

Wireless e-Security™

The issue of security dominates e-commerce and the enterprise. Whilst e-commerce enables businesses to operate on a global basis without physical presence, it also represents new challenges in assuring both customers and merchants that they are operating within a trusted environment. For Enterprises, the security of corporate assets is paramount. Information, intellectual property, documentation, content, networks, data, computers, employee and customer data can all be considered part of a corporation's assets. These should all be safeguarded by a corporation to ensure it remains competitive within its marketplace.

Cryptography offers virtually unbreakable systems for security on open networks. A wide range of new Internet security systems have evolved in the past few years which utilize cryptography for a wide variety of requirements. These systems secure other systems such as messaging, email, electronic payments, software applications and network communications.

The Telepathy WST extends the ability to create such security systems from the Internet to the WAP-enabled wireless world. Telepathy WST allows developers to build in Wireless e-Security. At a basic level security can be divided into a number of elements: confidentiality (privacy), authentication, authorization, integrity and non-repudiation. Telepathy WST enables developers of system to build in confidentiality, integrity and authentication. Authorization and non-repudiation can also be incorporated by integrating with more sophisticated systems such as Public Key Infrastructures (PKI). Telepathy WST has built-in functionality that makes this integration to PKIs a simple process.

Security in WAP

The WAP specification is a major achievement because it defines for the first time an open, standard architecture and set of protocols intended to implement wireless Internet access. Wherever possible, existing standards have been adopted or have been used as the starting point for WAP technology. Optimizations and extensions have been made in order to match the characteristics of the wireless environment.

The key elements of the WAP specification include:

- WAP Programming Model

Very similar to the current WWW Programming Model. See figure 1.

Figure 1: Wap programming model

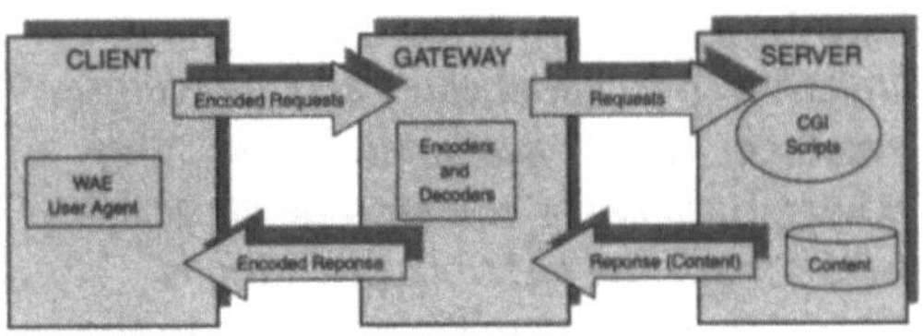

- Wireless Markup Language (WML) and WMLScript (WML Scripting Language)

A markup language adhering to XML standards that is designed to enable powerful applications within the constraints of handheld devices.

- Micro-browser Specification

A specification for a WML/WMLScript aware micro-browser in the wireless terminal that controls the user interface and is analogous to a standard Web browser.

- Wireless Telephony Applications (WTA) Framework

This allows access to telephony functionality such as call control, phone book access and messaging from within WMLScript applets.

- WAP Stack

A lightweight protocol stack to minimize bandwidth requirements, guaranteeing that a variety of wireless networks can run WAP applications securely, see Figure 2.

Wireless Transport Layer Security (WTLS)

Security within WAP is mandatory. It initially appears in the form of WTLS. WTLS provides the key security elements of confidentiality, integrity and authentication. WTLS is the wireless version of the industry standard Transport Layer Security (TLS), which is equivalent to the widely used Secure Sockets Layer (SSL) 3.1. TLS provides a secure network connection session between a client and a server, most commonly used between a web browser and a web server.

The transformation of TLS to WTLS is based upon the need to support datagrams in a high latency, low bandwidth environment. To operate within this environment WTLS provides an optimized handshake (initiation of a secure session) through dynamic key refreshing. Dynamic key refreshing allows encryption keys to be updated on a regular and configurable basis during a secure session. This not only provides a higher level of security, but also provides considerable bandwidth savings on the relatively costly handshaking procedure.

The additional security elements of verified authentication, authorization and non-repudiation are provided by integration into a PKI. Baltimore Technologies range of Telepathy WST products allow users to implement WTLS and take full advantage of PKI systems. The Telepathy WST range of products supports open standard APIs and protocols to allow the widest possible interoperability with

Figure 2: Security within WAP is provided by WTLS

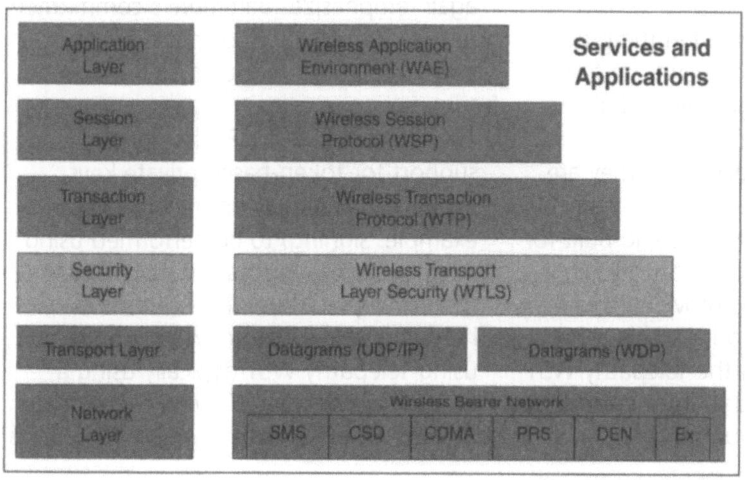

existing open standards based implementations.

Telepathy WST Overview

Telepathy WST is a powerful software development kit allowing application developers to create secure encrypted sessions between online networked applications. W/Secure contains an implementation of Wireless Transport Layer Security (WTLS) 1.1 allowing developers to build full strength security into their WAP V1.1 client and server applications. WTLS 1.1 is the mandated security layer within any WAP v1.1 compliant product.

Figure 3: WAP clients and gateways can incorporate security with the Telepathy WST

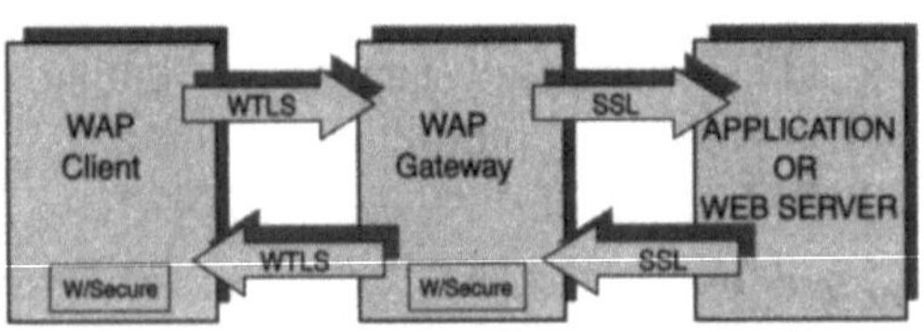

The high level API provided in the product, abstracts the developer away from the complexities of cryptography, and allows them to concentrate on the core functionality of the applications they are building. Underlying security can easily be integrated into applications being built for the wireless world. Basic certificate handling functionality is provided as a standard part of the product. This allows applications built using the Telepathy WST with an excellent foundation to enter the now widely adopted world of PKI.

The Telepathy WST allows the developer to integrate WTLS data encryption capabilities into any online networked application. This entails the ability to configure the security parameters to be used for authentication data security, and to initiate and receive WTLS-secured connections.

The Telepathy WST API includes fully configurable support for:

- Session caching
- Security re-negotiation
- Temporary key reuse
- Dynamic re-configuration during a session
- Integration into datagram layers defined in the WAP specification i.e. UDP/IP and WDP

Telepathy WST supports a number of standards regarding the secure storage of private keys and digital certificates. These include PKCS#1, #7, #8 and #12, allowing both private keys and certificates to be integrated into security applications from all major industry-standard formats. This again emphasizes Baltimore's commitment to fully follow and implement all relevant standards in our products.

The Telepathy WST API also contains support for token-based private keys, allowing private key operations (for example: signing) to be performed using any private key mechanisms available to the developer. Thus smart-cards and hardware tokens can now be integrated using Telepathy WST (typically using a PKCS#11 interface) allowing even greater security when establishing a WTLS session.

Telepathy WST supports a wide range of public key cryptographic algorithms, which can be configured within its cipher suites:

- RSA
- Diffie-Hellman
- RC5
- DES, Triple DES
- IDEA
- SHA-1
- MD-5

The key strength used by these suites is configurable to suit the level of security required.

Telepathy WST is available as a set of ANSI C libraries with an object orientated interface on the following platforms:

- Win32 (Windows 95/98, Windows NT)
- Sparc Solaris 2.5+
- HP-UX 10.2+
- Linux

session. A Telepathy WST support service is a utility object that provides support for Telepathy WST sessions; support facilities include providing security parameters, CA certificates, session caching facilities, etc.

Before you can establish a Telepathy WST session with a remote host, you must decide what cipher suites you will accept, what Certification Authorities you will trust, etc. Telepathy WST allows you to do this by creating and configuring a Telepathy WST Support object. A single support service can be used to support multiple Telepathy WST sessions. When creating a support service, you must initially specify whether your application is an WTLS server or an WTLS client. The distinction is usually obvious: A client makes outgoing connections to WTLS servers; a server receives incoming connections from WTLS clients.

Telepathy WST Architecture

There are two major concepts involved in the use of Telepathy WST: Telepathy WST sessions and Telepathy WST support services. A Telepathy WST session is a single WTLS-secured connection with a remote host; data can be securely read from and written to a Telepathy WST

Figure 4: Telepathy WST Architecture is based around Session and Support Service objects

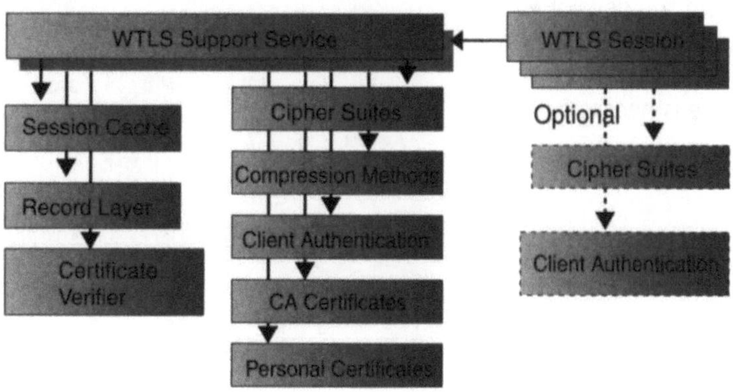

e-Commerce and PKI Integration

The WAP specification is an ongoing process. Some of the basic elements of security such as confidentiality, authentication and integrity have now been addressed with the publication of the WTLS specification. Baltimore Technologies is one of the first companies to market with products that meet this specification.

However full participation in e-commerce requires that the additional security elements of authorization and non-repudiation be addressed. In real terms this implies integration with PKI systems that have already been deployed and new systems for the future. In the wireless arena these systems will be defined in WAP. Interoperability with these different systems is a key design principle with any of Baltimore's e-security products.

Telepathy WST contains the seeds to begin this integration and will be enhanced in the future to make this as seamless as possible. Underlying protocol and format changes are hidden from the developer, so they can be easily introduced when the time is necessary.

Baltimore Technologies

Baltimore Technologies is a leading supplier of global e-security solutions for e-commerce and enterprise systems. These solutions are based on a family of products and services offered directly by Baltimore and through approved members of Baltimore's Trusted World™ selected channel partners.

Each product contributes to the overall vision of Baltimore's e-security framework, which is designed to offer full strength security for a variety of business contexts. This security enables companies to operate more efficiently and to offer new levels of customer service.

Telepathy WST is an essential component of Baltimore's vision for security. It provides developers with the necessary foundation to build to secure wireless applications with strong cryptography.

Figure 5: Wireless e-Commerce with complete security

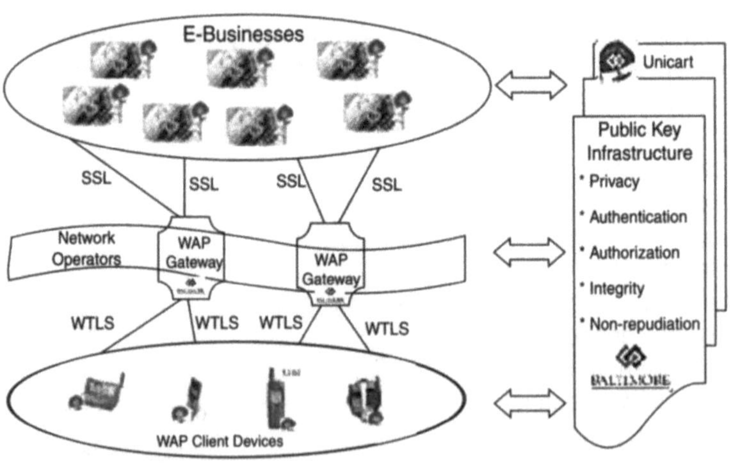

Baltimore Technologies is a member of the WAP Forum and provides regular input into the security working group tasked with defining e-security standards in the wireless world.

Email Connectivity for WAP Enabled Mobile Terminals

Title: *Email Connectivity for WAP Enabled Mobile Terminals*

Author: *Dialogue Communications*

Abstract: *The growth of mobile phone usage, particularly in the business sector has provided people with increased voice connectivity to their colleagues, customers and suppliers. The advent of mobile data has revealed a high demand for data connectivity for people working away from their office, and the use of the internet has shown how any business can operate worldwide. People want to access business information, and they want to do it from their mobile phone.*

Introduction

In 1998, 73% of European corporations were using some kind of mobile data solution and 91% of those who were not said, that they would within 1999. In addition, 60% of the mobile data users in the United States said they wanted to use a mobile phone for the mobile data instead of a computer or some other device. These results clearly show that there is a true demand for access to more information and services on mobile phones.

Until recently the only mobile data available on a GSM mobile phone was SMS (the short message service). SMS was limited to text messages of 160 characters in length, or binary messages of 140 bytes. This was limited but many organisations attempted to deliver a variety of value added data services using SMS, and SMS has been recognised by GSM operators throughout Europe as an unqualified success in terms of the growth of message volumes. But SMS on its own was not enough and gradually everyone began to see that the web-browser paradigm was the one most likely to offer long term success.

Some companies, particularly Nokia had already developed SMS encoding to enable information services to be delivered to so called 'smart' terminals. This encoding was known as ' Smart Messaging' and it used a markup language known as TTML (Tagged Text Markup Language). In parallel with this Unwired Planet developed HDML (Handheld Device Markup Language) which would enable small devices to operate micro-web browsers. These early attempts to deliver functionality as well as data to mobile phones showed some of the possibilities of this medium, but also highlighted some of the dangers. Competing technologies together with inappropriate bearer services could have compounded to stifle the growth of mobile connectivity.

Fortunately however the industry realised that the benefits of working together out-weighed the risks of sharing technology and the WAP Forum (Wireless Application

Protocol) was created with the objective of delivering a set of standards for device independent, and bearer server independent technology for delivering content and functionality. To date the WAP Forum has over 90 members and represents over 90% of all mobile users.

It is estimated that in 2005 there will be about one billion mobile phone subscribers in the world, and that a substantial portion of the phones sold that year will have WAP capabilities. The implication of this forecast has not been missed by those currently leading the IT industry. Nokia alone sell more mobile phones in one month than PCs are sold in one year, underlining the importance WAP based phones have to offer in connecting people to Internet and Intranet based services. Businesses that have benefited from a web presence are now focussing on reaching mobile browsers. No one is in any doubt that by 2005 people will be using their phones for a wide range of data activities, what nobody really knows is which suppliers of technology will be most successful in this new arena.

Integrating Two Powerful Environments

The email and the mobile phone have become two fundamental tools in millions of peoples' working and social lives. They still remain however, unconnected giving rise to separate places to check for messages and separate contact points for an individual. Together with fixed telephones and fax a complex communications environment has evolved.
The advent of unified messaging systems has gone simplified the communications environment by providing a single universa inbox where all messages are represented by electronic mail with attachments for fax and voice mail. Users require a view onto this message store when both when they are have access to their desktop PC, and also when they are mobile.

All operations performed on a message (e.g. read, delete, move folder, forward, reply etc.) need to be reflected in the message store. WAP based email connectivity gives the user the ability to view and act on messages and ensure that these actions are recorded in the central message store. Many existing ways of delivering email to mobile devices such as PDAs involve messages being downloaded and result in duplicating the message stor

Mobile Email – A new strategy

Broadly there are two categories of email users. Those that use corporate email systems, for example those based around Microsoft Exchange or Lotus Notes, and those that use internet email systems and have an account with an ISP. Dialogue aim to address both of these markets with the products and services strategy.

Figure 1

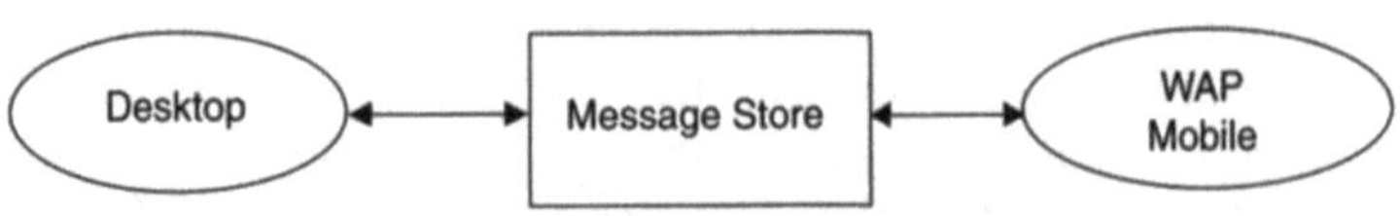

Mobile Email Products for Corporates – providing access to existing corporate email systems

Mobile Email Products for ISPs and Mobile Operators – providing carrier class solutions for provision of large scale services

Email System Access

Dialogue WAP based email connectors provide a mobile viewing portal onto existing email systems. A proxy application server translates calls from the mobile device into native calls to the email system across a LAN or WAN.

The functionality provided by the email connector may vary depending on the type of system being accessed. Features such as calendar or task management are for example offered by MS Exchange but not in POP3 based email systems.

Access to All Email Hosted Services

Email systems today have common features that enhance basic messaging:

- Calendaring
- Personal Address books
- Access to public Directories (e.g. LDAP)
- Task Management (to do lists)
- Forms and Workflow

WAP enabled mobile phones have similar features

- Calendaring
- Personal Address books
- Tasks and reminders

Clearly there is a need to ensure that these separate systems and message stores can be synchronised in order to maximise the value to the end user. This ensures that: The users desktop system can be used as an input device for tasks, appointments, address information etc.

The user has critical information replicated on the phone even when it is not connected to the server The user can browse and search address books and public directories on the server adding information to the local address book as required.

Figure 2

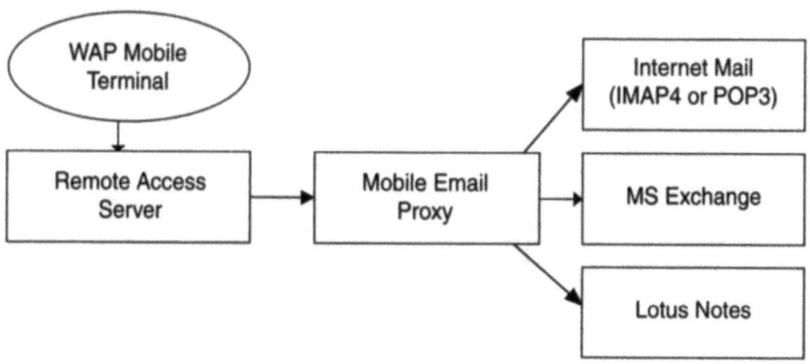

Figure 3

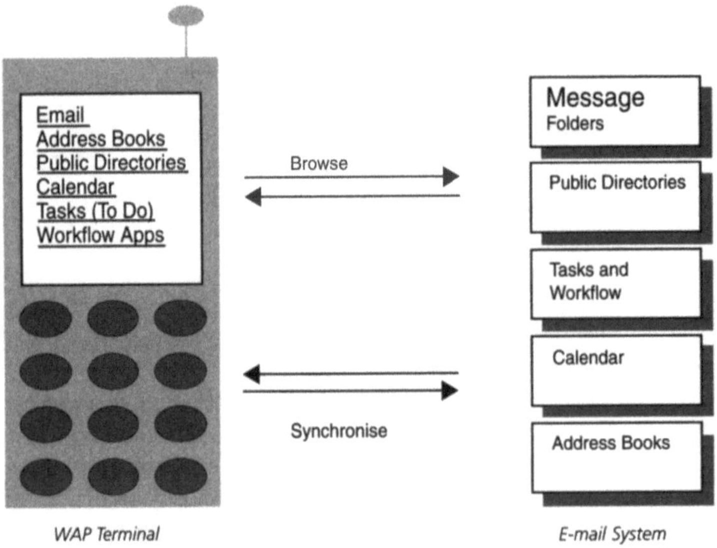

Increasing Mobile Bandwidth

The trend for mobile data means that more bandwidth will be available for mobile applications. Not only that, but a switch from circuit switched to packet switched operation and tariffing will further improve the speed and cost at which email can be exchanged with mobile terminals.

Figure 4

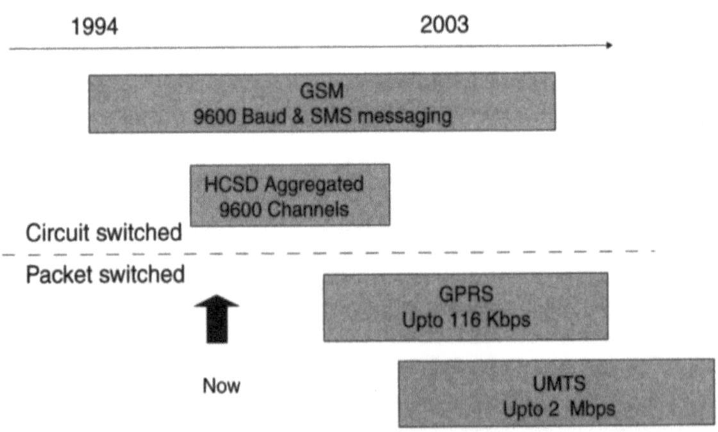

GPRS (General packet Radio Service) will start to become available in 2000 on GSM networks. GPRS terminals will support existing GSM service such as Voice, Circuit Switched Data and SMS.

The move to UMTS will improve bandwidth still further, making the delivery of email messages and attachments to mobile terminals.

User Interface

The user interface oı a mobile phone is limited by screen siz and also by the bandwidth that connects the terminal to the network. While both screen size and bandwidth will no doubt improve over the next decade, it is likely that there will always be a disparit between the mobile terminal' s ability to display information and that of the desktop computer or workstation. Information must be presented in a succinct way, and while there is limited capacity for images, much of the information must be transmitted by text. Below are some examples of sequences of user interface dialogues.

Logging into a Microsoft Exchange mailbox from a WAP enabled phone

There is no capability currently for terminals to be able to render graphical attachments or to receive streamed audio or video. While this may happen in the long term, short term strategies for handling attachments include:

- Document filters for delivering text to the mobile
- Integration with the terminals Telephony features (Through WTAI Wireless
- Telephony Application Interface)

Figure 6

Viewing and managing messages ...

Architecture over GSM

The diagram shows how WAP enabled terminals can access corporate email systems using GSM Circuit Switched Data and the Short Message Service for delivery of email notifications . The mobile email connector is called Expressway 2000.

Figure 7

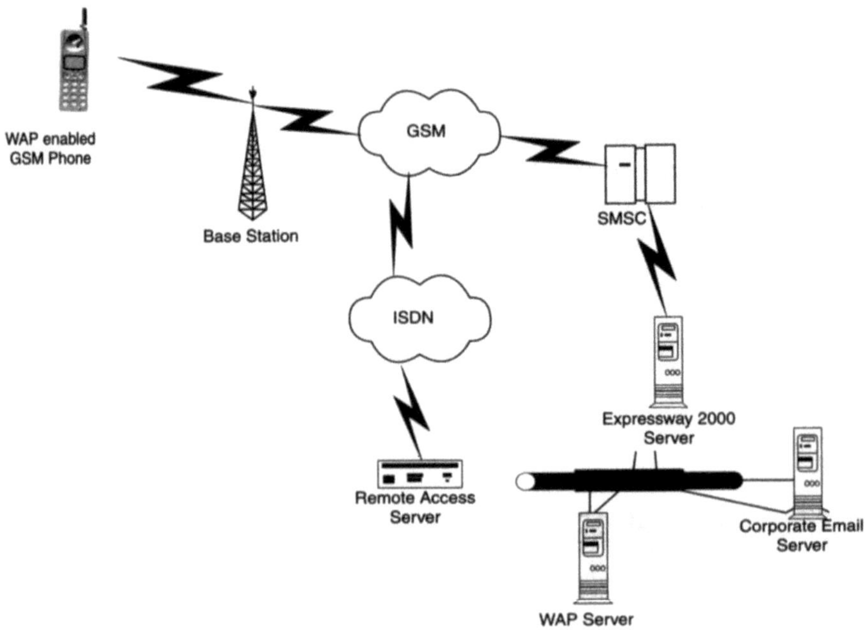

Figure 8

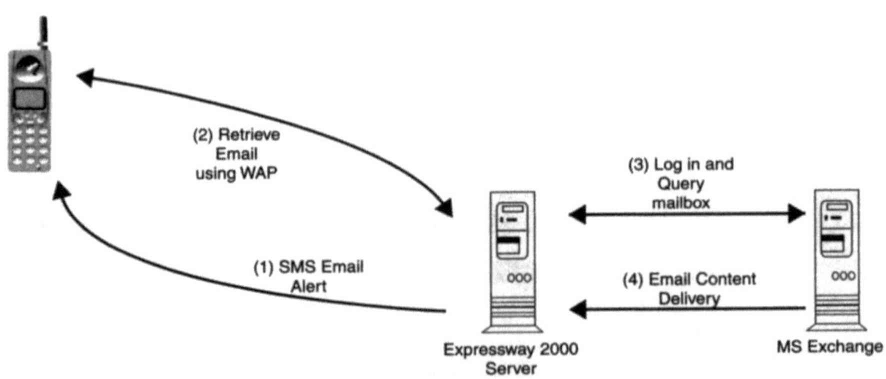

WAP Browser Compatibility Issues

While Mobile terminal vendors will produce handsets that conform to the WAP specification (Currently WAP 1.1), there can be differences in their implementation. The form factor of the device may give rise to radically different display capabilities, cache sizes and storage.

- Display size;
- Rendering capabilities;
- Cache size;
- Conformance to full WAP 1.1 functionality;
- Bearer services utilised;

When the browser or User Agent logs onto the email system, the browser identity is used to optimise the performance of the email application for that device

Email Session Proofing

The nature of mobile communications means that during the time that a user is connected to the email server the connection may be broken for the following reasons:

- Breaks in coverage;
- Timeouts over Circuit Switched Data connections, where the call is dropped. This may occur if the user is viewing cached content for more than the timeout period of the remote access server;
- Over Circuit Switched data the call can be dropped automatically when the user is engaged in offline activity e.g. composition;

The email server keeps the session open and stores the current state of the user' s session even when the connection is broken. On reconnection the user can continue the session from the point at which the connection was broken.

Figure 9

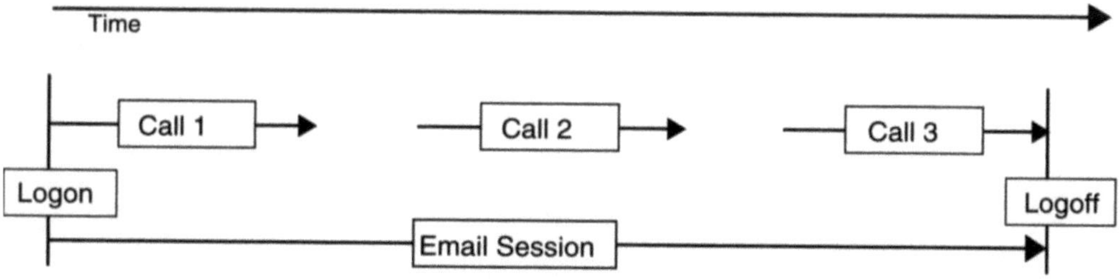

Mobile Commerce Report

Title: *Mobile Commerce Report*
Author: *Durlacher Research Ltd.*

Abstract: *Europe is about to experience an explosion in the uptake of WAP technology which will lead quickly to interim capacity constraints requiring investments in more bandwidth capacity: GPRS (Global Packet Radio Services) will be the first mainstream technology to bring the real advantage of mobile Internet to the user, through its provision of 'always on' connections. Increasingly, mobile operators will derive revenues from content and services, and will compete to develop value-added user experience. The initial killer application for mobile Internet services will be e-mail based on the current success of SMS. Instant messaging will substitute e-mail as GPRS arrives. The mobile phone incorporates indeal characteristics for ensuring secure electronic payment. Durlacher Research believes that it will become the de facto electronic wallet in Europe.*

INVESTMENT HYPOTHESIS

The mobile communications market is currently in a state of horizontal consolidation.

The era of partnering, which has been one of the telecommunication sector's defining characteristics over recent years, is now being supplanted by more aggressive acquisition-based strategies from major telecoms operators. Global One, Unisource and Concert are examples of some of the less successful transactions from the partnership era. The move to acquisitions is reflected by the recent take-overs of Airtouch by Vodafone, One-to-One by Deutsche Telekom, E-Plus by France Telecom and the current bid for Orange by Mannesmann. Today, mobile operators are increasingly becoming part of the consolidation picture. The mobile operators' current acquisition activities are targeted primarily at optimising the core processes for delivering mobile voice, creating synergy effects in terms of economies of scale and scope and generally reducing the large fixed costs of being mobile operator.

The emergence of mobile commerce provides a significant boost to valuations in the telecommunications sector, a point reflected in the latest valuations per subscriber for wireless operators. Mannesmann has offered Euro 9452 for each Orange subscriber and France Telecom paid Euro 4994 per E-Plus customer. This compares to Mannesmann (Omnitel), Deutsche Telekom (One-to-One) and Vodafone (Airtouch) all paying below Euro 4600 per subscriber in earlier take-overs during 1999. In comparing the 1999 revenue multiples of European (mobile) telecommunications operators (market capitalisation/1999 revenues), one finds that they generally lie in the range of 3 to 5. Similar multiples for traditional content companies vary widely between about 2 (e.g. Dow Jones, Disney) and 20 (e.g. Reuters).

Until now mobile operators have simply not experienced an internet type valuation, but with the arrival of mobile internet and mobile commerce, this might well change rapidly. Mobile operators will play a more active role providing portal services and content to their users. Multiples for major portals range from 60 (AOL) to 245 (Yahoo!). Mobile operators are likely to move in the same kind of direction. Large mobile operators do have the advantage that they already have a large number of users and an established billing relationship with those customers. Our investment hypothesis is that mobile operators will shift away from offering mostly voice services to become a true portal for the mobile terminal and beyond.

We believe that Europe is about to experience an explosion in the uptake of WAP (Wireless Application Protocol) technology which will lead quickly to interim capacity constraints requiring investments in more bandwidth capacity. GPRS (General Packet Radio Services) will be the first mainstream technology to bring the real advantage of mobile internet to the user, through its provision of "always on" connections. While GPRS/EDGE (Enhanced Data for GSM Evolution) technology will enhance theoretical bandwidths to match those of UMTS (Universal Mobile Telecommunications System), this technology will not provide any large scale capacity relief, rather it is expected to fill up available capacity even further. For this reason, we believe that the market will require significant investment in third generation UMTS technology, which will substantively solve capacity problems. From an investment perspective, we believe that Bluetooth technology will also emerge as a key enabler for a very wide spectrum of applications.

Increasingly, mobile operators will derive revenues from content and services, and will compete to develop a value-added user experience. In so doing they will become content aggregators and portal players allowing mobile to take its (very valuable) place in the internet jigsaw.

HIGHLIGHTS

The European m-commerce market is expected to grow from Euro 323 million last year to Euro 23 billion by 2003 and is currently about two years ahead of the US in development terms. Currently equipment vendors are creating over-hyped expectations on the development of the mobile commerce market. Mobile web browsing will not become a reality before 2002.

Broad market uptake of mobile commerce will be delayed until the main obstacles fo early market success are addressed. These obstacles include that little content and few applications are likely to be available initially, call set-up time is too long and fe WAP (Wireless Application Protocol) devices are in stores. At the beginning of 2002, mobile commerce in Europe will start to take off on a bigger scale, as GPRS (General Packet Radio Service) starts to become more widespread.

Mobile advertising will be the number one mobile commerce application (23%) by 2003.
The mobile device provides unrivalled one-to-one marketing capabilities, which the direct marketing industry will exploit moving forward. Mobile financial services, e.g. stockbroking, banking and payment (21%) as well as personalised, often

location-based mobile shopping services (15%) will also contribute significantly to market development. Mobile entertainment will become a major driver for mobile commerce only after 2003 using EDGE and UMTS.
The initial killer application for mobile internet services will be e-mail based on the current success of SMS (Short Message Service), which is necessary to pave the way for more transactional m-commerce services. Instant messaging from the mobile phone will start to substitute e-mail as GPRS arrives. Unified messaging will become mainstream technology by 2001. Mobile video telephony will not be an important application for mobile devices within the next 4 years.

Smartphones will become the standard mobile device from 2002 onwards. These devices will include a WAP microbrowser, which enables wireless internet access. The other main category of device will be so-called communicators (where Nokia has led the way), which have been derived from PDAs (Personal Digital Assistant) and which are equipped with or linked to a mobile phone. The borders between mobile phones, PDAs and consumer electronic devices will begin to blur after 2001. Phones with an integrated MP-3 player or a video player will appear in the market around this point.

Mobile operators are ideally positioned to lead the mobile commerce market as they possess comprehensive customer data, such as demographics, calling patterns and a detailed profile, as well as an existing billing relationship. Moreover, the operator owns information about the subscriber's geographic position, which facilitates the offering of location-based services, such as advertising, shopping, reservations and information provisioning. In the near future, mobile operators will have to undergo a major change in order to position themselves as mobile portal providers, content aggregators or WASPs (Wireless Application Service Provider). Considering the different business models for serving the increasing mass market demand for mobile phones and for building a mobile portal, a split-up of network operator organisations into mobile voice and mobile portal is likely.

We expect that, based on their unique customer relationship, the first mobile operators will move upwards in the value chain into the banking sector by acquiring a bank or a banking license in 2001. The mobile phone incorporates ideal characteristics for ensuring secure electronic payment and we believe that it will become the de facto electronic wallet in Europe.

METHODOLOGY

Research for this report commenced in May 1999, and in the interceding months much has happened in the communications market. Indeed, there are few areas where the words written on one day have not been superseded within days by further market developments. As will become evident through the report, there are numerous additional notes that illustrate these market changes, and explain how they affect its future.

Notwithstanding this volatility, we have published this report with the intention of providing operators, investors, mobile commerce service and equipment vendors, banks and others, with a pragmatic view and analysis of the m-commerce market in Western Europe today. We have also made our best efforts to forecast how and when this market will grow, and outline the applications that will drive adoption.

The report does not aim to provide an exhaustive overview of the m-commerce market place or the enabling technologies. In particular it should be noted that we explicitly exclude in-depth analysis of any of the equipment vendors or operators. Other market reports and technical documents are available that fulfil this role. In this report, we attempt to discuss important technological trends and mobile commerce applications that will be enabled through this environment, and evaluate them according to their ability to deliver business benefit in relation to other communications and commerce solutions.

Many of these observations are not unique to Western Europe and we believe that these regional trends provide some indication of the wider global market.

PRIMARY RESEARCH

Over the past seven months, we have conducted original research and interviews across Europe, and have exchanged ideas with many operators, banks, equipment vendors, service platform developers, application vendors, content providers and portal companies.

We have built a view on the market that cuts through the hype that has been created in the last few months around mobile internet and m-commerce in particular.

Our understanding of the relevant developments in this particular segment has been shaped through discussions with industry leaders.

However, the market is just being created, growing out of the earlier "mobile data" classification into what is being regarded a m-commerce today. In such a new marketplace, it is almost impossible to extrapolate long term trends from an early, developmental snapshot.

SCOPE

The report is divided into nine main sections, Today's Market, M-Commerce Market Drivers, M-Commerce Enabling Technologies, M-Commerce Enabling Applications, Consumer M-Commerce Applications, Business M-Commerce Applications, Market Sizing and Forecasts, Industry Outlook and Investment Opportunities. These sections are further complemented by our views on how thes technologies and markets will evolve to provide growing momentum to the m-commerce applications market, how these applications will shape over time and how the various players will react to such developments.

INTRODUCTION

DEFINITION OF MOBILE COMMERCE

The working definition of Mobile Commerce for the purposes of this report is any transaction with a monetary value that is conducted via a mobile telecommunications network. In this report, we refer to Mobile Commerce as M-Commerce, Mobile Electronic Commerce or Wireless Electronic Commerce, using these terms interchangeably.

According to this definition, m-commerce represents a subset of all e-commerce transactions, both in the business-to-consumer and the business-to-business area (m-commerce will not only expand its share of this market, but will expand the market overall, through the rapid uptake of m-commerce services).

Therefore, regular SMS messages from one person to another are not included in the definition of mobile commerce, while SMS messages from an information service provider, that are charged at a premium rate, do represent mobile commerce according to our definition.

Until now, the term "mobile data" has always been used for everything which is non-mobile voice. We believe that this terminology is slightly outdated, today's m-commerce is all about applications and services on the mobile phone. It is not about capacity, it is about content.

Europe (specifically, Western Europe) has been the primary focus of this report despite the fact that key e-commerce trends and business models usually derive from the United States. This is because, in the specific area of mobile

Figure 1

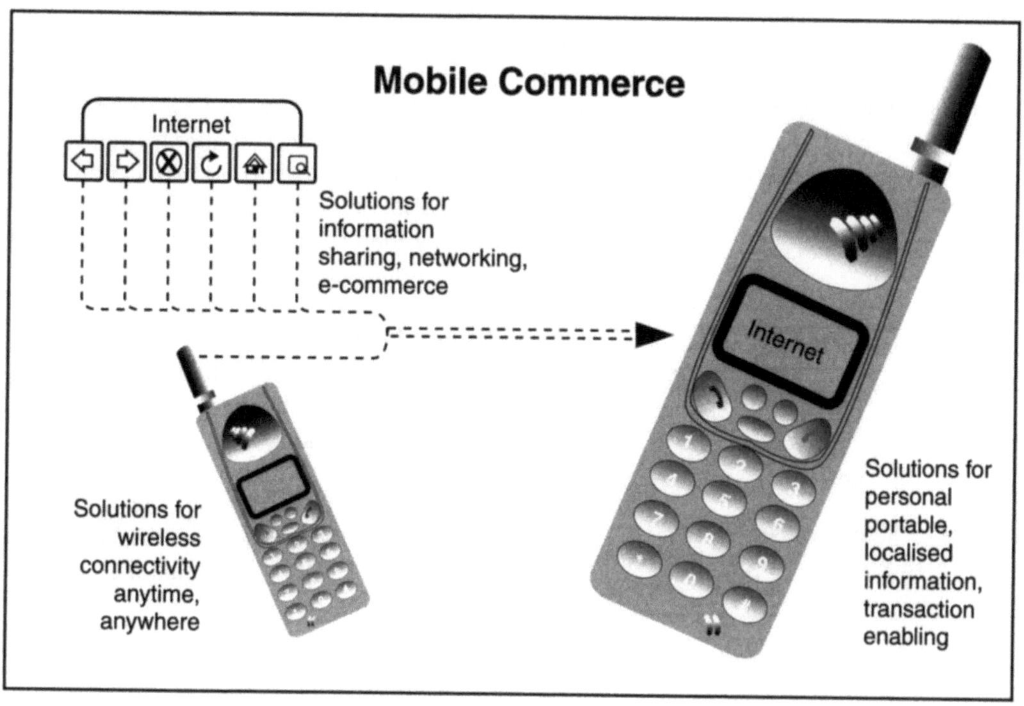

communications, Europe has adopted a clear lead in terms of usage and application development. Europe has a high penetration of mobile phones and has successfully adopted a single standard, GSM (Global System for Mobile Communications) which dominates the wireless world throughout the continent. The US has not been able to reach this single standard nor has it managed to settle on a generic type of terminal, thereby retarding the arrival of a critical mass of handsets in the open market for the introduction of new services. Instead, a wide selection of both analogue and digital devices as well as all types of pagers can be found in the US.

TODAY'S MARKET BACKGROUND

Mobile commerce applications that combine the advantages of mobile communications with existing e-commerce services will be very successful, but we will also see entirely new services built around the mobile. Some of the key drivers for the increasing sophistication of the mobile market are:

Figure 2

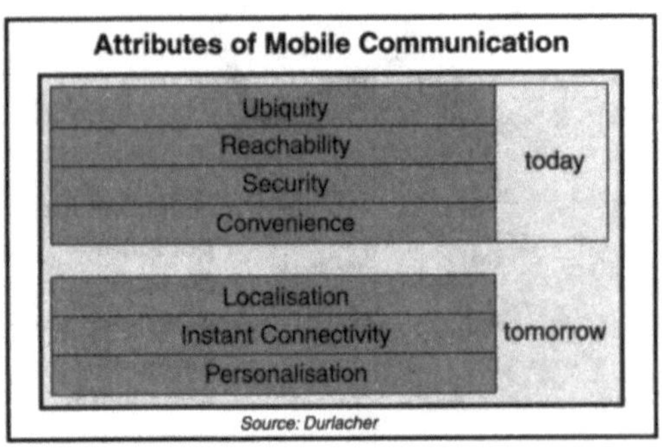

Source: Durlacher

Taking these drivers in turn:

Ubiquity

Ubiquity is the most obvious advantage of a wireless terminal. A mobile terminal in the form of a smart phone or a communicator can fulfil the need both for real-time information and for communication anywhere, independent of the user's location.

Reachability

Reachability is important for many people who want to be in touch and be available for other people. With a mobile terminal a user can be contacted anywhere anytime. The wireless device also provides users with the choice to limit their reachability to particular persons or times.

Security

Mobile security technology is already emerging in the form of SSL (Secure Sock Layer) technology within a closed end-to-end system. The smartcard within the terminal, the SIM (Subscriber Identification Module) card, provides authentication of the owner and enables a higher level security than currently is typically achieve in the fixed internet environment.

Convenience

Convenience is an attribute that characterises a mobile terminal. Devices store data, are always at hand and are increasingly easy to use. Enhanced functionality that will become available, based on technological advances, on tomorrow's devices will include the following:

Localisation
Localisation of services and applications will add significant value to mobile devices. Knowing where the user is physically located at any particular moment will be key to offering relevant services that will drive users towards transacting on the network. The mobile operator will soon know where the user is physically located, so for instance abusinessperson arriving on a plane into Helsinki can expect to receive a message asking whether she needs a hotel for the night.

Instant Connectivity
Instant connectivity to the internet from a mobile phone is becoming a reality already and will fast-forward with the introduction of GPRS services. With WAP or any other microbrowser over GSM, a call to the internet has to be made before applications can be used. Using GPRS it will be easier and faster to access information on the web without booting a PC or connecting a call. Thus, new wireless devices will become the preferred way to access information.

Personalisation
Personalisation is, to a very limited extent, already available today. However, the emerging need for payment mechanisms, combined with availability of personalised information and transaction feeds via mobile portals, will move customisation to new levels, leading ultimately to the mobile device becoming a real life-tool. So, returning to the businessperson landing in Helsinki, if she responds 'Yes' to the question regarding the hotel room then

Figure 3

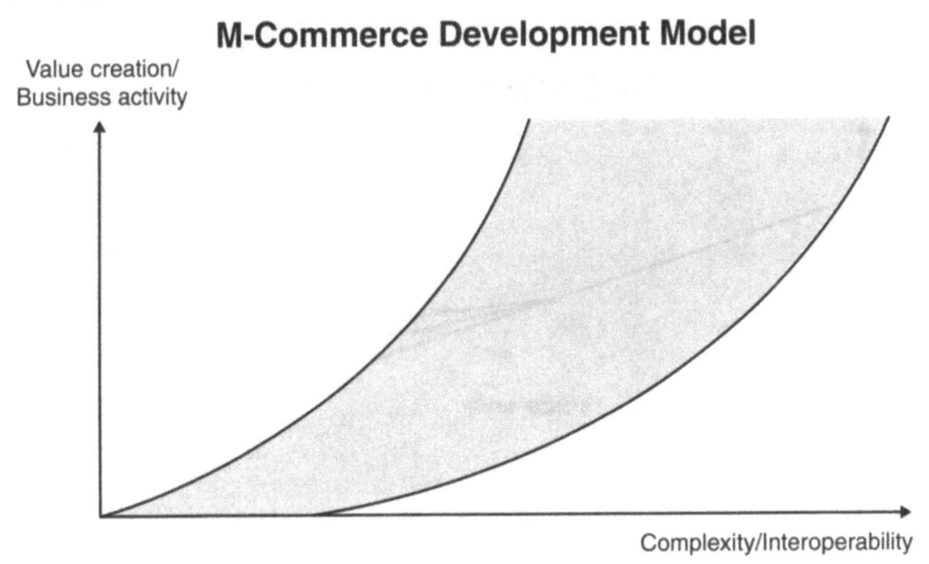

Source: Durlacher

the network will advise her what is available in her price range (and will match any other variables she may have input through her personalisation tool).
We believe that we will see the following path for m-commerce service deployment in Europe. Although mobile e-mail is not considered to be a commerce application (rather a communications application), it is featured in the chart in order to reflect the key role it has in developing the market.

MOBILE COMMERCE MARKET DRIVERS

The following principal drivers are responsible for the growth expectations of the mobile commerce market.

Mass Market Mobile
With mobile communications reaching the mass market, network operators are facing decreasing ARPU (Average Revenue Per User). Price erosion for mobile voice service is faster with 3rd, 4th and sometimes 5th mobile operators having entered the market in many European countries. There is a common understanding throughout the industry, that within a 2 to 3 year period mobile tariffs will come down to the same level as fixed tariffs. The network operators must continuously implement new services on their networks if they want to slow or turn around the trend of decreasing ARPU.

Mobile data and SMS services have not been very successful in the past, generating usually no more than 2-3% of an operator's turnover, although in Finland

Figure 4

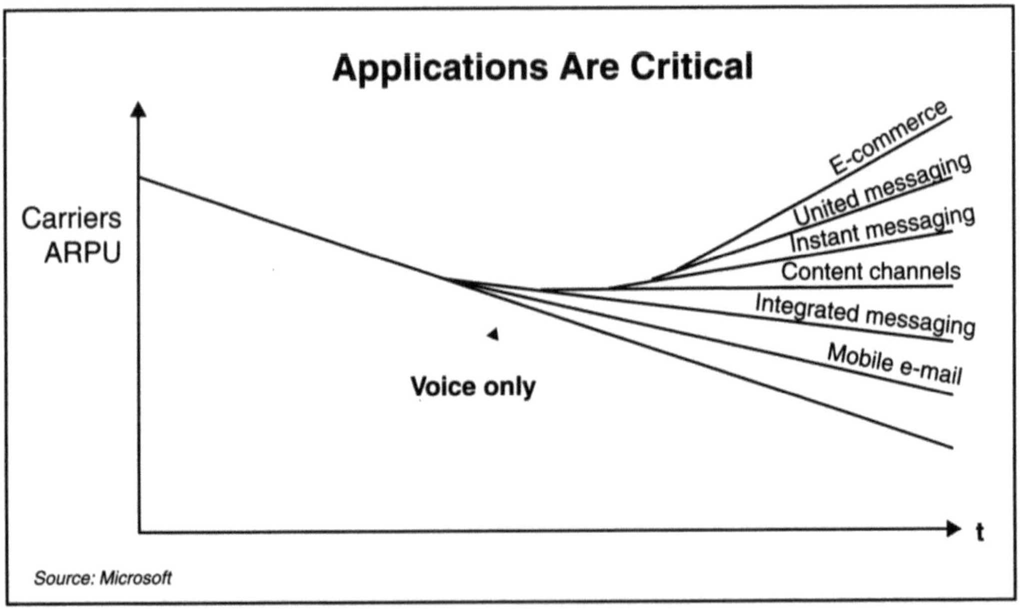

Figure 5

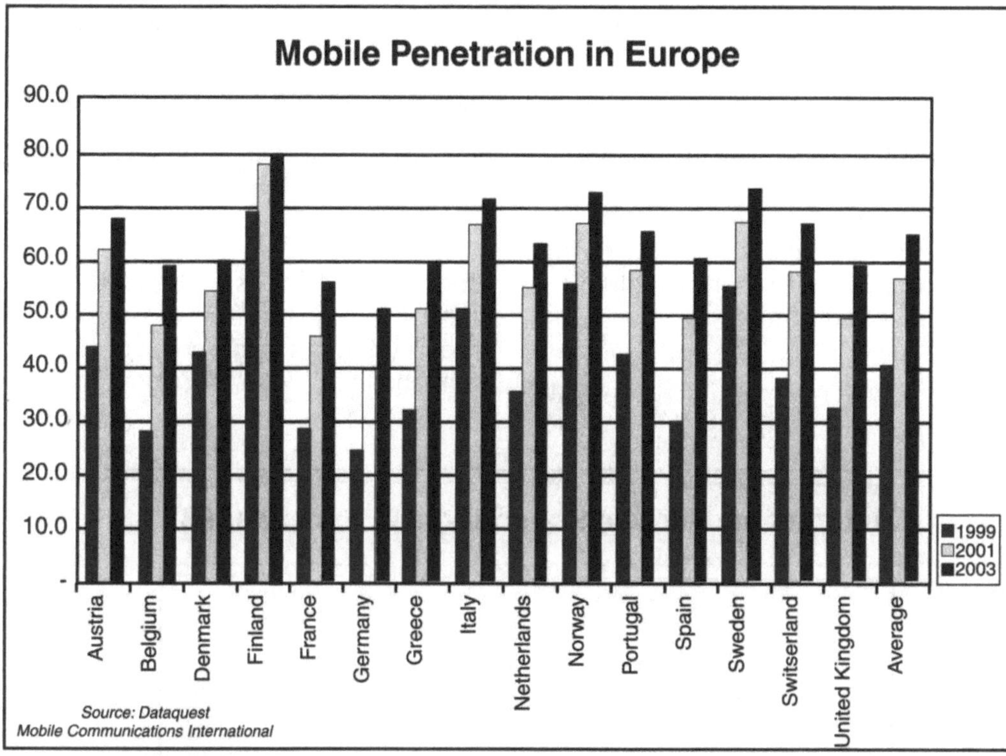

a 7% revenue share has been reached. Since SMS is mainly used for communication between people in the 15-25 year old age group, they have earned themselves the title "Generation Text". The power of SMS can be seen in countries such as the Philippines where it is a national phenomenon amongst the youth market.

The number of mobile phone subscribers is going to outnumber the number of fixed line telephone lines. The point of mobile dominance has been reached already in some countries in Europe, such as Finland, which reached a mobile penetration of over 65% during summer 1999. It is important to point out that although mobility of people is increasing, the need for content remains location-dependent according to the "business is local" principle. Thus, we believe that content offerings will change with the location of the user.

Mobile commerce is the strongest future potential source of revenues for operators once wireless bandwidth becomes more or less a commodity. The purchase of goods and services or the trading of stocks via a mobile device is no longer considered to be simply a wireless data or mobile value-added service. They are key commercial transactions, which happen to be conducted in a mobile environment.

Booming Wireline Internet

The current number of internet users is indicated below in comparison to the number of mobile phone owners. The gap between those numbers is particularly large in Southern European countries, such as Italy, Greece and Portugal.

E-commerce is growing rapidly throughout the world as more and more people are getting online. In 1999 European E-commerce has reached a volume of about Euro 8 billion according to Forrester Research. A larger part of the population is undergoing its first online shopping experience.

Most forecast numbers are based on the on-line population on PCs with internet capability, but this model does not withstand the arrival of new tools for

Figure 6

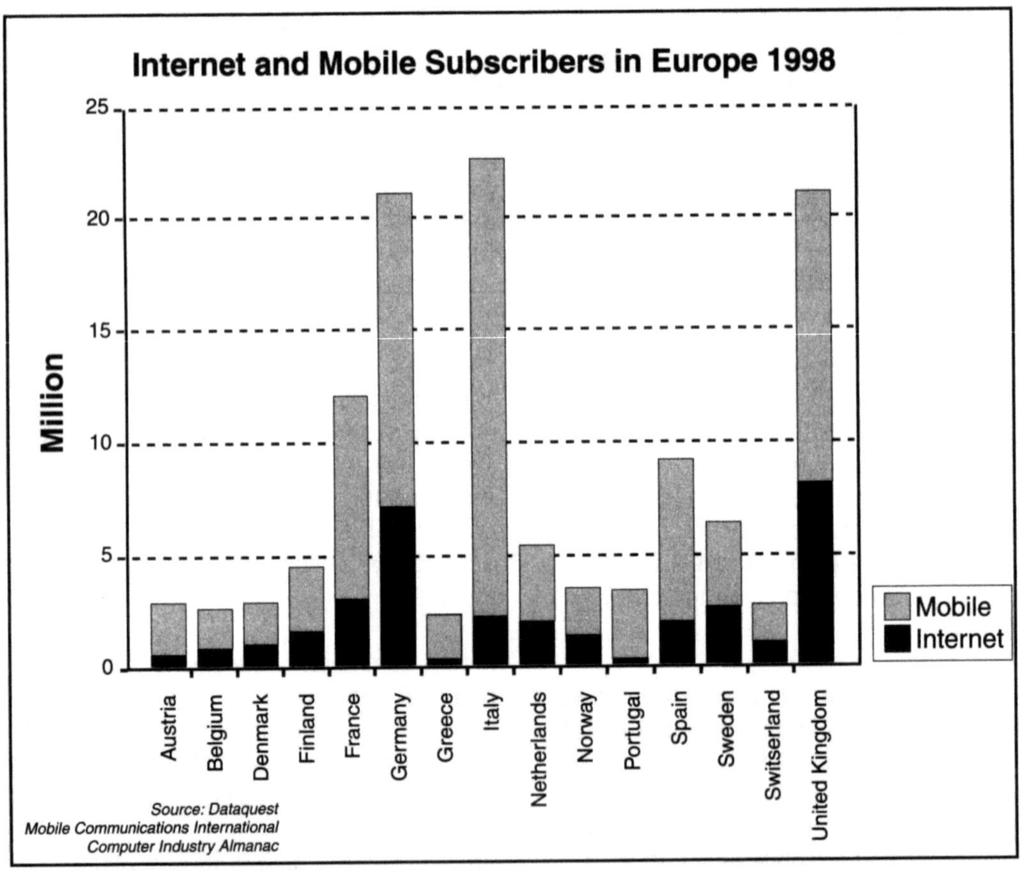

Figure 7

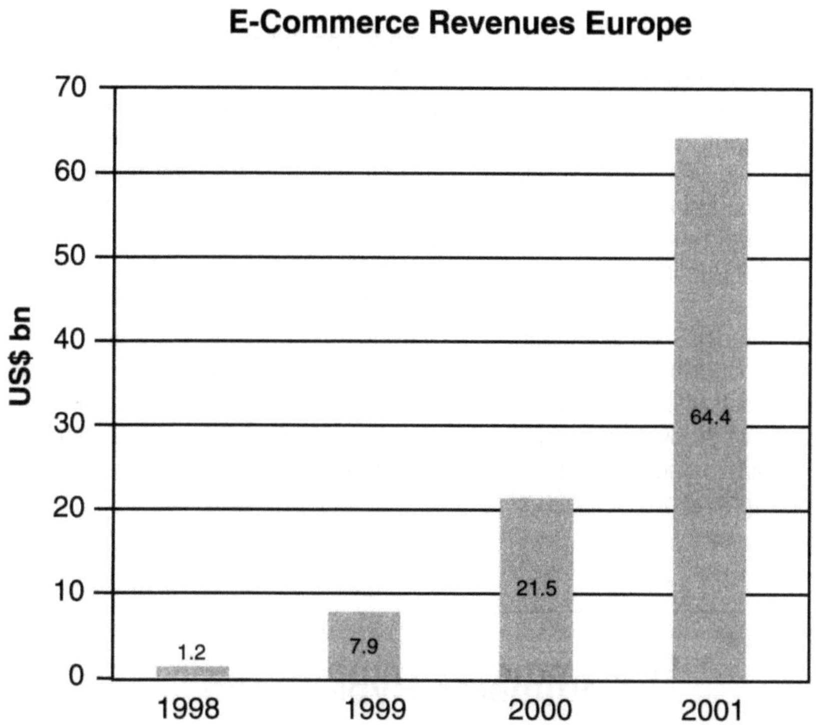

Source: Forrester

integrating internet functionality: WebTV, smartphones and communicators. Mobile commerce is per se not included in the traditional e-commerce market models. M-commerce will be able to increase the overall market for e-commerce, because of its unique value proposition of providing easily personalised, local goods and services anytime and anywhere. We believe that m-commerce will be adopted rapidly in Europe because of the high usage of mobile data services and because of increasing exposure to fixed line e-commerce.

The increased competition in regional markets (with up to 5 cellular players/market offering virtually identical services at often identical tariffs) puts a further pressure on service differentiation as each player tries to distinguish their position in the competitive landscape. Mobile commerce in all its shades will provide a key differentiator for a network operator, who must be forward thinking and innovative to move successfully from being a pipe (infrastructure provider) to become a content aggregator (mobile portal) and customer solution provider (systems integrator).

Supplier Push
The push from equipment vendors for WAP gateways and microbrowser-enabled smartphones is helping to drive the market for mobile commerce. Telecom99 has been the climax of the market hype so far, with mobile internet generating cover stories in all major economics publications.

We believe that the hype will eventually cool down, before the true uptake of m-commerce becomes a reality by 2002. Although the rate of innovation is very high within the mobile and internet industries, false expectations in terms of equipment availability and functionality (e.g. realistic network capacity) are created.

New Billing Principles
With the arrival of GPRS, per minute pricing of mobile services, as we know it today, will not be relevant any longer. Instant access to the internet requires an "always on" mode, necessitating some new form of pricing mechanism. This infers a commodisation of bandwidth. However, a shortage of bandwidth until the introduction UMTS in 2003 limits somewhat the extent to which mobility can be viewed as a commodity. In our view this is an issue which sits on the critical path in the adoption of new services.

Figure 8

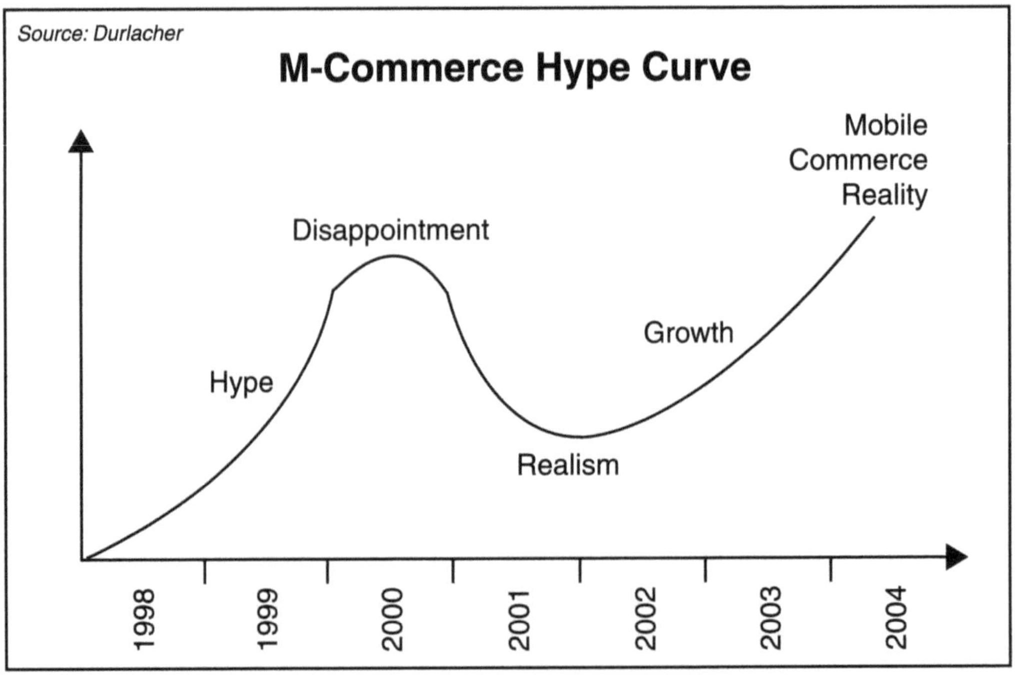

Figure 9

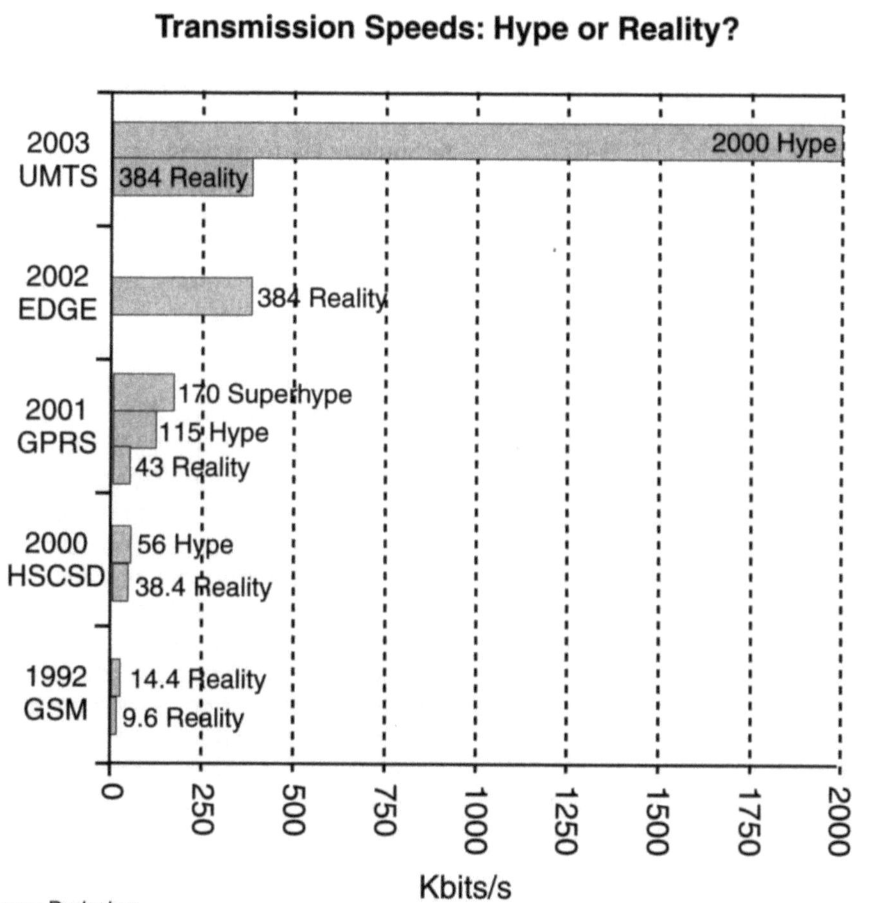

Source: Durlacher

The new pricing model favoured by the operators will likely be (in the initial instance at least) a series of flat monthly rates for a certain amount of traffic. The operator needs then only to roughly control the traffic volume to ensure that it is not too far over the allotted volume. This approach also takes into account the bursty (meaning significant intensive peaks of usage followed by periods of protracted troughs) nature of GPRS traffic. For the customer this would make life easier in one sense as they would have to calculate only one price. But on the other hand, no consumer will know how many packets they really use in a month, and indeed the billing metric itself is a somewhat technical one.

Alternatively, services could be charged based on value-based pricing. This would relate to various price tags for certain services whether they are stock trades, e-mail or maps for example. NTT DoCoMo has been using this model successfully since February 99 for their i-modeservice, the first operational mobile portal.

We believe that the acceptance of this second model, value based pricing, will ultimately drive the success of mobile commerce, because it is a model that customers can understand. It is likely that many services will be provided for free, such as news services, while transactional services will be paid for, maybe even by the selling partner. The packaging of these services over networks may be brokered by an m-infomediary rather than directly by the operators themselves.

UMTS Licensing

UMTS licenses are being awarded in most European countries during the year 2000. Only Finland has completed the licensing procedure already, by giving the UMTS licenses to holders of the current GSM licenses during March 1999. The award of limited packages of radio spectrum, which will enable IP (Internet Protocol) based services to be accessed using the same handset all over the world, will be mostly used to increase competition and add new market players. Companies from the media industry such as Bertelsmann and strong marketing organisations such as Virgin are planning to compete for UMTS licenses, as are all the existing mobile operators and a few of their foreign counterparts. The licenses are expected to cost some Euro 100 million in each of the countries.

MOBILE COMMERCE VALUE CHAIN

Technology Platform Vendors

The technology platform vendors are delivering the operating systems and microbrowsers for mobile devices such as smartphones and communicators. The battle for the dominating OS (Operating System) has been reduced to two major camps of players, Microsoft and its followers with Windows CEon the one hand and Symbian with Palmon the other hand. Symbian, the industry consortium now comprising Psion, Motorola, Ericsson, Nokia and Matsushita began by building upon Psion's EPOCoperating system, and has now agreed to collaborate with 3Com, owner of the popular PalmOS. The challenge will be to combine the PalmOS with EPOC. Nokia, from its position as the mobile industry's innovation leader, seems to be the most likely party to make this happen.

The microbrowser market is today largely dominated by Phone.com (formerly Unwired Planet), who have gained support from all major mobile phone manufacturer except Nokia and Ericsson, who are marketing their own microbrowser products.

Infrastructure Equipment Vendors

The leading suppliers for mobile network infrastructure equipment: Motorola, Ericsson, Siemens, Nokia and Lucent, have developed solutions for mobile data,

Figure 10

The Mobile Commerce Value Chain

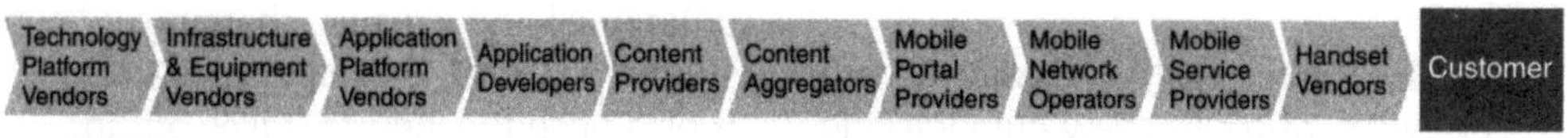

Source: Durlacher

mobile internet and thus for mobile commerce. They are creating significant hype around the entire topic and are driving the mobile industry with the speed of innovation and new technology developments such as WAP, HSCSD (High Speed Circuit Switched Data), GPRS, EDGE and UMTS. In this sense the technology is well ahead of the market since to a large extent applications are yet to be developed which utilise these developments.

Application Platform Vendors

A particular key driver for providing wireless internet applications is the availability of middleware infrastructure, i.e. WAP gateways at either the mobile operator's site or at the corporate customer's site. The companies, who have developed their own WAP stack include Phone.com (who also acquired Apion), Nokia, Ericsson and Dr. Materna.

In order to drive the industry and to formulate standards, interest groups have been formed in addition to ITU (International Telecommunications Union), ETSI (European Telecommunications Standards Institute) and the GSM MoU (Global System for Mobile Memorandum of Understanding), such as the WAP Forum, the Mobile Data Initiative, Bluetooth Special Interest Group, GAA (GPRS Applications Alliance) and the UMTS Forum. They are setting de facto standards by assembling the key players and agreeing workable development conditions much faster than the traditional standards bodies.

Application Developers

Applications for the mobile environment are currently being built primarily on Windows CE, Symbian's EPOC32, or PalmOStechnology platforms. Currently, most of these applications are used off-line rather than via the mobile network. However, WAP is receiving increasing support from developers, who are going initially after the smartphone market rather than the PDA market. This makes a lot of sense if one compares the number of PDAs sold in Western Europe (1998: 1.4 million) to the number of mobile phones (1998: 90 million).

There is an army of developers working on applications for SIM Toolkit and especially WAP. While there have been a number of applications built based on SMS applications by companies such as WAPIT (Finland) and Dr. Materna (Germany), there is an easy migration from SMS to WAP. This migration is likely to be realised over a period of time, during which time both platforms are likely to be offered. Revenues are generated today with SMS-based information services, while WAP service revenues are likely to come on stream during 2000, dependent on the arrival (in volume) of the long awaited handsets.

As the CEO of a Finnish WAP-start-up puts it: "Everyone who can chew gum walking down the street, is now doing WAP". Although this might be true for Finland and Sweden, where a lot of development activity is taking place, we believe that the number of available applications is still very limited and many more players are going to enter this market.

Content Providers
Technologically advanced content providers are moving into the mobile space to be ready for when mobile commerce will happen. They are using a variety of distribution channels for their products according to the axiom "when content is king, distribution is King Kong". For example, Reuters is delivering its information via partnerships with Ericsson and Nokia as well as via existing portal sites, such as Yahoo! and Excite, who are building mobile portals as well. Additionally, Reuters is building its own mobile portals in a number of markets, having recognised the emerging importance of mobile as a distribution channel.Other early content providers include Consors, the online brokerage, Multichart, the stock exchange information provider and Webraska, the traffic news company.

Charging for content is difficult in a mobile commerce world, even though users are accustomed to paying for value-added mobile services. The easiest way to create revenues for mobile information providers is by taking a share in the call revenues. However this model is classic first generation and it is our view that differential and dynamic charging structures (based on value) will rapidly evolve once the industry takes off. In future, advertising, sponsoring and subscription models will also be realised. In Germany, for example, it is expected that there will be more than 1500 mobile services, many of these subscription based, available on WAP by the end of 1999.

Content Aggregators
A new category of content aggregators is starting to emerge that repackages available data for distribution to wireless devices. The added value is in delivering content in the most appropriate package. One example is Olympic Worldlink, which has developed a solution called Mobile Futures, that provides real-time information from the futures and options markets as well as financial market, company, political and general interest news. It also adds trade data from exchanges and clearing houses the world over. Another UK company, Digitallook.com, is providing a service that lets PDA (Personal Digital Assistant) users download share prices and news headlines from the BBC, CNN and AFX. However, it is not yet dynamically updated via the mobile network.

Mobile Portals
Mobile portals are formed by aggregating applications (e-mail, calendar, instant messaging etc.) and content from various providers in order to become the user's prime supplier for web-based information that is delivered to the mobile terminal. Mobile portals are characterised by a

greater degree of personalisation and localisation than regular web portals, since the success of m-commerce applications is dependent on ease of use and on delivering the right information at the right moment. This is something which we at Durlacher refer to as the value-for-time proposition and, moving forward, will be a key dynamic in determining the success or otherwise of mobile (and indeed other) services. It has been estimated that every additional click-though which a user needs to make in navigating through a commercial online environment with a mobile phone reduces the possibility of a transaction by 50%. MSN Wireless and Yahoo! Mobile are among the first portals to offer service for the mobile community, but they are still very much focussed on the US. Mobile operators across Europe have put first portals out, e.g. BT Cellnet with its Genie, Telia with its MyDOF (My Department of the Future), Sonera with its Zed or Deutsche Telekom's T-Mobil subsidiary with T-D1@T-Online.

Mobile Network Operators

Operators, such as Mannesmann, Orange, Telia or TIM (Telecom Italia Mobiles), are best positioned to benefit from the introduction of new m-commerce services, because they already own a billing relationship with the customer and they control the portal which is pre-set on the SIM card when it is distributed. The operator's intention is to position itself in a key role for mobile commerce by owning the portal and participating in the revenues accrued by services over its network. Those revenues will be significantly higher than the sheer increase in call minutes or volume, particularly as the incremental price per minute falls to zero. The mobile operator has the opportunity to become an ISP (Internet Service Provider) in the sense that the mobile network is going to be built on IP technology with UMTS and that the operator will provide a transport pipeline for content services. Therefore, numerous operators are trying to move up the value chain.

Mobile Service Providers

The phenomenon of mobile service providers as an intermediary for faster marketing and sales of mobile phone contracts and terminals has been seen in many European markets. The service provider has the contract and billing relationship with the customer, but does not own any infrastructure. They are buying the services at a discount of typically 20-25% and can sell them under their own brand. The mobile network operator determines the functionality of services and therefore dominates the information displayed on the screen. However, control over the billing relationship puts the service provider in the position to offer m-commerce applications by charging goods and services directly to the phone bill, if the network operator has provisioned for it.

The influence of service providers is slowly decreasing. While there is currently overall strong subscriber growth, service providers have been growing less than the overall market. Most valuable service providers have by now been acquired by larger telecom operators (Talkline by Tele Denmark, debitel by Swisscom), enabling these operators to gain a customer base without deploying costly infrastructure.

Handset Vendors

Handset vendors are critical in the value chain. Generally, customers do not shop for a particular service provider or network operator, but rather for the handset brand. The emergence of the mobile phone as not only a consumer electronic device, but also as something personal such as a pen or watch, has created lots of value for the handset brands.

In mobile commerce, the handset vendors are a bottleneck in bringing new devices to the market, that support not only SIM (Subscriber Identification Module) Toolkit, but more importantly WAP, GPRS and W-CDMA (Wideband-Code Division Multiple Access). Innovation cycles are becoming continuously shorter, but significant m-commerce will not take place before the right end-user terminals are widely available. The handset vendors have to develop a wider variety of products, as future applications will require different combinations of features. Handsets, optimised for music download and listening, video streaming and watching, computing, game playing or just managing one's life will become possible choices.

At the same time, mobile handset manufacturers are coming closer to the traditional PDA manufacturers, as they are both offering smartphones and communicators with combined functionality. Production capacity problems with respect to the expected large quantities of phones needed by the market are likely, considering that the total demand for mobile phones is going to almost triple over the next 5 years.

Customer

For consumers, mobile commerce will be a new experience, since thus far most of them have used their mobile phone primarily for voice, and more recently for SMS messages.
According to a Nokia study on mobile VAS (Value Added Services), the primary target markets for m-commerce consumer services are:

- Teens (18 years and under)
- Students (19-25 years old)
- Young business-people (25-36 years old)

The business market can be divided into three main categories of organisations that possess distinct m-commerce needs:

- Sales-driven organisations, such as manufacturing companies and banks
- Service-driven organisations, such as consultancies and system houses
- Logistics-driven organisations, such as taxi companies or courier services

Depending on which segment it falls under, a company will become more likely to use a specific mobile commerce application, such as CRM (Customer Relationship Management), fleet management or integration of mobile devices into corporate ERP (Enterprise Resource Planning) systems. Finally, it should be pointed out that payment agents play an important role as an enabling force in the m-commerce value-chain, although the dominant mode of payment for m-commerce services has yet to be determined. Banks have been traditionally the natural providers of payment agent services. Now they are becoming increasingly concerned about the future role of mobile operators, who allow their subscribers to charge purchased

goods and services to their telephone bill (e.g. Sonera). Therefore, the banks themselves are becoming front-runners in mobile commerce in order that they do not become disintermediated.
Merita Nordbanken has for example established its own proprietary SOLOpayment system for e-commerce and m-commerce payments, while Visa, Cartes Bancaires and Barclays Bank are all experimenting with potential payment solutions for mobile commerce services.

M-COMMERCE ENABLING TECHNOLOGIES

We will not see significant growth in the mobile commerce market until the necessary enabling technologies are developed and deployed. We analyse below the various technology enablers we feel will contribute to the development of the market, and assess their impact on m-commerce in Europe.

NETWORK TECHNOLOGIES

Mobile protocols are all very similar and are ultimately chasing the same applications. All of the protocols are client-server based and involve new functions on the mobile phone and new servers connected to the mobile phone network. Although there are several overlaps, one needs to analyse these protocols to identify the potential winner (the protocol that will gain the strongest support in the market from an application point of view). The protocol that delivers the strongest commercial value at any point in time will be supported by the largest number of attractive applications, and therefore, will jockey into the lead position.

GSM

GSM (Global System for Mobile Communication) operates in the 900 MHz and the 1800 MHz (1900 MHz in the US) frequency band and is the prevailing mobile standard in Europe and most of the Asia-Pacific region. GSM is used by more than 215 million people (October 1999), i.e. representing more than 50% of the world' s mobile phone subscribers. North America has only about 5 million GSM users in late 1999, while the majority of subscribers are using a variety of technologies for mobile communications, including pagers and a high percentage of analogue devices. Additionally, the North American mobile market development is handicapped by the "Called Party Pays" principle, which has led to a low usage of mobile phones. In Europe, the common GSM standard provides the critical mass to make it economically feasible to develop a large variety of innovative applications and services. Thus, we believe that Europe and Asia will be at the forefront of the development in m-commerce and about 2 years ahead of the US. However, an increasing number of very innovative solutions is coming out of the US for the m-commerce market, e.g. from Spyglass, W-Trade and Aether.

HSCSD

HSCSD (High Speed Circuit Switched Data) is a circuit switched protocol based on GSM. It is able to transmit data up to 4 times the speed of the typical theoretical wireless transmission rate of 14.4 Kbit/s, i.e. 57.6 Kbit/s, simply by using 4 radio channels simultaneously. HSCSD services are being launched during Autumn 1999 by operators such as E-Plus and Orange. In total there are only 18 GSM operators worldwide who intend to offer HSCSD service, before they introduce GPRS.

The key problem in the emergence of this market is that there is currently only Nokia who can provide PCMCIA modem cards (CardPhone 2.0) for HSCSD clients, which offers a transmission speed of 42.3 Kbit/s downstream and 28.8 Kbit/s upstream. The typical terminal for HSCSD is a mobile PC rather than a smartphone. Call set-up time is still 40 seconds needed for the handshake of the modem. Application usage is more like existing mobile connections to the internet and intranet, which are used especially for accessing e-mail services. The frequent business traveller seems to be the primary target market. Other vendors are expected to hit the market soon with alternative HSCSD terminals.

We believe that HSCSD is an interim technology and that it will mainly be used for speeding up existing mobile data applications. The opportunity window for HSCSD is limited as GPRS services (which offer instant connectivity at higher speeds) will be going into operation by late 2000. The impact of HSCSD on mobile commerce will therefore be very limited.

GPRS

GPRS (General Packet Radio Service) is a packet switched wireless protocol as defined in the GSM standard that offers instant access to data networks. It will permit burst transmission speeds of up to 115 Kbit/s (or theoretically even 171 Kbit/s) when it is completely rolled out. The real advantage of GPRS is that it provides an "always on" connection (i.e. instant IP connectivity) between the mobile terminal and the network.
Network capacity is only used when data is actually transmitted. It will be available in the second half of 2000 in the first GSM networks in Europe. The actual speed of GPRS will be initially a lot less than the above dream figures: 43.2 Kbit/s downstream and 14.4 Kbit/s upstream up to 56 Kbit/s bi-directional some time thereafter. GPRS will be the first transport mode to allow full instant mobile internet access and will become the enabler for a wide range of applications. In this sense it truly may pave the way for UMTS.

Pilot GPRS networks are already in place today in many European markets. However, GPRS will require new terminals that support the higher data rates, and these seem to be the bottleneck to the early adaptation of the technology. So far no handsets have been released, but they are scheduled for Q3/2000. In order to push the development of applications for GPRS, the GAA (GPRS Applications Alliance) was founded in October 1999 by Ericsson, Palm, IBM, Lotus, Oracle and Symbian.

We believe that the availability of a large number of applications is critical for the take-up of GPRS as a bearer technology for internet access. Implementing innovative billing and pricing systems will be necessary to make the shift away from a per minute charge.
According to our evaluation, since both are packet-based technologies, GPRS will be

widely installed by operators as a step in the evolution towards the UMTS world.

EDGE

Enhanced Data Rates for Global Evolution (EDGE) is a higher bandwidth version of GPRS permitting transmission speeds of up to 384 Kbit/s. It is also an evolution of the old GSM standard and will be available in the market for deployment by existing GSM operators during 2002. Deploying EDGE will allow mobile network operators to offer high-speed, mobile multimedia applications. It allows a migration path from GPRS to UMTS, because the modulation changes that will be necessary for UMTS at a later stage will already be implemented.

While a number of mobile operators are considering implementing EDGE as an interim data technology between GPRS and UMTS, the success of EDGE depends very much on the timely availability of the products and applications. We believe that the opportunity window for EDGE will be very short, unless major delays occur during UMTS deployment.

3G

3rd generation (3G) is the generic term for the next big step in mobile technology development. The formal standard for 3G is the IMT-2000 (International Mobile Telecommunications 2000). This standard has been pushed by the different developer communities: W-CDMA as backed by Ericsson, Nokia and Japanese handset manufacturers and cdma2000 as backed by the US vendors Qualcomm and Lucent.

After long negotiations the intellectual property rights were cross-licensed between Ericsson and Qualcomm in June 1999, and it looks as though there will be a "peaceful co-existence" of standards. The goal of being able to have one single network standard (CDMA) and use one handset throughout the world seems to be capable of being reached. But within the one standard there will be 3 optional, harmonised modes (W-CDMA for Europe and the Asian GSM countries, Multicarrier CDMA for North America and TDD/CDMA for the Chinese). The first 3G network is expected to be in operation by NTT DoCoMo in Japan by late 2001.

UMTS (Universal Mobile Telephone System) is the third generation mobile phone system that will be commercially available from 2003 in Europe. First licenses have been granted in Finland, but the rest of Europe will award the licenses starting from 2000. Although many people associate UMTS with a speed of 2 Mbit/s, this will be reached only within a networked building and indeed only with some further development to the technology. Realistic expectations suggest a maximum capacity in metropolitan areas of 384 Kbit/s, at least until 2005.

This is in fact the same transmission rate that can be realised much earlier with EDGE.
In fact, some mobile operators are currently reconsidering their UMTS roll-out plans due to concerns relating to the capacity differentials and the cost/benefit of migrating from one to the other. Durlacher believes that, considering capacity problems are a major challenge

for GSM operators (because of higher penetration rates and the expected boom in mobile data and m-commerce traffic), it makes sense for mobile operators to invest early in UMTS in order to win a share of the available UMTS spectrum.

The business case for operators to provide nationwide UMTS coverage (comparable to that achieved with GSM) is still negative. Therefore, initially only metropolitan areas will be covered by UMTS networks. Forced sharing of infrastructure, as demanded by the UK regulator OFTEL, is likely to be imposed in many markets in order to foster the operator focus on new services rather than solely on building the network.

SERVICE TECHNOLOGIES

SMS

Since 1992 Short Message Service (SMS) has provided the ability to send and receive text messages to and from mobile phones. Each message can contain up to 160 alphanumeric characters. After historically finding it tough going in the GSM markets, during the year 1998 SMS started suddenly to explode. In October 1999, there were about 2 billion SMS messages sent per month within the GSM world, doubling the number six months earlier. The latest figures show that SMS has taken off, with exponential growth experienced in many markets once the 20% penetration threshold was reached.

Figure 11

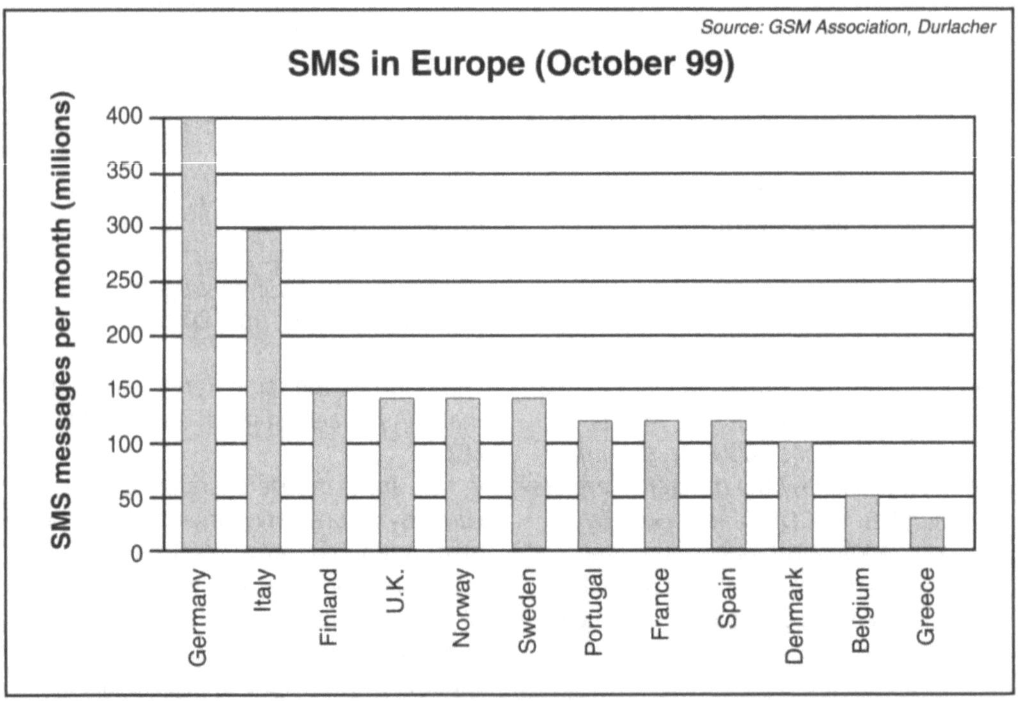

About 90% of SMS messages are voice mail notifications or simple person-to-person messaging. The rest is mobile information services, such as news, stock prices, sport, weather, horoscope, jokes etc. Additionally, SMS e-mail notification, SMS chat and downloading of ringing tones has been offered recently in some markets. SMS services can be customised using SIM toolkit technology. SMS is an ideal technology for pushing information from one-to-one or one-to-few.

We expect that SMS will show rapid further growth, with the number of messages doubling every half a year. SMS will benefit from the hype created with the arrival of WAP. It will continue to be available until at least 2005, but it will lose importance and become part of an advanced messaging solution built around instant messaging via GPRS and e-mail. Many m-commerce applications will be initiated using both SMS and WAP as a platform, until the latter has sufficient support on the handset side.

USSD

Unstructured Supplementary Services Data (USSD) is a means of transmitting information via a GSM network. It is to some extent similar to SMS, but in contrast to SMS, which is basically a store and forward service, USSD offers a real-time connection during a session. The direct radio connection stays open until the user or the application disconnects it. A USSD message can have up to 182 characters. It is relevant for real-time applications, such as mobile stocktrading, where a confirmed information transmission is needed. USSD is a WAP bearer service.

We believe that USSD will grow with the further market penetration of WAP. We see it used mainly for mobile financial services, shopping and payment.

Cell Broadcast

Cell broadcast (CB) is a technology that is designed for simultaneous delivery of short messages to multiple mobile users within a specified region or nation-wide. CB is similar to SMS, but it is a one-to-many service rather than a one-to-one or one-to-few. At the moment, only those users that are within the broadcast area when the message is sent can receive the messages. It is a mass distribution media mainly for news and generic information. The user has to turn on the particular channel in order to receive the news from a selected provider, so that spamming can be avoided.

Usually, cell broadcast services are distributed to the consumer on a no cost basis. The business model works in such a way that the network operator charges the content provider for sending the messages and the content provider will try to make money on follow-up services. For example, a free news service could turn into a paid for service after six months, when the customers grow accustomed to it. Alternatively, follow up revenues could be realised with a premium rate phone service, where additional detail is provided.Platform vendors include Logica Aldiscon, Telecast, CMG and Sema.

We believe that cell broadcast might become a technology to be used in convergent offerings for internet communities or followers of sticky local 'passion centre' content such as football, music, or cars. It could also be used to provide a city information service fed by a local newspaper.

SIM Application Toolkit

SIM Application Toolkit (SAT) technology allows network operators to send applications over the air as SMS or as Cell Broadcast message in order to update SIM cards with changed or new services.

SIM Toolkit applications are built in Java for a client server environment. SAT servers have been built by smartcard specialists, such as Gemplus, Giesecke & Devrient and Orga, as well as independent developers, such as Across Wireless of Sweden. SIM Toolkit handsets have been developed by all major cell phone manufacturers. But because there are many different classes of the protocol, although all claim to be built on the GSM standard, not all handsets allow all applications. In contrast to SAT, WAP provides a more web-centric/thin client environment, that is easier to manage and to maintain.

SIM Application Toolkit is targeted at phones that do not yet fall into the smartphone category. Small programs can be fairly simply created by the network operator. For example, SAT defines how the mobile phone talks to a bankcard, which is inserted into a dual slot phone. This is also the same technology being used to allow users to download new ringing tones.

Security is a key feature of SIM Toolkit, since data confidentiality and integrity are already included in the standard. Mobile banking has been the trial application with the strongest demand for SAT, but mobile e-mail and mobile information services have been also helping the demand for it.

We are of the opinion that, although SIM Toolkit is being heavily pushed by the smartcard industry, it will be an interim technology and will not be able to survive once GPRS terminals hit the market, since WAP will be the GPRS-supported protocol. WAP 2.0 will include SAT.

However, SAT is available now and it enables numerous trial applications today that can be tested for demand and impact in the market. SIM Toolkit helps to create the market, awareness and business models for mobile commerce, but many operators are directly implementing WAP.

WAP

WAP (Wireless Application Protocol) is an open, global standard for mobile solutions, including connecting mobile terminals to the internet. WAP based technology permits the design of interactive, real-time mobile services for smartphones or communicators. The WAP Version 1.1 specifications were announced by the now over 200 member strong WAP Forum on June 30, 1999. The primary goal of the WAP Forum is to bring together companie from all segments of the wireless industry value chain to ensure product interoperability and ultimately growth of the market. The wireless application protocol with its different protocol stacks compares to the internet protocols as follows:

Table 1

Wireless Application Protocol (WAP) vs. Internet Protocol	
Internet	**Wireless Application Protocol (WAP)**
HTML Javascript	Wireless Application Environment (WAE) WML
HTTP	Wireless Session Protocol (WSP)
	Wireless Transaction Protocol (WTP)
TLS-SSL	Wireless Transport Layer Security (WTLS)
TCP/IP UDP/IP	Wireless Datagram Protocol (WDP) WCMP; User Datagram Protocol (UDP) SMS, USSD, GPRS, CSD, CDPD, HSCSD, ETC

Source: AU-System Radio, Durlacher

There is little doubt about the future success potential of WAP and even companies like 3Com and Microsoft have recently joined the WAP Forum. WAP is compatible with GSM 900, GSM 1800 and GSM 1900, CDMA and TDMA (Time Division Multiple Access) wireless standards as well as the proposed 3G communication systems.

In our view in order for WAP to make an impact on the market, the following three criteria must be met:

1. The penetration of WAP terminals must be sufficiently high (Status: first WAP phones to hit the shops only by year end 1999)
2. Relevant WAP applications, which really provide added value should be made available (Status: applications are limited and not location based, personalisation has to improve)
3. WAP gateways must be installed at operators or corporates in such a way that users can access WAP based

services (Status: gateways are only being installed slowly)

The big advantage of WAP is that it makes it easy and user friendly to receive and react to information on the mobile telephone. Therefore, WAP is expected to lift the entire area of mobile information services to a new plane, one, which the SMS world is only a poor approximation of. WAP has been able to gain support from all major players in the market. Therefore, we are convinced that WAP is likely to succeed. WAP based information is also optimised for GPRS, so that the transition will be very smooth to the "always on" mode.

We believe however, that as more advanced services, such as mobile broking and banking, mobile advertising and mobile shopping, are offered, increasing value is added to WAP. After all it is the applications that will make it successful. We believe that after 2001 no mobile phones will be shipped that are not AP enabled.

Web Clipping
In the United States the web clipping service for 3Com's Palm handheld device has been very successful. The Palmhas a 75% market share of PDAs in this market. Web clipping is a Palm proprietary format for delivery of web-based information to Palm devices via synchronisation or wireless communication to the Palm VII. Avant Go is the primary content aggregator of web clipping services. A number of prominent content providers including AOL Instant Messenger, Amazon.com, UPS, Fedex, Yahoo! and others have developed real time content for wireless delivery via this service. This service does not have the same recognition in Europe, primarily because of lack of availability of the Palm VII device and its wireless network in Europe. 3Com has no plans to launch the Palm VII in Europe. Web clipping may co-exist with WAP in the fragmented US market. However, in Europe it is likely to be superceded, even on the Palm platform, by WAP based services.

MExE
The Mobile Station Application Execution Environment (MExE) is, essentially, the incorporation of a Java virtual machine in the mobile phone. The purpose of MExE is to provide a framework on mobile phones for executing operator or service provider specific applications. It allows full application programming. The protocol is integrating location services, sophisticated intelligent customer menus and a variety of interfaces, such as voice recognition. MExE will incorporate WAP, but also provides additional services exceeding the WAP functionality.

We believe that MExE might be built into future UMTS phones, which will have the processing power to run the Java programs. For application developers it will be increasingly important to develop their products for more than one protocol, since many of the above protocols will be in the market at the same time. MExE will be the next logical step after WAP.

MOBILE MIDDLEWARE

A number of middleware platforms are emerging with the arrival of m-commerce:

Mobile Portal Platforms

Oracle is offering a Portal-To-Go platform, which was known earlier under the name Project Panama. It allows mobile operators to translate web pages into WML (Wireless Mark-up Language) format pages, so that they can be read by AP enabled smartphones. Spyglass' Prismis a similar platform.

In contrast, IBM has linked its MQSeries Everywheremiddleware technology to let mobile workers with smartphones and communicators exchange data with back-office systems.
It is a key component of IBM's "pervasive computing" initiative under which the company develops technology, products, and services for portable devices and embedded systems. Thus, the same software is run on mobile devices as on back end systems. Oracle is providing links between company databases and database software embedded in wireless devices as an alternative.

@Motion of the US has also announced the development of a wireless portal for carriers or traditional portal providers that lets them launch an internet voice portal including text to speech and browser entry. The German company Dr. Materna has also developed a packaged portal solution for mobile operators.

Mobile Commerce Platforms

A variety of vendors are currently positioning themselves in the market for middleware solutions for m-commerce. For example, the middleware m-commerce server from Logica provides an interface between a retail bank and the mobile operator, so that bank information, bill payment and electronic value download can be supported from the mobile phone. HP has come up with an integrated m-commerce platform, that is a hardware, software and services bundle based on WAP and which integrates HP technologies like e-speakand third party software. Intershop has also developed its a sell-side m-commerce platform with Danet from its former e-commerce product, which is used for example by T-Mobil in Germany.

Oracle has developed its Oracle 8i Lite 4.0database for the use of data and applications by a mobile workforce. They have been first to support smartphones and communicators based on PalmOSand Windows CE, but IBM and Sybase are already following along the same lines.

Mobile Payment Platforms

Mobile operators are generally deploying proprietary billing solutions from vendors such Kenan (acquired recently by Lucent), Logica or LHS. These platforms have been developed for per minute charging of standard voice, SMS and premium rate calls only and not for charging for particular content. Kingston-SCL is providing its billing solution to France Telecom Mobile and is charging for weather, news and traffic information on WAP.

Payment solutions targeted especially for the mobile market have been developed for example by start-up More Magic

Software, which is financed by Siemens' Mustang Ventures. This Finnish company has developed MBroker, a micropayment platform that lets mobile operators bill for diverse content and services rather then on a per minute basis using a variety of payment methods. Brokat is offering one of the leading e-payment solutions with its Twisterplatform.

Mobile Banking Platforms

Mobile banking and trading platform or solution providers are evolving very quickly and include the following players:

In the US, Aether Systems are the main suppliers of mobile banking technology. Aether, partly backed by Reuters, IPO-ed in October 1999, with a share price rise of 200% on the first day of trading. They have been tasked to bring Charles Schwab the world's largest on-line broker, into the mobile world.

Table 2: Overview of the key mobile banking and broking technology players

Category	Suppliers
Mobile financial service providers	Aether Systems, w-Trade and EmailPager (all US), Multichart, Teledata
Mobile data service providers	GIN (acquired by Saraide.com), Research in Motion, Multichart
Mobile software developers	Brokat, Yellow Computing, Netlife, Aspiro, DataDesign
Mobile system integrators	IBM, HP, Logica
Mobile communications software/gateway companies	Apion (acquired by Phone.com), Phone.com, Nokia, Digital Mobility, 724 Solutions (Citibank with Sonera), Sonera SmartTrust, CMG
Mobile operators	Mannesmann Mobilfunk, T-Mobil with T-Online, Cellnet, Cegetel/SFR, NTT DoCoMo, Swisscom, Telia

Source: Durlacher

MOBILE COMMERCE TERMINALS

Operating Systems

The operating system for mobile terminals is not standardised and currently there is an ongoing battle to become the technology standard of the future. Each of the operating systems is gathering a number of application developers around them, who mostly develop their products for one OS only. There are three major players, who have each developed their own operating system.

Microsoft has developed a lighter version of its Windows operating system, called Windows CE, that has been created especially for small palm-size, hand-held PCs and other consumer electronics devices. A large number of handheld computer/PDA manufacturers mostly coming from the PC industry, such as HP, Casio, Philips and Compaq, have developed their devices around CE. However, CE has faced problems surrounding ease of use, robustness, synchronisation and memory requirements. Philips was first to announce the production halt of its NinoPDA, weakening further the position of the Microsoft Windows CEcamp. Instead Philips sees the future in AP enabled phones that will take over some of the palmtop's functionalities.

Symbian is a consortium of leading mobile handset manufacturers Nokia, Motorola, Ericsson, Matsushita and UK PDA manufacturer Psion, and was established in June 1998. Together these handset manufacturers produce more than 58% (Dataquest 1998) of the world's mobile phones. The operating system, which is based on Psion's earlier software, is called EPOC. It is especially designed for two types of wireless information devices: smart phones (mobile phones with add-on applications and PC connectivity) and communicators (handheld computers with connectivity to or built-in mobile phones). They have the market power to push through EPOC for the smartphone category, which will outnumber the PDA segment by far. Therefore, we believe that Symbian can succeed against Microsoft if it is able to manage its shareholders' (potentially divergent) interests.

1. Symbian's microbrowser supplier is STNC of the UK. STNC was bought by Microsoft in May 1999 and this was viewed as a strategic acquisition striking at the heart of its competition. Similarly, Microsoft acquired Swedish Sendit AB in July 1999, for its Internet Cellular Smart Access server and also its mobile communication protocol, which is licensed by Symbian for use in EPOC. This two hits are likely to harm Symbian's time to market. Some believe Microsoft have understood internally that Symbian is the top competitor and primary threat to its future extension plans in the mobile market.
2. The EPOC Release 5 comprises an application suite that includes a wide variety of communication tools (e-mail, fax, SMS, synchronisation), PIM (Personal Information Manager), office functionality and utilities. It is used in the recently released Psion 5mx.Nokia is currently still using its self-developed GEOSoperating system in its Communicator 9110 product until the end of 2000, but is moving over to

EPOCfor its new developments.
3. 3COM is the smallest player for mobile terminal operating systems, but it is the global market leader in the PDA market (72% according to IDC in 1998) with the Palm Pilot product and its proprietary OS. The operating system is regarded to be inferior to its competitors', but the Palm is much simpler to use in both software and hardware terms. 3COM is intending to spin-off its Palm division in 2000 and IPO it. The PalmOShas a particular wide acceptance in the US, where the Palm VIlwith its wireless connectivity and web clipping technology has hit the market already. However, in Europe there will be a different type of device based on WAP.

Under increasing pressure from Microsoft, Psion and 3Com as well as Nokia and 3Com have decided to work more closely together in developing a common standard. Psion and 3Com have agreed to make their operating systems compatible. Earlier Qualcomm had lined up with Palm to create the PdQ communicator, despite its joint interests with Microsoft in WirelessKnowledge.

In addition to its existing product developments, Nokia decided not to await further discussions, to license PalmOS and to produce a combined OS of both Symbian's EPOC (as a base) and the PalmOS(running over it). The goal is to produce a top-of-the-line hybrid communicator consisting of the palm pen-based UI (User Interface) and a wireless earpiece.

This alliance was necessary and makes a l of sense for all parties creating as it does win-win situation. Microsoft will be the likely loser to this powerful group, which brings the US and European market leadeı in the PDA market together with the thre global mobile phone market leaders. A much closer co-operation or a merger between those parties seems necessary in order to create a true combination of thei operating systems.

Whoever takes over the Palm division will determine whether the market goes towards CEor EPOC.Microsoft might be very interested in eliminating another competitor in order to clear the way for CE.

Physical Terminals

There is going to be a large variety of mobile devices/interfaces in the market that will provide a fit for the various consumer segments. We distinguish the following categories, but note that the borders between them are blurring:

1. Mobile phone: today's mainly dumb devices with voice only capability (e.g. Motorola StarTac)
2. Smartphone: a mobile phone with added applications and PC connectivity (e.g. Ericsson R380, Nokia 7110, Alcat OneTouch)
3. Communicator: a PDA-type equipment integrated with or attached to a mobil phone for data and voice (e.g. Ericsson Mobile Companion MC218, Nokia Communicator 9110)
4. Laptop PC: this includes all the sub-notebook sized equipment (e.g. Sony Vaio)

There are a few AP enabled smartphone products, which are commercially available between end 1999 and early 2000. Although the number of products with internet connectivity is initially very limited to a few types of terminals, by 2001 the majority of mobile phones produced will be AP enabled.

In order to easily enter information for SMS, e-mail and internet within a mobile phone environment, two different approaches have been chosen. Nokia is using Tegic's T9 software in its 7110smartphone that recognises the word to be typed when putting in the first letters. Motorola is using its own Lexicus iTAPsystem, with the same functionality on its Timeport L7089 phone. Ericsson has developed an external keyboard, which can be clipped on to an existing phone.

The smartphones coming with GPRS and 3G will also have all kinds of future functionality included which de facto lets them become a multifunctional consumer electronics device, such as a smartphone MP3 player, a communicator video viewer or a smart gameboy phone.

The Siemens S25has been developed as a AP enabled terminal. However, the product has been developed according to the WAP 1.0 standard, which is not used in Germany. Germany is Siemens largest market for mobile phones.

Laptops can be connected to mobile networks to receive and send e-mail, access corporate intranet data and browse the internet with speeds up to 14.4 Kbit/s on regular GSM networks and on 38 Kbit/s on HSCSD networks. They are increasingly becoming smaller and lighter, and the borders between these and top of the line PDAs are blurring.

Microbrowser

The microbrowser is a software product that is used to access the web from a handheld device. Content that has been created using WML is thereby accessible over mobile devices and networks. The two possible locations to position the browser are either in the phone or on the SIM card. However, the goal of microbrowser developers is to sell their own WAP gateways to mobile operators.

As noted already the microbrowser technology market is dominated by Phone.com and its UP.Browser, which is licensed for free to 90% of the world's mobile handset manufacturers and to 3Com's Palm. The 2 major exceptions are Nokia and Ericsson. Nokia is using its own version of a WAP microbrowser in its smartphones and communicators (as in the 7110), which it licenses to other terminal manufacturers. It also distributes the software via Spyglass, the US-based web solutions provider, to other handset vendors.

Microsoft is using its own proprietary product, which was developed especially for Windows CE. Microsoft acquired STNC in July 1999 mainly for its advanced microbrowser product, which is supplied to Symbian, Microsoft's strongest competitor for PDA/communicator operating systems. Microsoft is continuing to sell the microbrowser, but Symbian is urgently considering an alternative supplier, in order that they do not become wholly dependent. Across Wireless (formerly AU-System Mobile) of Sweden, which is backed by Schroder Ventures, has worked with Finnish SIM card specialist Setec to put the WAP browser on the SIM card, so that it can be used by most standard GSM phones. This alternative is especially useful until WAP handsets are widely available.

German smartcard specialist Giesecke & Devrient (G&D) have developed their own wireless internet gateway browser, STARSIM, which is based on SIM toolkit and which can be used with GSM phones already. It is a temporary solution as it provides the potential to offer mobile access web applications today, which can be later transformed to WAP services.

The microbrowser needs to be on each handset in order to access WML content while the user is on the move. Durlacher believes that the microbrowser in itself provides no real competitive edge moving forward.

Bluetooth

Bluetooth is a low power radio technology that is being developed to replace the cables and infrared links for distances up to ten meters. Devices such as PCs, printers, mobile phones and PDAs can be linked together to communicate and exchange data via a wireless transceiver that fits on a single chip. We estimate that the unit cost will drop from Euro 20 today to below Euro 5 in a few years.

There are more than 1000 companies world wide supporting the technology through the Bluetooth Special Interest Group. After considerable delay, Bluetooth equipped devices will now be available on smartphones from approximately 2001 onwards.

Key applications of Bluetooth are the synchronisation of different pieces of equipment, e.g. mobile phone, PDA and PC, which will make it possible to perform only one single entry with any of the devices used. Additionally, data exchange (for example, with POS (Point Of Sale) terminals), ticketing or e-wallet application for mobile commerce might also boost th success of Bluetooth. Since it has a throughput of about 1 Mbit/s, Bluetooth might also be used in wireless LAN applications.

Using Bluetooth, it will be possible to separate the transceiver unit of the mobil phone from the earpiece and the display. Thus, the transceiver unit could be in the belt buckle and the display in the watch with no wires needed.

We believe that although mobile commerce would be possible without Bluetooth, the technology is adding convenience for both mobile payment an security. Bluetooth provides some of the key functionalities to change a mobile phone into a "lifetool".

Smartcards

Smartcards, i.e. chip cards with a small microprocessor such as GeldKarte, Proton or Mondex, can have credit/debit functionality as well as digital signature or electronic wallet functionality. They are als capable of being used as a loyalty card or as a health record card. The SIM (Subscriber Identification Module) cards used within the GSM phone are (miniature) smartcards as well. Their size and compatibility with the magnetic stripe card theoretically makes the smartcard an ideal carrier for personal information, such as secret keys, passwords, customisation profiles and medical emergency information.

Although many smartcards have been delivered to customers for other reasons (for instance as ATM (Automatic Teller Machine) cards) and not as a debit card for direct payments, there is ongoing speculation about the success of smartcards as an electronic purse. So far, there are about 30 million Proton-cards and 55 million GeldKarte circulating in the market. However, the smartcard for micropayments has not had any real success. In Germany, only 30% of all owners of a GeldKarte know that they actually possess one, because the small processor functionality is included in a normal bankcard and many customers have never used the electronic wallet functionality.

A common standard for smartcards is still absent. The 20 member strong OpenCard organisation grouped around IBM, Sun, Visa, Gemplus and Schlumberger have tried to push for interoperable smartcard solutions based on Java across many hardware and software platforms that are based on Java, but they do not seem to be overly committed to make it fly. Visa, for example, has also developed a proprietary solution, called Open-Platform that it is pushing independently into the market.

Additionally, there is an ongoing effort to develop the Multos-card, a multi-application card, which is claimed to be compatible with most international standards, such as GSM and EMV (Europay-Mastercard-Visa). So far the Multos card has not been compatible to the Javacard, yet another widely supported smartcard technology. The basic compatibility between the two has been agreed and we are awaiting the results.

However, too many standards have been created by the various vendors for both the smartcard operating system and the POS-terminal. It is our view that in the current situation, no one standard will gain a leading position. Technology is not the key differentiator; rather it is about business issues, such as customer take up and the availability of a common card reader.

Microsoft has positioned itself as a unifier of card operating systems by pushing its Windows for Smartcardsto be launched in late 1999. The SIM card running on Windows will come with applications and downloadable applets such as Windows 2000 log-on, internet access, e-mail, calendar, small address book, home banking, encryption, payment transaction, electronic signatures as well as the standard GSM requirements authentication, identification and portable storage.

We believe that, when WAP has reached critical mass in terms of penetration, applications and services will move away from the client and towards the network, so that there will be a thin phone client. It is very likely that the client's SIM card will continue to host the phone number, authentication, digital signature, identification, offline synchronisation, some favourite web sites and encryption.

Smartcards will continue to be a difficult market to forecast until a standardised true multi-application card is developed, which allows the incorporation of a number of different applications, including GSM-SIM and security applications, such as PKI (Public Key Infrastructure), on the same card. Ultimately, the mobile phone companies might create this kind of environment to strengthen their vision of the mobile phone as a "lifetool".

PKI

Security is a key enabling factor in m-commerce. Although GSM provides some improvements through the PIN (Personal Identification Number) when turning on the handset, through an authentication protocol between handset and network and through SSL (secure socket layer) encryption of voice and data, it is not sufficient for ensuring highly secure wireless commerce.

It is widely believed that smartcards will be the preferred way for gaining access to a secure system. The smartcard can be in form of a credit card or in the form of a miniature card, like the SIM card. It is possible to run a variety of applications on one single small SIM card.

Encryption is used to ensure confidentiality through a secret key in association with an algorithm. This produces a scrambled version of the original message that the recipient can decrypt using the original key to retrieve the content. The key must be kept secret between the two parties.

There are two basic methods, which can b used to encrypt a document: symmetric and asymmetric. With the symmetric method the same key is used for encryption and decryption. The problem is that the key has to be transmitted to the recipient of the message, and a third part could gain access to the key during this transmission.

Using an asymmetric algorithm, also known as public key methods, a set of tw keys is used: a private and a public key. Information encrypted using the public ke can only be retrieved using the complementary private key. With this system the public keys of all users can be published in open directories, facilitating communications between all parties. In addition to encryption, the public and private keys can be used to create and verify digital signatures.

Today, symmetric encryption such as DES (Data Encryption Standard) or 3-DES is most common. With the symmetric encryption method both parties have the same key, typically 40-128 bits. Asymmetri encryption becomes more relevant as it is statutory requirement in some countries, e.g. in Germany by the Digital Signature Law. It is set to become more widespread in the future driven by the EU directive on electronic signatures of March 99, which has to be further translated into national laws. Within asymmetric encryption, each party has a key pair, i.e. a public and a private key with typically 1024-2048 bits.

The market leader in security for mobile commerce is Sonera SmartTrust, who has offered PKI (Public Key Infrastructure) for cell phones since early 1999. Sonera is co-operating with GTE CyberTrust, the US PKI specialist. PKI uses certificates, certification authorities, asymmetric encryption and digital signatures. Sonera has implemented PKI on a SIM card (manufactured by Finnish Setec) within the GSM phone, so that no additional smartcard reader is required. Other PKI solution providers include Baltimore of Ireland, who have a co-operation agreement with WAP gateway vendor Apion (acquired by Phone.com in October 1999). Full security is reached in PKI through:

- Digital signatures for authentication of customer and merchant
- Non-repudiation of the involvement in the transaction
- Strong encryption
- Integrity of the message
- Confidentiality

A new initiative, the Radicchio alliance has been launched in September 1999 by EDS, Gemplus, Sonera and Ericsson to promote PKI (Public Key Infrastructure) as the standard for secure wireless commerce transactions. The success of the initiative will largely depend on how well they can market themselves to the public and on how many other players they can get on board. 50 supporters by early 2000 is considered by them to be critical mass.

Security solutions are becoming increasingly integrated into application or platform offerings in order to increase reliability and decrease bandwidth-intensive add-on cryptographic software. JP Systems of the US is integrating Certicom's asymmetric encryption and SSL Plus technology in its wireless e-mail systems. Similarly Puma Technology, the synchronisation software provider, is also partnering with Certicom.

Brokat's Twister based electronic services delivery platform is also using electronic signatures, which are on the SIM card. Brokat of Germany, who is using a symmetric 128 bit encryption, has also created a PKI solution that includes mobile phones and is piloting it in Singapore and with Deutsche Telekom. HP's VeriFone division has also adapted its VeriSmarte-commerce software to support secure sessions with WAP based handsets.

We are convinced that after many years of PKI being 'just around the corner', fast take up or otherwise in the mobile market of Sonera's best-in-class encryption solution will determine if the technology will be able to finally reach critical mass.

Synchronisation

Synchronisation is a key technology enabling mobile commerce, because there will be demand for both web-centric and local applications on a PC or any type of mobile device. Synchronisation is the process by which identical versions of applications and data are maintained wherever and on whichever device the user chooses. So far in corporate purchases, management and synchronisation issues for mobile devices have been largely handled by individuals rather than the corporate organisation, because of the relatively low penetration rate of PDAs.

However, in order to boost employee productivity, more companies will integrate communicators or smartphones into their corporate IT/communications environment. Mobile business applications, such as ERP, CRM, KM (Knowledge Management) systems or fleet management, are also demanding that particular data be resident on the client at any point of time, making mobile devices an extension of the corporate network.

For example, Puma Technology's Intellisync Anwheresoftware allows corporations to use remote and LAN (Local Area Network)-based synchronisation of PalmOS devices with MS Exchange and Lotus Domino as part of the business infrastructure. Motorola has initiated the Starfish TrueSynchinitiative to advance synchronisation technology. The UK's Paragon software has developed FoneSync software that allows users to update the mobile phone's directory from a PC.

FusionOne has developed an internet synchronisation technology (Internet Sync) that facilitates anytime/anywhere information demand. The software provides the possibility to maintain files and contacts on the office PC, which are then converted and forwarded to a storage space on a web server. The space on the web server, however, is limited to 25 MB and only available in the US. From there, the data is automatically downloaded to other specified devices, such as a communicator, a smart phone or a home PC, when they are connected to the internet the next time. Additionally, a number of companies, such as Jump (acquired by Microsoft), Visto's Briefcase and desktop.com, are going to make the user's applications, files and data centrally available via the internet and accessible via any mobile device.

MOBILE LOCATION TECHNOLOGIES

The ability to locate the position of a mobile device is key to providing geographically specific value-added information that stimulates mobile commerce. Mobile location services may either be terminal or network based. The largest push for this technology is coming from the US. There, mobile telephone operators have been forced by the FCC to provide emergency 911 services by Octobe 2001 in such a way that the location of th caller could be determined within a radius of 125 meters in 67% of all cases. ETSI ha standardised the first three LFS (location fixing schemes): GPS, TOA and E-OTD in 1999.

GPS

GPS (Global Positioning System) is a systerr that consists of 24 satellites circling the earth in a particular constellation to each other so that several satellites fall within line of sight for any GPS receiver on Earth. Because the satellites are continuously broadcasting their own position and direction, the GPS receiver can calculate its position very exactly. Anybody can use the GPS system for free with an appropriate receiver.

GPS has been developed in the US for military use, but from the beginning of the decade it has been usable (with lower resolution) for civilian purposes. GPS requires additional equipment or some modification in the mobile device, so that i can become a GPS receiver. This technology is used in car navigation systems. GPS technology for mobile phones is being currently developed for example by SnapTrack and SiRF. GPS is already used in Benefon dual mode GSM/GPS handsets.

TOA

TOA (Time Of Arrival) technology requires larger network modifications and is therefore not very cost-effective. Rolling out TOA for an entire network is estimated to cost as much as 10 times the price of an E-OTD system.

E-OTD

The E-OTD (Enhanced Observed Time Difference) system works by using the existing GSM infrastructure to determine the mobile phone's location. When a user calls selected service providers, E-OTD simultaneously sends data indicating the phone's position.
It works by comparing the relative times of arrival, at the handset and at a nearby fixed receiver, of signals transmitted by the underlying mobile network base stations. The E-OTD system overlays the existing mobile network. Suppliers for E-OTD solutions include CPS, Ericsson and BT Cellnet.

COO

COO (Cell Of Origin) can be used as a location fixing scheme for existing customers of network operators, but it is not as exact as the three other methods. It requires no modification to the mobile terminal, but the network operator has to do some significant upgrade work. While in urban areas COO might be sufficient to determine location fairly accurately, because the cell size is very small. In more rural areas, where the cell radius is larger, it might not be exact enough.

LFS Independent

SignalSoft has developed its own proprietary Wireless Location Services solution that can use any or several of the above methods to determine position. The software is installed on the operator's network and is able to combine the position with relevant content. SignalSoft has also developed a tool for provisioning the service and fixing the latitude against a defined zone.

CellPoint (formerly Technor), a recently Nasdaq listed Swedish start up, has developed yet another approach. It is handset based, and it works on the standard GSM network without any modifications. The solution needs only a proprietary server, and works on triangulation between the handset and the nearest base stations from both ends. Thus, the system is very quickly installable and cost efficient when compared with competing technologies. CellPoint's system provides a precision of 100 m in urban and 200 m in rural areas.

We believe that GSM positioning is a key technology, which will permit the distribution of highly valuable, localised information. It is very early to determine which of the technologies will lead the pack and dominates the market, because only one real commercial project has commenced so far (November 99), where CellPoint's technology has been supplied to Tele2 in Sweden. Because of the cost and time advantages we favour CellPoint in the short-term as a technology solution for operators, while in the medium to longer term, E-OTD might capture market share. However, this might only occur if CellPoint is not able to develop a sufficient granularity based on its existing approach or by integrating GPS.

The missing link will be to bring companies who provide the content together with geo-coded information, to make use of the technology. Applications using mobile location service technologies include fleet management, vehicle tracking for security, tracking for recovery in event of theft, telemetry, emergency services, location identification, navigation, location based information services and location based advertising.

MOBILE PERSONALISATION TECHNOLOGIES

Personalisation technologies are needed in order to develop a user profile, which can be recalled when the user is identified via cookies or username and password information.

Proactive applications are expected in the mobile environment. The mobile user needs services that are able to learn from him or her either explicitly, i.e. through entered preferences, or implicitly, i.e. through behaviour.

Broadvision and Vignette are globally the leading vendors of personalised e-commerce applications. Both have only very recently (in October 1999) announced their interest in the m-commerce market. Broadvision is partnering with HP, Nokia, Phone.com and Brokat to create packaged applications and solutions around its One-to-Onem-commerce platform. It minimises the need to navigate with your mobile phone and is configurable by either the network operator or by the subscriber.

Vignette has developed StoryServer 4, a system to manage customer relationships throughout the user's lifetime. It consists o a content management system and deliver a combination of profiling, personalisation and reporting services to create unique user experiences. These individual user experiences will be key in providing portal services to the mobile commerce user. Vignette is a member of the WAP Forum.

We believe that while personalisation technologies are already very successful in the e-commerce environment, they will be absolutely crucial in the m-commerce arena, where every additional click require from the user reduces the transaction probability by 50%.

CONTENT DELIVERY AND FORMAT

XML

The eXtensible Markup Language (XML) has the potential to be the standard language of e-commerce. XML is very like HTML in application and origin (in fact HTML becomes a sub-set of XML). They are lightweight meta-languages, languages used to describe the content and the structure of the data contained within. The primary difference between the two is that HTML is used to visually depict content on the web, whilst XML is designed to communicate the meaning of the data through a self-describing mechanism. If companies' information systems are XML compliant, data e.g. purchase orders, can be exchanged directly (computer-to-computer) even between organisations with different operating systems and data models. The drawback to XML is that it is not a data description language, rather it is a specification for creating data description languages. Partners must still agree on the meaning of the data they exchange.

XML allows information to be presented differently depending on the device used to access it, e.g. PC or smartphone. XML based microbrowsers are heavily promoted by the Microsoft camp, while the larger part of the market is supporting WAP products and services using WML instead.

Microsoft promotes XML as an easier means to access data from HTML based applications, because only 10% of the data has to be changed.
WirelessKnowledge, the joint venture of Microsoft and Qualcomm, has developed a product for converting HTML and Exchangeapplications to XML.

WML
WML (Wireless Mark-up Language), which is a rough subset of XML, has been developed especially for WAP. WML is basically to WAP what HTML is to the internet. WML is the format in which

Figure 12

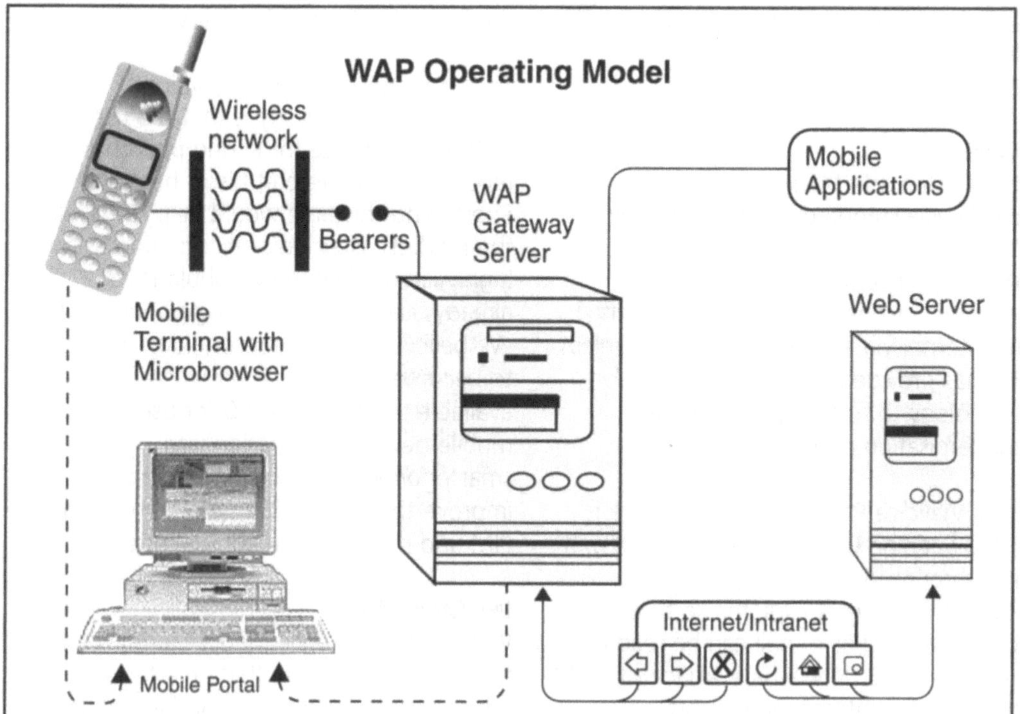

Source: Nokia, Durlacher

information can be read with a AP enabled phone and the built-in microbrowser.

If a request in WML is made, it is passed to a WAP gateway, which in return contacts a web server with the content in standard HTML or preferably WML format. In case the content is in HTML format, a filter in the WAP gateway will try to translate the content into WML. Argo Interactive has developed ActiGateexactly for this, because the vast majority of content on the web is today only available in HTML format. While the HTML to WML translation is an interpretation of the site, most commercial solutions will require the site to be rewritten using WML from the ground up. The requested information is then sent back from the WAP gateway to the WAP client on any bearer service (e.g. SMS, USSD, HSCSD, GPRS).

VXML

Voice eXtensible Mark-up Language (VXML) is a standard being pushed by the VoiceXML Forum, driven by Motorola, AT&T and Lucent. VXML is aimed at enabling voice recognition for accessing the internet via a phone, wired or wireless. IBM has provided its speech recognition technology to further enhance the standardisation process.

IBM's Viavoice Pro Millenium software is already adding voice-activation to internet sites through AOL, Netscape and Microsoft's Internet Explorer, so that web content as well as e-mail can be read out loud by the PC. Nokia have allied with IBM to use their joint experience and technology in order to bring voice recognition to mobile devices. Microsoft has also invested heavily in speech recognition technology, snapping up Entropic and Lernaut & Hauspie. Motorola has linked up with Unisys to integrate its VoxML gateway with their Natural Language speech assistant in order to encourage developers to build voice access applications to the internet.

The frontrunner operator in applying voice recognition technology in a commercial application is Orange of the UK. Orange has introduced its Wildfireservice in Augus 1999. Wildfire provides a humanised electronic assistant whom you can talk to to set up and access your voice-mail. The service talks back to the user and to external callers, and is being extended into a fuller unified messaging service.

The common problem with voice recognition technology is that the user has to train the device for a long period to recognise and interpret his or her voice. Shortening this time would help to develo the market faster. Voice recognition using highly limited learned vocabularies is already used in voice dialling phones today We believe that voice recognition will start to become a broader commercially available product by 2000 for use with mobile devices. Integration into smartphones and communicators will improve the usability for communications, PIM and m-commerce applications.

WWW:MMM

WWW:MMM(Mobile Media Mode) is a trademark to indicate that an internet based value-added service and a mobile device will work seamlessly together. The goal is to promote content that has been optimised for access via smartphones and communicators and the mobile internet overall. The trademark is jointly owned by Nokia, Ericsson and Motorola, who will license it to third parties for WAP-compliant products, content and services. It is possible to run a site and browse other users' sites with a mobile phone under WWW:MMM. The user's homepage is his business card with all the personal details, address and contact information. Additionally, he can publish classifieds, express opinions or maintain a PIM (personal information manager). WWW:MMMhas the strongest backers in the mobile phone industry one could wish for, who do have a considerable interest in ensuring applications for their handsets are developed. However, WWW:MMM has not got so much attention yet as the market is flooded with new abbreviations all centred around the arrival of WAP. We believe that the trademark will need a strong marketing push, if it is to succeed.

M-COMMERCE ENABLING APPLICATIONS

E-MAIL

In our view e-mail is the killer application for wireless internet usage for two reasons: people generally understand the application and they need e-mail often as their prime communication link to stay in touch with their organisation or their friends and family. So far, e-mail access has been the key application for wireless data usage in Europe (together with internet access and SMS). There are broadly two different types of e-mail users:

Figure 13

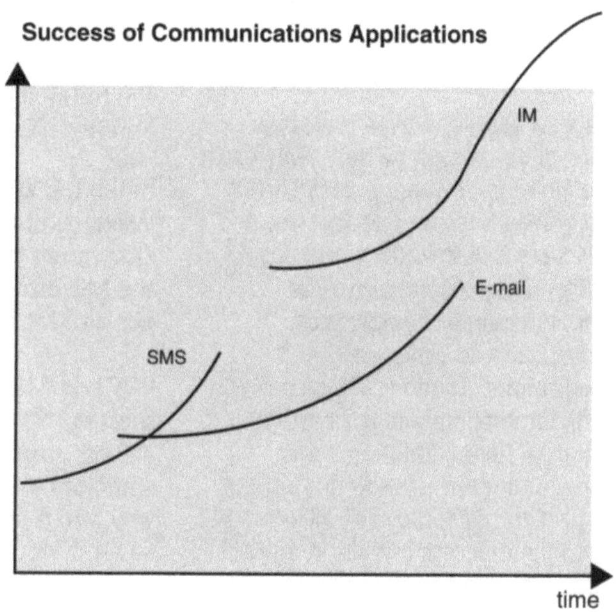

Source: Durlacher

Corporate Users

Until now e-mail has been accessed on the road by corporate users using a laptop, a GSM modem and a connected mobile phone. They are remotely dialling into the corporate network's mail server. This is by far the most frequent application used when remotely connecting to the internet/intranet. Current corporate remote access e-mail solutions are based typically on either Microsoft Exchange or Lotus Notes/ Domino platforms, the dominating collaboration tools.

Alternatively, e-mail messages can be accessed by using a web-browser from other locations, such as hotels or internet cafes, to access (forwarded) e-mail from the corporate e-mail account. However, the highest convenience and accessibility would require a solution in which the e-mail service would be device independent and available on an anywhere/anytime basis.

E-mail access solutions from a mobile terminal will be offered by both, Microsoft with the Platinum Exchange 2000server building on the Internet Cellular Smart Access(ICSA 3.5 in Q1/2000) and Lotus with its Dominoproduct, as well as European independent application developers, such as Dialogue Communications, Dynamical Systems Research, Satama Interactive (through acquisition of Seiren Solutions) and Peramon. Peramon is providing a unique solution, called Lexicos, which allows the user not only to access e-mails in inbox and outbox, but to conduct a keyword search (using AltaVista technology) on her entire, fully indexed e-mail account. Thus, the e-mail system becomes effectively a personal knowledge management system, accessible from any web browser including a WAP terminal.

From Autumn 99, BT and Microsoft will be piloting (under the name " Project Nomad") access for business users via Exchangeto e-mail, calendar and corporate intranet from a wireless, CEbased phone o PDA in the UK and Norway. There are 2 second-tier phone makers, Samsung and Sagem, who are offering phones running on CE for the trial, but BT/Microsoft are also entering the hardware market and are going to jointly develop CE based smartphones and communicators. Thus, they are moving to a new position in the value chain and are redefining the rules of the wireless industry. A future alliance partner is likely to be AT&T for the US market, being BT's international partner. Microsoft's entry into the wireless market i very aggressive and it is targeted to where the future market will be created: in Europe.

In the US, Microsoft is using its WirelessKnowledge joint venture with Qualcomm to run similar trials. However, the Microsoft solution is not using WAP, but an XML based browser.

POP3 or IMAP 4 web based e-mail system such as HP's OpenMailare also used, but are less established in the business community. Web based e-mail requires on very thin clients and thus provides for easier it would provide easy WAP access.

Residential Users

Residential users are generally accessing e-mail either at one of the big portals (MSN Hotmail, Yahoo! Mail, Lycos Mail, AOL mail etc.) or at their local or preferred ISP. Choices for free e-mail accounts are virtually endless. E-mail makes the portal a particularly sticky application, as users are going back there to check their e-mails frequently. Some mobile portals such as Sonera or Mannesmann D2 are offering e-mail services straight from the service launch.

Server-based e-mail via WAP would generate a key competitive advantage for any ISP or mobile operator at the current point of time. For example, Telfort Mobile has installed an e-mail gateway using SIM Toolkit from Microsoft (which was Sendit before the acquisition) in their network – this allows e-mails to be sent directly to the mobile phone as (phonenumber@carriersname.nl). UK based Genie(from Cellnet) offers similar services for SMS today.

In the US, on the other hand, the dominance of e-mail as a driver for wireless data is reflected in a variety of companies focussing only on that element. For example RCN Corporation, a regional ISP with over 500,000 subscribers, is promoting access to corporate e-mail on MS Exchange via the BlackBerry mobile e-mail terminal with PDA functionality, from Research in Motion. Paul Allen bought a quarter of the company, which is in total valued at about Euro 7.5 billion.

Other examples are the wireless ISPs GoAmerica or Alerta, which send e-mails to the wireless phone using an assigned e-mail address, as well as a copying them to a different e-mail address for free. Saraide offers wireless e-mail service based on Critical Path's platform. Intelligent Information Incorporated and Infinite Technologies have both developed solutions based on e-mail for the corporate and the carrier market.

MOBILE INSTANT MESSAGING (MIM)

Using mobile instant messages (similar to ICQ, MSN Messenger or AOL's Instant Messenger service on fixed networks) while being connected to the net (i.e. anytime, when using GPRS), will provide the opportunity for mobile users to check instant availability of people, communicate instantly on a non voice basis and use a chat-like environment for exchanging ideas and information. If a smart phone is used as the terminal to communicate it will probably feature some standard "canned messages", as it will be still complicated to enter text. Intelligent text recognition software, such as the T9function in the Nokia 3210 or 7110 handsets, will ease the difficulty of entering words, by remembering earlier entries. Mobile messaging is a logical extension of SMS (Short Message Service), but it will not be limited in length to 160 characters and the message will be transmitted within a fraction of a second, i.e. near "real-time".

As a first, hefty push into that new market, Microsoft announced its wireless messaging product in September 1999, while AOL and Motorola have agreed in October 1999 to integrate AOL's Instant Messengerinto Motorola's new smartphones and communicators. Thus, from Q1/2000 new Motorola phone users will be able to communicate with e-mail users at PCs instantly. In North America, AOL, Yahoo! and others are already integrating instant messaging into the wireless Palm VIIdevice.

Messaging applications that are currently being developed include e-mail, chat, message boards, and access to internet telephony software, video conferencing, and on line games. These have the opportunity to become the dominant universal messaging tool.

The real advantages of mobile instant messaging will be utilised once the always on, always connected mode using GPRS network technology becomes a reality in the second half of 2000. It will be possible to send messages any time to the mobile device and check if the device is switched on. There are already companies developing applications around instant messaging technology, i.e. trading at a stock/options exchange. A very advanced application has been developed for example by Olympic Worldlink, who have developed a product, called Mobile Futures, which can transmit real-time prices and information to a PDA and provides instant communication/messaging capabilities under GSM.

We expect that mobile instant messaging will be very successful within exactly the segments of the market in which SMS usage has its stronghold today: the youth segment aged 15-25. Pricing of the service will be critical in order to convert existing SMS customers to MIM.

UNIFIED MESSAGING

Unified messaging systems (UMS) are a ke application for mobile usage in future. Unified messaging is an emerging service that is crossing the boundaries between different communications media and is focussing on customer need rather than any particular technology. Theoretically, voice mail from fixed and mobile phones, SMS, e-mail, fax messages and instant messaging messages will all end up in the same mailbox. Unified messaging will enable the user to access the various messages with any one single interface, i.e PC, PDA, mobile phone or fax machine, independent on the original medium, which has been used to send the message This requires the availability of text-to-speech and speech-to-text transformation technologies, which at this moment still seems something of a problem. The qualit of the speech output is key to the success of UMS, and so far it has simply not delivered.

UMS has been offered for a number of years for both, corporate and residential customers, but only including a selection o the media described above. In the mobile world UMS is being offered by a variety of vendors, such as Comverse, Nokia, 2Communications and Dr. Materna and therefore comes in different flavours. Integration of WAP is key to the further roll-out of UMS.

UMS have been deployed by some operators, such as Singapore Telecom with OneM@il, or Telia with its DoF (Departmen of the Future) mobile portal based on the Oracle Application Server and Oracle web-based application software. Telecom Italia

Mobile has launched its Universal Number product in November 1999, which is an advanced UMS product that sits on the Intelligent Network platform. It allows the user to send e-mails directly to a mobile phone without a PC. It is also possible to send audio as WAV files as attachments to e-mails in order to listen to voice mails from a PC. In Italy the penetration of mobile phones is about double the penetration of PCs.

The very advanced IP service application iPulse from OZ.COM and Ericsson, instantly connects computer, PDA and fixed or mobile phone. Additionally, it provides PIM and web collaboration functionality. We believe that unified messaging is set to become a mainstream application within two years.

MOBILE CHAT

Mobile SMS chat has already been offered as a service for about one year in some European markets, such as Finland, Germany and the UK. Chats are based on communities of interest between like minded people. The number of possible communication links increases by an order of magnitude when the number of participants double. Thus, the stickiness of a chat group theoretically increases significantly as user numbers grow.

Chat is a very popular application in the PC based internet world especially among the 15 to 25 year-old user group. This same segment of the market also constitutes the bulk of SMS users. Thus, linking those two applications is a natural development. Dr. Materna of Germany has developed a chat application that allows the combination of online, videotext and SMS into the same service and the user can use any platform to participate. Another vendor of an SMS chat platform is WAPIT of Finland, key supplier for Radiolinja, the Finnish GSM operator. With the arrival of instant messaging via mobile devices, we expect that chat will move to this new platform, and will be both faster and more cost efficient.

MOBILE VIDEOTELEPHONY

The video telephony concept has been developed by handset vendors as a potential application of mobile internet despite the unsuccessful experience in the fixed telecommunications market. In no European market has the penetration of videophones ever taken off and reached critical mass. Correspondingly, it is in our view very unlikely that we will see this as one of the key applications in the mobile business. We believe that videotelephony will be only a niche market product for the first years of the next decade.

MOBILE PIM

Some form of scheduling functionality, such as calendar, address book, tasks and journal, is included in every PDA and in any Office type or groupware software package. Most of these are device-oriented, while the Personal Information Management (PIM) evaluated here does basically the same, but runs on the net and is thus device independent.

The advantage of the net version is that it can be accessed from any web-browser, whether it is a PC based or a AP enabled smartphone. Changes on the PIM are always made in real-time mode, thus helping work-groups to stay closely in touch, even when out of the office. The operator can actually host the PIM functionality for both corporate and private customers.

Mobile accessible PIM-solutions have been developed for example by TimeSystem, the former European market leader of the paper-based day planner, GIN (acquired by Saraide.com), Future Internet Technologies, and Seiren Solutions (who was also acquired during 1999). In the US, there were about 10 PIM players, out of which 5 were bought up including When.com (by AOL) and Jump (by Microsoft).

We believe that a mobile accessible PIM is an interesting addition for a complete portal, useful for those people who do not always carry their PDA or communicator with them. Availability, security and ease of synchronisation with existing legacy systems (MS Outlook, Lotus Notes) and existing PDAs is key so that the data has to be maintained only once.

CONSUMER M-COMMERCE APPLICATIONS

In this report, we analyse m-commerce applications first in a consumer context and thereafter in a business context. This section covers consumer applications. The following diagram indicates when applications will be commercially available

Figure 14

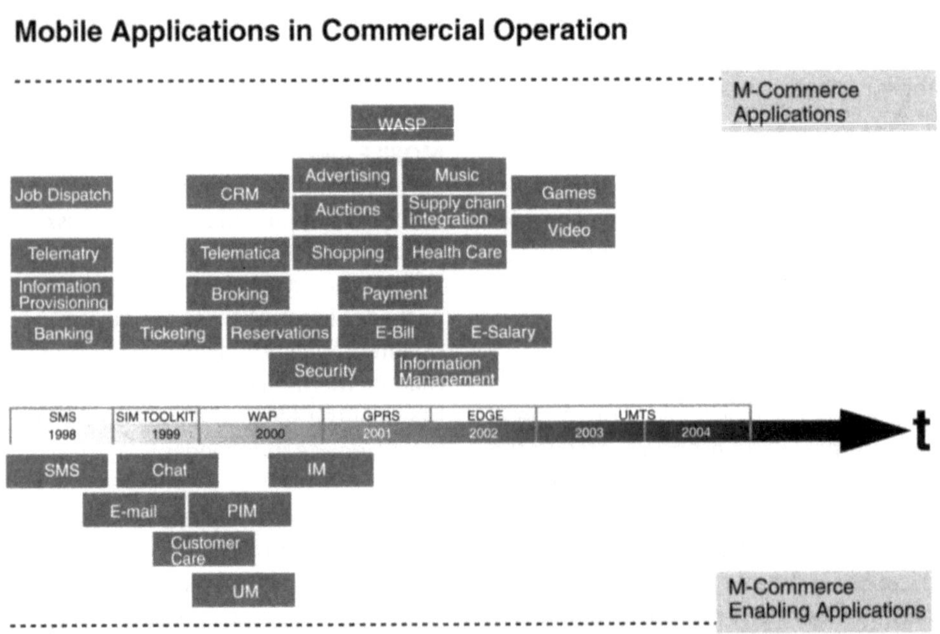

Source: Durlacher, Veba

in Europe and which technology is the particular enabler.

MOBILE FINANCIAL SERVICES

Financial services are a key commercial driver for the mobile commerce market in Europe and beyond. Retail banking and stockbroking markets are in the midst of major industrial restructuring resulting in significant M&A activity and the online dimension is accentuating this. Taking their online activities to a mobile device opens a new service channel for the financial institutions. In a recent survey by Nokia, looking at which application types various parts of the market might demand and when, Mobile banking was the top application demanded by more than 85%.

Mobile Banking

Mobile banking is a subset of online banking, a service that is being offered by 94% of all banks in Europe in 1999. The motivator for mobile banking from the bank's perspective is to have an additional distribution channel and to further cut costs, as every transaction on the internet, fixed or mobile, is saving money on the bricks and mortar operations side.

Based on an existing back office online banking operation, mobile banking can be deployed in a straightforward manner. It requires the elements in the form of a private information service, SIM Toolkit or WAP support and security.

The services mainly considered for offering through mobile banking are:

Public Information

- check exchange rates
- check interest rates

Private Information

- check account and credit card balances
- administer credit lines
- check interest earned on deposits
- check last transactions

Transaction

- transfer funds
- pay invoices
- apply for credit line

The simplest mobile banking solutions can be pull based via a voice call to an IVR (Interactive Voice Response) system or via an SMS request. Alternatively, the service could be push type, thus sending information on an event basis, depending on a certain time or value. All information is currently typically sent via SMS over GSM networks. Interestingly, mobile banking services can generate some additional revenues where banks are sharing in operator revenues generated by the SMS enquiries.

There are more than 30 mobile banking services already in place world wide, and a few have reached mass market status, such as Czech operator Paegas with Expandia Bank and Merita Nordbanken of Finland/Sweden.

The first mobile banking service was started by Merita, the innovation leader in mobile banking, which allowed their customers to make bill payments via GSM as early as 1992 and SMS-based balance checking and transactions from 1997.

In the UK customers of Barclaycard, Barclay's Bank and Cellnet have conducted mobile banking since 1997. A special handset with a bank direct dial " Barclay button" allows the users to receive mini statements, check credit card limits, balances and next payment days. Apparently this led to more than 150,000 very loyal customers. The Woolwich Building Society has also announced mobile banking plans in the UK.

Paegas has developed multi-banking support, but each subscriber can only use one specific bank. The SIM Toolkit solution from Giesecke & Devrient, which is based on point to point SMS service, uses symmetric encryption (triple DES). The customer can actually download a bank's specific menu structure to the phone's SIM card. The mobile banking solution is the main driver leading to the current 20% penetration of customers using mobile commerce.

Elsewhere SIM Toolkit based mobile banking solutions are now offered by the DVG, a German savings bank association, German direct and online bank 1822direkt.

On arrival of WAP phones, Merita launched a commercial WAP banking service in October 1999 after completion of an 8 month trial period in partnership with Nokia.

Deutsche Bank, the world's largest financial services organisation, has also decided to move to WAP with the help of Nokia as a full solution provider, i.e. also as a systems integrator. Svenska Handelsbanken has teamed up with IBM to launch a WAP based mobile banking service in Q1/2000. Thus, we believe that the future is WAP banking rather than SIM Toolkit banking.

Mobile Broking

Mobile broking is a killer application for mobile commerce. Location independent, real-time information about a share price reaching a particular stop mark and the possibility to act on it provides a very high value to many stock traders, private or professional. Shares exceeding certain pric points could trigger messages asking whether to buy or sell. In the US, many online brokers are already offering mobile trading via cell phones or Palm VII.

Traditional stockbrokers are moving increasingly online and new online brokerage firms are being established all across Europe, such as the First E-Bank (which is teaming up with Wit Capital to combine banking and broking in Europe). Competition is becoming fiercer in an online world, but mobile broking provides a new differentiator and increases customer loyalty. The online trading commissions will still be charged, but trading volumes are expected to be higher as clients have more opportunities to trade.

Mobile broking provides the following key functionalities:

- receive alerts about price-movements
- receive message when order is executed
- check quotes
- manage portfolio
- buy and sell stocks, options, mutual funds, other financial instruments
- browse and delete existing orders

Merita-Nordbanken of Finland and Sweden is already experiencing increased trading volumes through online brokerage. Today, more than 57% of their stock orders are done online. Therefore, Merita has prepared well to launch in November its WAP brokerage service. Also Leonia Bank and Oko-Bank, the two main competitors of Merita, are betting on WAP. And the leading online brokerage in Finland, eQ-online, is involved in a pilot with Sonera.

Delivering financial information regarding the stock market in a real-time or offline mode is a very useful aspect of mobile broking, but it still is missing the transaction capability. In order to conduct a transaction based on the information, the investor would still have to call his broker. For example, in Germany Dr. Materna in partnership with Teledata real-time share prices, offers exchange rates and lists of share prices, portfolio management tools and so on. Multichart is offering its KISS(me)Broker product, which is augmented by an additional news service. In the US, Stock Smart is offering more than 400,000 web pages with financial information that can be customised and sent on an event basis to mobile devices.

When it comes to transactions however, many mobile brokers seem to rely solely on the standard GSM SSL encryption, without any advanced features and guarantees. PKI must be standard to ensure secure stock trading.

While an average PC online share trade takes about 5 minutes, Fraser Securities of Singapore is experiencing an average trading time of only 2 minutes for mobile devices.

The expectations of the mobile broking service are high, but it is believed that the payback period (in terms of ROI (Return On Investment)) for the S$500,000 investment is only a few months. The take up of the service within the first 4 months of operation was so high, that more than 20% of all the online trades were done via mobile phone, although only one particular mobile phone type could be used with a microbrowser equipped SIM card.

We believe that mobile broking will be a regular way to trade shares in the whole of Europe within the next five years. Many more people will own a mobile phone than a PC by that time, and a new level of convenience and timeliness of decision making can be reached. Thus, the application will help to drive mobile commerce as a whole.

Mobile Cash

Mobile electronic cash refers to loading cash onto a stored value card via the wireless network. It is also referred to as mobile ATM or mobile phone cash machine. The solutionsso far being piloted all use Motorola's dual slot StarTac Dmobile phone to authenticate and verify transactions. The phone is thicker than the regular model, because it has space for a full sized smartcard reader at the back. The other manufacturers have not developed a dual slot phone, but some are experimenting with a smartcard reader somehow included in the mobile handset.

In one of those trials, Visa in the UK has been experimenting with a debit smartcard called Visa Cash, to which the cash can be loaded, while Barclaycard has supplied the card and is providing the banking infrastructure. The trial was conducted over the BT Cellnet network in early 1999. Smartcard manufacturers, such as Gemplus and De La Rue (acquired by Oberthur in September 1999), have been involved using SIM Toolkit to ensure the authenticity and integrity of the transaction. In a trial in Singapore that will be launched commercially at the end of 1999, local debit cards can be reloaded with the same technology.

We believe that these trials are an interesting playground for seeing how the different technologies work together, but dual slot terminals are unlikely to become widespread in the market, for several reasons: the phone would be thicker and heavier, and smartcards for micropayments have not worked so far. Moreover, a smart card reader on a mobile phone might not be robust enough to survive external impact on the handset. BT Cellnet has already officially announced that dual slot phones will not be introduced commercially for the foreseeable future.

Mobile Payment

In August 1999, France Telecom launched its Iti Achatservice for purchasing via the mobile network on a trial basis. It was launched as a mobile commerce pilot, but is in fact more a mobile payment service. It is based on a dual slot Motorola StarTac phone with a smart card reader that allows payments via Cartes Bancaires (CB), the bank card supplier group. While they have already signed up some interesting merchants, the process is still far from a viable commercial mass-market initiative.

After the customer has ordered the product by phone, internet or Minitel, the merchant will send an SMS message with the relevant price. Then the customer must insert his CB card into the second slot, enter a password and the transaction data is then transmitted to one of the participating banks. The service is planned for commercial launch during Q1/2000.

While the service provides some additional security in comparison to just giving the credit card number over the phone or via the internet, we do not believe in the future success of this type of solution. Dual slot phones are unlikely to become mainstream products and the solution does not fulfil the basic requirement of making the user's life easier. A more advanced mobile payment solution is realised by Sonera, formerly Telecom Finland. It all started with the famous Coke vending machine, which has been converted into a Pepsi automate since, from which a soft drink could be purchased with a GSM phone. Currently, it is possible to use Pay-By-GSMfor a variety of low-priced products and services such as passport photos, copies, golf balls, jukebox, toy car ride, shoeshine machine and car wash.

There are three ways in which customers can purchase a product or service from a vending machine or the internet using a mobile phone:

- Dial a premium rate number (0900) which has a call charge equivalent to the product price.
- Dial a prefix plus a premium rate number to indicate that the product should be charged to a different bill (important for users of corporate

mobiles).

- A pre-standing agreement for credit card payments is put in place. For authenticity, a PIN has to be entered at the time of purchase.

The role of Sonera is that of a clearinghouse, (which collects the money from the users and credits it to the service providers) and that of a network provider, who is generating additional traffic. Ericsson and Unisource have also piloted the use of vending machines in Stockholm, but these did not become as popular as Sonera's trial.

The first showcase Bluetooth vending machine has also been developed by Sonera in 1999, which allows the purchase of soft drinks, CD covers and candy bars with a Bluetooth-equipped mobile phone.

We believe that mobile payment will have an enormous potential, especially when using Bluetooth technology. But this technology must be pushed by the Bluetooth technology and handset vendors, since mobile operators have no additional call revenues, since no calls are established via the network. However, the operators might take over the clearinghouse and billing function instead, thereby switching their role in the value chain towards content.

Finland has produced another innovative payment solution, which is today being used for "traditional" internet purchases, but which will be also available towards the end of 1999 via WAP terminals. It is an SSL-based solution (thus only standard GSM security), which is offered by Merita-Nordbanken using its established Solo brand to pay directly over the (mobile) internet to 700 merchants from your bank account just by entering your username, password and a TAN (i.e. a one time transaction number). Key for the success of this payment method is that the technology functions according to users' habits, it is only adding an additional distribution channel at no additional cost for the user and it does not try to change the user's habits. Finnish banking customers have been using this method since 1984, when online banking was first introduced to the market.

Again, Merita-Nordbanken, Nokia and Visa are experimenting with the EMPS-phone, the Electronic Mobile Payment System mobile phone, which is based on a dual SIM concept for the Nokia 7110 WAP phone. A second, semi-permanent, multi-application, minisize smartcard, that is a Visa credit and debit card as well as a Merita bankcard, is inserted into the mobile phone. One of the major difficulties Visa has with this solution is that the logos are only displayed on the screen. Bluetooth shall be used to make local payments via the shop POS equipment, while other payments can be made over the internet. Major problems retarding the uptake of the solutions are that there are no dual SIM WAP phones on the market and that there are no POS terminals for smartcards with or without Bluetooth capability available.

KLELine, a subsidiary of French bank Paribas, has taken up a role as trust centre for online payments. They provide the user with a virtual wallet for their mobile phone that can be loaded from his Visa, Mastercard or American Express card up to USD 100. The wallet can then be used to pay for mobile commerce transactions, such as premium content or any other goods offered.

Confinity, a Silicon Valley start-up, is building a service called Pay-Pal, that allows users with a handheld device to make payments to anyone else with a similar handheld device – allowing the settling of debts, borrowing of cash or splitting of bills without exchange of cash. In order to use the service the user of Pay-Pal has to register his credit card details on the Confinity web page and download the software. The free service allows the beaming of money from one mobile terminal via infrared to another. Only when the sender goes back online again, the amount is actually dispatched. The business model is based on having the money sent into Confinity's account for a few days and Confintiy using the interest from it. The company received backing among others from Nokia Ventures and Deutsche Bank Tech Ventures. Currently it is available only in the US and only with a PalmPilotPDA. Pay-Pal can be seen as a pre-Bluetooth application, as it only uses infrared. It might be upgraded in 2001, when Bluetooth chipsets are widely available. Bluetooth would facilitate payment at the point of sale, but it requires the technology at both ends and the capability to load mobile phones with cash or to direct debit a bank account.

Solving the payment issue in mobile commerce is key for the future uptake of the industry. So far no standards have emerged, but, for the time being, it seems most feasible to use existing payment methods such as credit cards, direct debit, etc., and transferring those to the m-commerce market place. However, security issues remain, as do the demand for convenience and micro-payments. The ideal solution to us seems to move the payment application to the mobile device itself and use the handheld to pay either at the POS viaBluetooth or over a distance via the internet involving some kind of security measures such as PKI. Mobile operators will try to position themselves as key providers of the mobile payment solution, which will in our view lead to partnering with and ultimately acquisition of banks or a banking license.

Mobile e-bill

In Finland you can already receive electronic bills to an e-mail address or to a mobile phone, e.g. from your telephone company, which can be paid via semi-direct debit from the handheld terminal. Thus, no paper invoice is sent any longer.

This will cut costs significantly for the bill issuer saving in both production costs and postage. For the user, mobile e-bill will significantly reduce the effort required to pay bills to trusted parties. The security issue in respect of the digital signature must be solved in order to roll out this service to the entire market.

Mobile e-salary

Sonera employees have to choose today whether they would like to receive their monthly payslips via e-mail or via SMS to their mobile phones. Paper payslips will no longer be produced and distributed to employees by default. The cost cutting effect is significant, because the administration work for physically distributing 9,000 payslips every month as well as printing are no longer necessary.

MOBILE SECURITY SERVICES

The mobile phone with its integrated SIM card is an ideal bearer for the private key digital signature of a PKI system. Thus, the mobile device can become a security tool, for example for secure payment in e-commerce and m-commerce.

The wireless terminal can function as a security device for gaining access to buildings in at least two different ways. First, access could be via using the GSM part of the mobile phone and second, Bluetooth technology could be used as the authentication mechanism.

Sonera in Finland is using GSM already today to open the door to the corporate parking garage. The employees can dial the garage door when they are still some meters away, so that the door will open as they arrive. A back end database administers the mobile phone numbers, which are permitted to use the service. GSM is used in this case instead of a purpose-built infrared sender, which would create significant hardware costs, if all the employees were equipped with such a device.

As soon as Bluetooth technology becomes available in the first handsets in 2001, it will be an ideal way to establish entrance permissions for a given phone. Similarly, the mobile device can be used as an identifier for permission on Pay TV systems.

MOBILE SHOPPING

Mobile extends your ability to make transactions across time and location and creates new transaction opportunities. It is important to note that only a part of the purchasing process is conducted with the mobile terminal. The basic point is that you need to know what you want in advance of making a mobile purchase. Moving forward, it seems most likely that a shopping list might be created with a web interface, which may then be executed from a mobile.

At the current stage of technological development the customer must ideally be faced with a one-button purchase experience for mobile shopping. The purchase suggestions will often be based on the user's past behaviour patterns.

Mobile Retailing

There have been a number of network-based services with respect to mobile retailing already available but there has been little success so far. A value-added GSM service, such as D2Blumen, a flower ordering service of German Mannesmann Mobilfunk in conjunction with Fleurop, the world wide network of flower shops, makes it possible to get connected to a call centre and order via credit card.

It is an interesting application to use a smartphone to order pizza from a delivery service; this might be even more appealing than ordering the pizza via internet, because it takes a long time to boot the PC or a PC might not be available. The hurdles for a fast uptake are of course that microbrowser phones are not spread in the market and applications are missing. For example, pizza delivery services have so far not been fast to move to a web-based model. Mobile commerce combined with location identification creates new value, for example, when ordering a taxi or a pizza the vendor can automatically know where the service is to be delivered.

However, we believe that there will be a large space for e-retailers to become m-retailers, when the personalisation and location issues are well addressed. Books, CDs and groceries are often items, which the user knows well and where he needs just a tool to make a purchase. The purchase will be made when the user has spare time, independently of the shop opening hours and physical location.

Mobile Ticketing
Mobile electronic purchase or reservation of tickets is one of the most compelling proposed services, because ticket reservation/purchasing is hardly a pleasant expertise today. Either one has to go in person to a ticket booth, or has to call an agency or the outlet. Calling outside opening hours means having to go through a lengthy IVR (Intelligent Voice Response) system.

It is clearly more convenient to select and book tickets for movies, theatres, opera and concerts directly from the mobile device, because often the decision to purchase is made while outside or on the move among friends.

This is one of the first WAP applications being seen in many markets. It will take some time until the process is fully automated, because even if today many movie theatre schedules are on the web, this does not mean that it is possible to make a purchase or a reservation there and then. In most cases, due to lack of back end integration, one still has to call to book the ticket. A first step on the way to full automation of the transaction would be to offer one-button dial to the ticket issuer.

The travel market and especially the frequent business traveller market is likely to be an early WAP growth market. Using a WAP handset, train, plane, bus and boat tickets could be booked in a similar manner to movie tickets. The argument goes that mobile commerce will be the driver of market growth in this arena, rather than a phenomenon, which lags behind wireline commerce growth.

The mobile device must be intelligent enough to be able to learn that I go home every weekend to my family and I always need a ticket for the same train. Thus, the mobile portal should suggest this automatically as a default option.

A mobile ticket shop installation is already in place for commercial operations in Norway by Telenor Mobil with Across Wireless' WAP platform for both, cinema and theatre tickets. The subscribers can even make payments for the tickets with their GSM phone. A ticketing application trial has also been developed in Germany by Intershop and Danet consultants (part of the Deutsche Telekom group), which includes reservations in the first phase, but is likely to be extended to film reviews and mobile payment as well.

We believe that ultimately, the tickets will be downloaded onto the mobile device and the device will communicate with the check-in counter at the movie theatre or at the airport via Bluetooth or infrared. Some airlines (such as Lufthansa, BA and SAS) already provide their frequent travellers today with the possibility of electronic ticketing, that is they use their frequent traveller smartcard to identify themselves at the airport to get their boarding pass without ever having held a real ticket in their hands. Why shouldn't the mobile phone also be used to identify the owner via Bluetooth or infrared? The loyalty application, i.e. the entire frequent flyer program details, should be placed ideally in a way that it is easily accessible via the WAP terminal.

The problem is that a lot of transport operators and airlines have legacy systems in operation that do not support electronic ticketing and therefore mobile ticketing either. Thus, the existing IT-infrastructure is one of the greatest handicaps in the move to optimised m-commerce business processes.

Bluetooth will be able to take over many functions and it seems likely that it will be also used for example as a touchless ticketing system for public transport ticketing or for paying toll on streets, since it allows communications to take place directly from machine to machine.

We believe that the airlines will make the first significant push into the mobile ticketing space, because they do have already some experiences in electronic business and they have lots of costs to cut. However, it will take years of passenger education before we see mass uptake of the service. Taking the example cited above, e-tickets make up probably no more than 1-2% of tickets, which are issued e.g. by Lufthansa.

Mobile Auctions

Auctioning is gaining significant page views on the internet as more and more auction sites are popping up for B-to-C, C-to-C and B-to-B. Moving into the mobile environment seems to be a natural extension of the existing business models. The base proposition is that bidders want to continue to participate even when they are not in front of a PC.

A AP enabled smartphone, a communicator or a regular SIM Toolkit mobile phone could be used to receive an SMS message about the latest bid at a pre-set stop mark combined with the questions to bid higher or not. Thus, the bidder does not even have to be online, she could still participate in the process in an almost real-time mode. However, SMS is not too time sensitive, such that the delay might be at times higher than acceptable. We believe that a solution might be the use of USSD technology over WAP terminals, because the interactive session will be in an online mode.

Once instant messaging and instant access to the internet over GPRS have been fully implemented into wireless devices, participation in the auction could be within a real session, but with costs to the user being generated only when data is transmitted.

As one of the first attempts to move into that space, Xypoint and Wireless Services Corp. have joined forces to create WebWireless, a mobile commerce platform. It works in conjunction with a standard web browser, where a notification parameter is turned on, so that an SMS alert can be received from the user for online auction prices. Then the user must actually call an IVR to articulate his intended action. This service will cost up to Euro 10 per subscriber per month. BTCellnet has announced that it will offer mobile participation in auctions on its Geniemobile portal from 2000.

We believe that the auctioning model will expand into the mobile space very quickly, once eBay (who is launching already a co-branded auction pager in the US), QXL and others begin seriously integrating the additional distribution channel, because wireless devices provide the ideal environment for auctions. Mobile devices can overcome the hurdle of being tied to a fixed location, while the user is able to be responsive in an almost real-time mode.

Mobile Reservations

Mobile reservations for restaurants and hotels has been one of the most featured applications in mobile commerce, since the prospect of easily finding a restaurant or hotel that suits personal taste and fits the relevant criteria at least is intuitively very appealing. Especially as a location based service, mobile reservations become a valuable application for the business or leisure traveller.

In the future scenario, using the smartphone's or communicator's microbrowser, the restaurant item can be selected from the mobile portal and particular choices can be made, if the user profile is not already making a suggestion based on past choices and the pre-set parameters. Once down to the preferred restaurant, the user connects directly to th web site of the restaurant (in WML format ideally) to make his reservation.

The problem, currently, is that although there are a few online restaurant guides operating, not too many restaurants or hotels have a web presence. The second best choice would be to get their telephone number and address from a Yellow Pages directory or restaurant guide for the next few years, call the number an gradually move to the web-based model. In order to reverse the payment structure so that the call initiator is the vendor rathe than the customer, RealCall and Argo have recently announced a strategic alliance to develop a solution that permits mobile phone users to access a web site from the phones. The users can receive simultaneou live telephone calls via digital call-back from the relevant vendor.

Thus, the restaurant can be called up for a reservation or, using geo-coded location information to indicate the exact location of the place, the mobile handset can tell how to get there. A push se-vice is also likely to be available soon, v hich is basically a form of mobile advertising. Once a person enters a particular area, an SMS or IM (Instant Messaging) might be generated to indicate that there is a particular type of restaurant close to his or her actual position. The phone would ideally indicate directions from the current location of the user to the restaurant using services such as those piloted by CPS: "Where am I?" and "How do I get there?" based on positioning technology.

At the back-end, the information from the restaurants and hotels would be available in different formats. It could be in the online Yellow or White Pages or in some form of online City Guide. However, in order to leverage the location information fully, the data must be in a geo-coded format. Companies such as Saraide.com (through the acquisition of Dutch GIN), InfoSpace (US) or Swedish start-up CitiKey are supplying or intend to supply their information in this format.

Notwithstanding the limitations as indicated above, we believe that a usable location based service could start to be available by the second half of 2000, once smartphones begin to appear with integrated positioning modules deployed by the operators and vendors.

Mobile Postcard

This wireless picture messaging service lets the users send digital images from a regular mobile phone, a WAP phone or a mobile portal on the internet. Sonera has developed this service, which lets the receiver obtain a real postcard (via regular mail), a virtual postcard or an e-mail postcard. A digital camera must be used to make the photo. Small digital cameras are already being integrated into Sony handheld computers.

Users of this technology could be private customers, who want to send their own photos as postcards to friends and family, photo journalists, builders, surveyors or other professionals who need to send images back to the office. We see this application as a perfect use of mobile technology with integrated consumer electronics, but the impact on the market will be rather limited.

MOBILE ADVERTISING

There is a widespread opinion, that mobile internet will not be as dependent on advertising revenues as the wired internet. Among the justifications for this view are that mobile phones currently have a very small user interface and that graphical visualisation on the screen is very limited.

Up to now, mobile advertising has been carried out only on a very limited level, for example by offering a free cell broadcast channel delivering news – the name of the content provider is always distributed as well. However, we argue that advertising on mobile devices, whether a smart phone or a communicator, will continue to have a strong business case, because it is the dream environment for every marketeer.

It is possible to mass customise a mobile phone for particular user requirements, which then in return would allow one-to-one marketing. It is business critical to market to the individual consumer through very pinpointed and localised messages.

The conditions for one-to-one marketing are ideal using the mobile device. The mobile operator or service provider has not only all the demographic data of the subscriber, but also has been able to build a data profile with lots of information about that user's calling patterns. And there are going to be plenty of mobile subscribers around the globe, about 1 billion by 2003. Additionally, by providing a mobile portal the network operator can get even more information on the subscriber, as he is requested to input his or her preferences and information needs, so that he will receive personalised, and thus more valuable, information. Finally, with the use of mobile positioning technology, the network operator can identify fairly exactly what is the location of the subscriber.

All of these factors combined would create the ultimate marketing tool. Since most business transactions are local and the mobile device is the only tool that enables location-dependent services so far, personalised advertising via the mobile device seems to make sense. Vendors can reach their target customers when they are near the actual outlet.

For example, people close to an Argentine steakhouse receive a message about a lunch special. Those people have been selected because they are physically near to the location of the restaurant and because they have selected Argentine food as one of their favourites on their mobile portal.

Today, mobile advertising is carried out using the short message service. In the near future, the RealCall/Argo solution may provide a way for the advertiser to actually call the mobile subscriber, while he or she is near their site. As we move from simple mobile phones via AP enabled to GPRS and W-CDMA smart phones and communicators, the potential for advertising that includes audio, pictures and video clips on a colour screen is increasing. Monochrome pictures can be displayed and sent on a Nokia Communicator 9110 today, while Samsun has integrated an MP3 player into one of their phones and Sanyo has already developed a mobile videophone.

As the functionalities of the phone are extended with the arrival of the new protocols, so the bandwidth available over mobile networks increases. Looking at GPRS with an "always connected" mode, very targeted ads could be sent easily. Moreover, advertising is also possible on the mobile portal, which is on the web an managed usually from a real PC, but accessed by a mobile device. The revenue model to be applied is likely to be slightly different to a regular internet advertising model. It is more likely to be like Cybergol or Webmiles, where the viewer is paid for accepting and watching an ad. Technology providers of mobile location services, e.g. CellPoint (formerly Technor) of US/Sweden, are keen to get themselves into a revenue sharing agreement with network operators because of the sheer size of the market. Sc far the concept of using the mobile device as a media channel has not been exploited perhaps because web advertising is driven from the US while mobile communications is driven from Europe.

Sunday (the mobile service rather than the day of the week) in Hong Kong is one example of a mobile advertising service which is already operational today, but it works in a pull rather than push mode. Fo example, Sunday'ssubscribers can call a number and receive special offers by phon from shops in a particular shopping centre when they are actually inside the shopping centre in question. Many of Sunday'scustomers are happy to receive an ad via the mobile phone as a voice or text message for the chance of getting a bargain. It would be interesting to discover whether local cultural dynamics have any impact on the choice of technology deployed when building a commercial mobile advertising service.

Over the next 12 months during the period when the mobile market for advertising is emerging, it is important to understand that this is primarily a pre-marketing tool. Content partnerships should be built to help develop the new media distribution channel. Highly targeted messaging can follow after the users have got used to the medium for receiving information services and alerts. Once the basics have been identified and secured one can then roll on the bells and whistles.

The key question to be solved in the future remains: who will be delivering the advertising – the mobile operator, the content provider or the mobile portal owner?

We believe that if the mobile operators succeed in becoming the portal player of choice, the advertising revenues will remain with them. It still has to be seen whether mobile operators, who have hitherto focused on network roll-out and voice might become serious value-added content aggregators, i.e. portal players. The existing portals have the know-how of how to run a portal, but they do not have all the subscriber data, such as billing address, demographics, calling patterns and locations. This is information that only the operator has accumulated.

However, it is also possible that the content provider will use a direct channel to the customer, for example via SMS. In Finland there is a trial from eQ Online, the online brokerage, who advertises using SMS messages with stock prices or exchange rates from Kauppalehti, the leading business daily. The solution has been put in place by 24/7 Europe, the advertising professional network. It is using a telephone number and a direct link for WAP users to respond to the ad. While no WAP phones were on the market during the trial period in October 1999, the hotline response was overwhelming.

Yet another move into mobile advertising has been made by Wired Digital in the US. It has placed banner ads for Hilton Hotels to users of Palm PDAs via AvantGo's PalmPilot network, which lets people download news. The reach was no more than 6,000 users, but the demographics were almost ideally fitting the mobile executive. Content managers, such as AvantGo, are just starting to open up to advertisers, but their business model is likely to evolve into sponsoring and advertisements as revenue sources.

We conclude that mobile advertising will become an important pillar of mobile commerce in the next three years, when larger parts of the population will have adopted the mobile device for far more than just voice telephony.

MOBILE DYNAMIC INFORMATION MANAGEMENT

The area of what we call dynamic information management is actually related to the mobile device as a secure storage tool for important information, that must be updated on a continuous basis.

Mobile Membership

Instead of using as a membership card a magnetic stripe or smartcard, club memberships could be stored on the mobile device, e.g. on the SIM card. Using Bluetooth in the phone and at the POS, you could be automatically checked in at your sports club, without having to carry the card with you.

Mobile Loyalty Programs

Loyalty or affinity programs, such as airline frequent flyer programs, require a card as well, which could just as easily be substituted by the smartphone or communicator. The device can also store the user's latest point levels for instant reference.

Danet in Germany has developed a WAP solution, which can be used to check your Lufthansa Miles & Morefrequent flyer account from your mobile.

Mobile Medical Records

The mobile terminal would be ideally suited to store a patient's entire medical records or to identify the patient enabling the records to be accessed via the web, so that they would be available whenever needed at a physician's office. This would not only add convenience, but it could significantly reduce costs to health insurance providers and patients alike.

Mobile Passport

Storing the passport information electronically on a secure device might in future replace the existing paper copy. Digital signatures or biometric fingerprints could replace today's hand-written signature. However, this is not likely to happen within the forecasting period of this report.

MOBILE INFORMATION PROVISIONING

As the web has shown, the range of information that can be provided is unlimited. It can be of very general, personalised or localised nature. Obviously, while the value of the information increases for the user, the more it is personalised and localised.

Until the arrival of WAP, mobile information provisioning has been based only on SMS. While WAP slowly penetrate the market (to reach a penetration of 85% of all mobile users by 2003, as predicted b Durlacher), SMS is already available in almost all GSM handsets in Europe. Information can be pushed or pulled to th mobile device. In Finland the user types in a keyword, sends it to the SMSC (Short Message Service Centre) with the adjacent content server and receives an SMS back (pull). In Germany, however, mobile subscribers have to go to a web site and subscribe to certain information services that are consequently pushed to their handset.

The following information categories are already offered today, each by at least one European mobile operator. SMS based information services are actually generatin very interesting revenues with high profit margins today, since the information acquisition is mostly quite simple and virtually no additional traffic is loaded to the mobile network.

For general news, content providers are supplying the input, such as CNN (who have an agreement with Nokia), Reuters (who have agreements with Ericsson and Nokia) or Handelsblatt Interactive (who have an agreement with Mannesmann in Germany).

A unique information set is provided by Omnitel in Italy, who have explanations to more than 100 tourist attractions throughout the country available in various languages.

General News

- What's happening today
- Text TV
- Headlines of the day
- Temperature conversion (°F to °C)
- DHL tracking information
- Shoe size conversion (UK to US to European)
- Speed, weight, power, volume conversion
- Events, festivals
- Attractions/sights opening hours
- Directory services

Sports News

- Football results
- Football league
- Football club
- Football game program
- Formula 1, Hockey, Basketball, Volleyball, Tennis, Rugby, Cricket,...
- Golf dictionary
- NHL, NBA results
- Horse races results
- World/European/Country Records
- Ski slope information

Financial News

- Stock prices
- Stock limit
- Interest rates
- Index prices
- News of the day
- Exchange rates
- Currency converter

Entertainment News

- Horoscopes
- Birthdays
- Today's history
- Jokes
- Gossips
- What's on in town
- Nightclub listing
- Band tour dates
- Today's slogan
- Museum opening hours
- Recipe of the day
- Music top 10
- Wine information
- Lotto, Bingo, Toto results

Program information

- Cinema/theatre
- TV
- Radio
- Radio frequency information
- Education

Travel information

- Train, bus, flight schedule
- Bus arrival and waiting time
- Parking space
- Hotel
- Lost credit card

- Taxi
- Tourist information

Each of these services generates at least one SMS on the network for the operator and the content provider or aggregator, who might make 25% of the revenues with this service. In Germany, Dr. Materna has supplied and managed the SMS services for the Mannesmann D2 network as well for E-Plus and Viag Interkom, and has even employed a 7 people editorial team, in order to create the content. Similarly, WAPIT creates Radiolinja's content in Finland, while Brainstorm is supplying information services to Vodafone in the UK.

MOBILE ENTERTAINMENT

Mobile Gaming

Currently there is virtually no wireless gaming existent. Because of the limited capacity of GSM, there are no multi-player games for use over the mobile network on the market. Within GSM phones, only very simplistic single-player games are available, such as Snake (on Nokia 6110), Mobiletrivia, Navystrikeand Stockmarket(all on Nokia 7110).

Nokia has also developed a mobile entertainment solution that allows users of a 7110 terminal to play interactive games that are located on a server, such as traditional board and adventure games. A pilot service will be running on Singapore's M1network in Q1/2000.

Mobile games could in future also be developed for PDAs, such as Handspring's Visor, which is equipped with a slot for an external Flash RAM (Random Access Memory).
Nintendo's Game Boy Advance, Sony's Pocketstation or Sega's portable device which connects with the Dreamcast controller are also likely to become linked to wireless networks within the next 2 years. Actually, Nintendo has announced that they plan to use the Game Boy to download games over communications networks from the first half of 2000. Moreover, users will be able to exchange game data with other users and send e-mails from the device.

We believe that the advance in mobile gaming might be driven through the success of NTT DoCoMo's iMode portal o the packet data network and by the fact that Japan will have their world's first UMTS network up and running in late 2001.

Mobile Music

The availability of portable MP3 players w soon lead to music devices integrated wit a mobile phone; Samsung has already developed an MP3-phone. Music titles will be stored locally at the mobile device. Korean consumer electronics company HanGo just announced a portable MP3 player that can hold 4.86 GB or up to 81 hours of music. Streaming audio files fro radio stations, record companies and othe web sites will become possible. Orange of the UK, soon likely to be a part of Mannesmann, have announced that they will introduce radio phones to the market in early 2000.

It could be possible that the user has a set of music licenses stored in his device, which permits them to download more titles. With the arrival of GPRS and new billing mechanisms, it will be possible to pay for titles rather than for a minute of downloading time.

We expect to see further mergers between mobile phone vendors and customer electronic companies, who might be able to further advance the combined device.

Mobile Video

Packet Video Corporation of the US is currently the only real-time streaming provider that delivers via wireless networks. Up to 5 frames per second can be delivered over the existing CDMA network. A new standard, MPEG-4, is emerging for video decoding.

We believe that store and replay technology from Replay Networks and TiVo is becoming integrated into GSM wireless devices, so that videos could be downloaded overnight or at a pre-set time and played from a phone-video player whenever wanted.

We do not expect wireless video reception to be shipped in quantities before the year 2003, although first showcase products have already been out for some time. We believe that there is a big demand for watching video content on transport, but the content must be short and of low bandwidth type, such as news pieces, sport highlights, weather, entertainment, horoscopes and the like.

Mobile Betting

Mobile betting will become a very interesting application for m-commerce, because it is time-critical, as with horse races, or it involves a lot of money, such as Lotto. In Germany, the first online Lotto company, fluxx.com, has announced that they are intending to offer their service via mobile terminals.

More interesting perhaps are the opportunities which will emerge for spread betting, such that for instance odds can be given on how many corners David Beckham might take in a particular football match or indeed how many he might take in the last 15 minutes of the match, or, where the match has already begun, in the next five minutes. The type of bets offered over the network would correspond to the profile submitted by the mobile betting service provider.

We are convinced that mobile betting will become a very interesting proposition and a variety of applications will be seen over the next two to three years.

MOBILE TELEMATICS

Driving directions provisioning is a very useful m-commerce application, since only relatively few luxury cars are equipped with a GPS-based in-car navigation system. In-car navigation has been realised so far mostly with GPS technology and CD-ROMs, which are inserted into the system inside the car. Motorola has developed a GSM mobile receiver system including GSM receive-only module from UbiNetics that receives updated traffic information from Trafficmaster via Cell Broadcast.

France's Webraska is providing real-time mapping, street-by-street guidance services and proximity services, that locate the nearest gas station from any address entered on the telephone keypad, for WAP mobiles. German CAS provides a WAP based route-planning service, which gives directions to the selected destination. Benefon of Finland has also launched a mobile map service, called Arbonaut, in co-operation with Geodata and a combined GSM/GPS phone, called Esc!

Debis Systemhaus in Germany, a subsidiary of DaimlerChrysler, has developed the PTA digital assistant for business travellers as a prototype. The PTA collects all traffic information, makes reservations of hotel rooms and concerts, selects the cheapest rental car and guides the user through the traffic. It is based on intelligent agent technology and is linked via GSM to a mobile phone or communicator.

Nokia has set up a business unit focussing solely on telematics, called Smart Traffic Products. They believe that every vehicle will be equipped with at least one IP address in the year 2010. The following applications are included within telematics: self-diagnostic service check for trucks and cars before break-downs occur, break-down-service when the vehicle has an immediate fault, emergency call when the car breaks down in a deserted area and, of course, positioning information about the exact location of the car. Similar services are currently offered via a satellite network for high end Chrysler customers in the US.

Durlacher is convinced that telematics will be an integral part of the mobile industry, once the above mentioned new services and devices hit the mass market during the next 12-18 months.

MOBILE CUSTOMER CARE

Current mobile service provider overhead costs for customer care are on average more than Euro 10 per subscriber per month (according to Phone.com). Therefore, mobile electronic customer care (m-care) becomes an economically interesting alternative to reduce costs by providing automated, unassisted operation directly from the handset. Mobile custome care has to be part of an integrated solution, which also includes access via th web, an intelligent voice response system and a call centre. There could be some extra revenues created by using m-care, but primarily it provides the possibility to reduce operational costs. In addition, mobile customer care could increase customer satisfaction because it puts customers in a better informed role and allows them to instantly change settings.

Interesting new technology in this space is Universal Registration, as developed by Davinci Technologies, i.e. single registratio for interactive services, which works on both fixed and wireless web access interchangeably. Using this service, a user profile of all the key data and preferences will be developed once only and wheneve needed sent to the particular web service provider.

BUSINESS M-COMMERCE APPLICATIONS

A variety of business processes could be streamlined by integrating mobility. Addin mobile devices as a choice of interface wil create more easy online information acces and data entry, extend the availability of key personnel for decision making and make processes more dynamic and real-time.

Below we indicate a selection of some areas of e-business, where we believe wireless will have a significant impact.

MOBILE SUPPLY CHAIN INTEGRATION
Integration of business processes along the supply chain is a key issue in wireline business-to-business e-commerce. We believe that as these become increasingly time sensitive and participants become increasingly mobile, smartphones and communicators must be integrated into the information exchange as one possible distribution path.
The integration of mobile could take place on the buy side as well as on the sell side of Enterprise Resource Planning (ERP).

Unified messaging, for example, could be at the centre of those communications, making the user's choice of information access device and technology less of an issue. Moreover, it will be possible to make mobile reservations of goods, order a particular product from the manufacturing department or provide security access to obtain confidential financial data from a management information system.

By integrating the mobile terminal into the supply chain, it will be possible for e.g. a pharmaceuticals sales representative to check from the motorway or the customer premises whether a particular item is available in the warehouse. Already today, a SAP application has been developed for deployment on 3Com's Palmdevice.

3Com and Aether Technologies have joined forces to create OpenSky and to offer a service that gives smartphones and communicators mobile access to database applications. OpenSky will provide remote wireless access via personal digital assistants to establish secure connections to applications, such as Lotus Notes, Microsoft Exchange, ERP and CRM.

TELEMETRY/REMOTE CONTROL
There are very many applications, in which telemetry is being used today, but with falling mobile rates, new ways of using GSM technology seem to develop.

Maintenance and service needs of costly static machines and industrial equipment, such as a copy machine or a large drilling machine, can be metered with a sensor from the distance via a regular phone line. Instead of using fixed PSTN (Public Switched Telephone Network) connection, the performance can be checked using GSM modules as well. This is especially relevant for machines in remote locations, where it would be very expensive and time-consuming to deploy a wireline telephone infrastructure, or for locations, where there are not sufficient dedicated phone-lines available for all the machines.

Preventive maintenance is applied, so that if there is any kind of problem with the machine, a warning will be sent via SMS or via modem sending a data call to a central supervisory system. In the latter case, the communication is purely machine to machine, while in the former, maintenance personnel will be able to access information about a particular component's performance and assess need for maintenance. If sensors are connected to an intelligent device such as a PLC (PROGRAMMABLE LOGIC CONTROLLER), the telemetry functionality can be extended to remote control, so that the service engineer can actively intervene at the remote site via a mobile connection. In Finland, these applications have been used, for example, in the water industry to remotely monitor and control pumping stations via NMT (NORDIC MOBILE TELEPHONE) since 1992.

Telemetry also offers strong applications for cars, for example as a remote vehicle diagnostics tool. Field trials have been conducted by DaimlerChrysler and Volvo to install GSM chipsets in cars to monitor performance and to provide an early warning system, which sends a message to the manufacturer indicating the problem occurring, e.g. high temperature in the engine, brake problems or "out of oil" alarm. The manufacturer's system will be able to analyse the various data and provide a fix via a software tool to be sent to the car or by asking the vehicle owner to go to a service station. Thus, developing faults can be found early and the continuous operation of the car can be ensured. It is likely that in future all cars will be equipped with mobile communication links, starting from the top of the line models first, but moving down the line fast. GSM chipsets have become cheaper and continue to fall in price. Compared to the incremental value create by the wireless link, the additional cost of installing a communication link is low. Communications costs can be negligible a calls are purely event-based. It is far more difficult to establish the processes around that application on the manufacturer's sid

NTT DoCoMo, for example, estimates the following connection figures for its mobile network by 2010:

Table 3

NTT DoCoMo's Customers 2010

Source: The Economist

Connected via Mobile	Number (millions)
Humans	120
Cars	100
Bicycles	60
Portable PCs	50
Motorcycles, boats, vending machines, pets, etc.	30
Total	360

JOB DISPATCH

Mobile phones and communicators are increasingly becoming an integral part of groupware and workflow applications. For example, non-voice mobile services can be used to assign new jobs to a mobile employee. A service technician could be assigned a new task together with detailed information about the customer's problem, while she is on the road.

The target application areas for mobile field, delivery and dispatch services are:

- Transportation (delivery of food, oil, newspapers, cargo, courier services, towing trucks, taxis)
- Utilities (gas, electricity, phone, water)
- Field Service (computer, office equipment, handymen)
- Health Care (visiting nurses and doctors, social services)
- Security (patrol, alarm install)

A dispatching solution allows improved response with reduced resources, real-time work order tracking, increased dispatcher efficiency and reduction in administrative work. An interesting solution is delivered by eDispatch.com. With a web-based dispatching solution using smartphones, it is possible to save about 30% of communication costs, and efficiency of the workforce can increase by about 25%.

FLEET MANAGEMENT

Fleet management, a subset of intelligent transport services, is predicted to be one of the biggest growth markets for the next 15 years. Ericsson provides solutions, for example, together with Scania, who will integrate new features in its Infotronics subsidiary.

Start-up Aspiro of Sweden has developed a WAP based fleet management solution, which just needs a smartphone to be placed with the car driver. This is targeted at the professional driver, but is much cheaper than existing systems.

MOBILE CRM

Web-based CRM is already forecast to become the leading application software with a CAGR of over 50%. Thus, in addition to the well-established CRM vendors Siebel and Vantive, companies like Microsoft, SAP, Oracle and Baan are entering the market as well.

We believe that business applications like mobile CRM will be able to gain quickly support (for example from companies in the IT and telecoms environment) and thus market share.

MOBILE SALES FORCE AUTOMATION

The current sales force automation tools, e.g. from Update, are already integrating a software architecture that is aimed for mobile commerce applications. The sales force on the road will be equipped with AP enabled mobile phones in order to have easy access to customer data at the central office. Key data, which can be retrieved, would include contact management information, order entry, product and spare parts availability and deal tracking. If the WAP device is a communicator type, sales forecasting and opportunity tracking could be done as well.

The travelling salesperson is able to check the latest status of his customer, just before she is going into her office, and she will be able to enter a successful business win immediately. One could argue that the demand for this kind of mobile tool, to get access to the office and valuable information fast, is latent with all travelling salespersons. We believe that we will see increasing applications in this space.

WIRELESS APPLICATION SERVICE PROVIDER (WASP)

We are seeing the first signs that the Application Service Provisioning (ASP) model for ISPs in wireline is moving to a Wireless Application Service Provider (WASP) model in mobile.

We see two basic WASP models arising in the context of this report. The first model is based on mobile operators providing applications services to corporate customers who wish to offer mobile commerce to end-users. The second model looks at providing wireless application services to the mobile operators themselves.

The mobile operators are in a prime position to host mobile applications for corporate customers, because many companies, especially SMEs, do not have the resources or the know-how to build and run them. Thus, the mobile operator turned WASP might not only provide a WAP gateway and the application server, but he might also develop the application and provide system integration services. Many mobile operators already manage the entire m-commerce platform for their (trial) customers in mobile banking applications.

The wireless application service Provider will offer not only the mobile commerce platform but also the implementation tool for both internet and intranet applications, or a services platform. Mobile operators today generally do not have sufficient expertise with system integration or with development of specific applications for the mobile environment. Therefore, they will need to co-operate with current SI (System Integrator) players in the market who bring competence for particular applications, such as ERP (mobile SAP), CRM, a mobile broking solution or a mobile commerce shop.

On the other hand, HP is positioning itself in the WASP space with a service called Mobile E-Services on Tap for mobile operators. These are advanced wireless capabilities based on infrastructure equipment owned and operated by HP an delivered to its customers, i.e. service providers and network operators, who pa for them on a subscription or pay-per-use basis. HP is offering the complete WASP solution.

With the Mobile E-Services on Tapmodel HP will host and operate the infrastructure for day-to-day business operators enabling the service provider to focus on its core competencies. This model includes all the technologies, management processes and global capabilities required for service providers to run and rapidly scale the business.

Mobile E-Services are value-added, internet-based services that undertake complex tasks including financial transactions and provide access to personalised information via the mobile network. In order to get their Mobile E-

Businessoff the ground, HP has established a Mobile E-Services Bazaar, that will provide a trading community and a developers program for wireless operators, service providers, enterprise application providers and technology partners. HP has signed up 30 partners to date and will provide each with an additional marketing and sales channel. Whether this offering gets beyond the press release/prototype stage remains to be seen.

Phone.com is providing a WASP service with its MyPhone service that lets Phone.com operate a mobile portal for carriers under the carrier's own brand. It is considered to be more vertical than mainstream traditional portals with a special focus on the wireless device. We believe that both WASP models have potential and will find a place in the value chain, since the entire mobile commerce market is still at the very beginning and infrastructure in terms of GPRS networks and (GPRS) WAP terminals is only now being put in place.

MARKET SIZING AND FORECASTS

The European market for m-commerce is driven by the increase in the number of mobile subscribers, by the availability of new equipment, by the amount and quality of applications and by push of new terminals via their price. Both, the mobile

Fugure 15

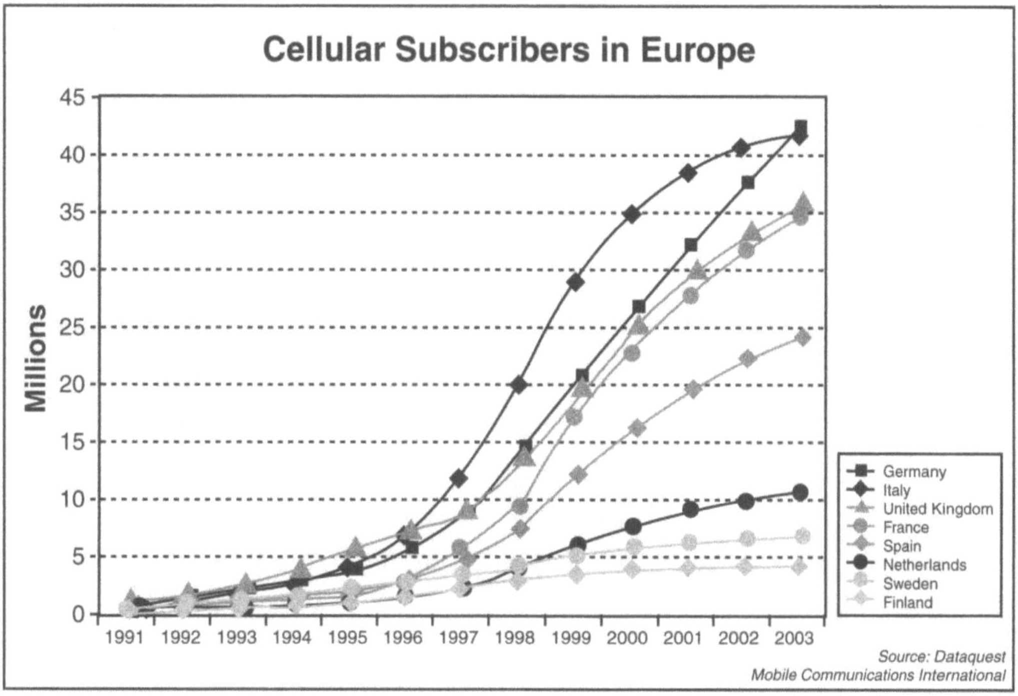

as well as the e-commerce market, have been traditionally always underestimated when forecast.

The number of mobile subscribers in Western Europe will increase from almost 90 million at the end of 1998 to over 237 million by the end of 2003 according to Dataquest forecasts. This represents an average penetratior of about 64% across the continent, with a minimum of 50% in any country. We believe that all equipment will be based on the GSM standard (or the derived protocols WAP, GPRS, EDGE) or compatible with it as UMTS.

At the beginning o October 1999, there were more than 300 million wireless users and more than 200 million internet users world wide. Globally an estimat of 1 billion subscribers for bot the internet and mobile communications seems to be realisti by 2003. There will be large overlap between both customer groups.

By 2003, we believ that there will be

Fugure 16

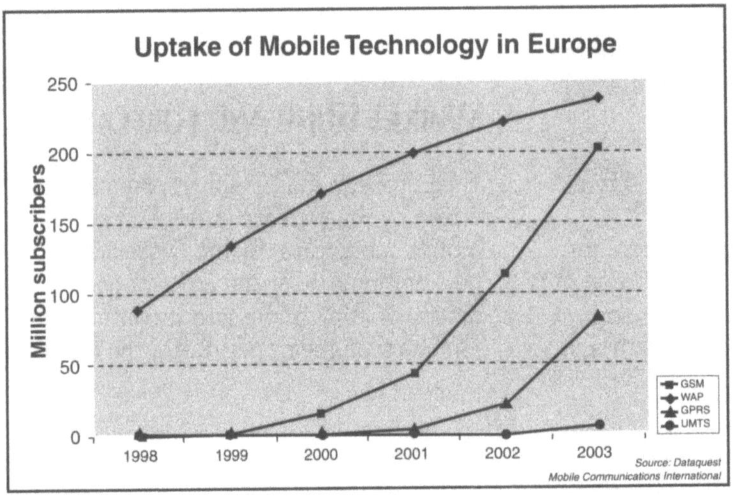

Fugure 17

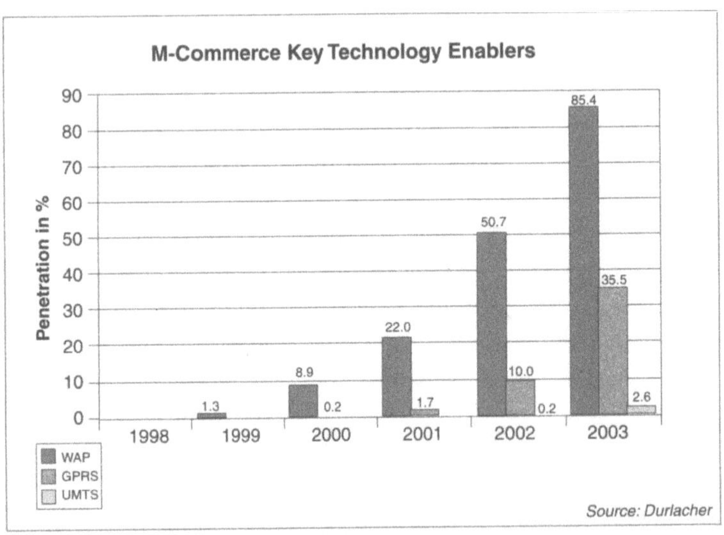

over 200 million subscribers in Europe accessing information from the internet with a mobile device. This represents more than 85% of the mobile phone owners. Based on statements from several vendors and operators, we have assumed that after 2001 no more phones will be shipped by the vendors that are not internet-enabled via a microbrowser.

While GPRS will have its big year in 2003 (being used by 35% of the mobile users) and also fostered by the arrival of EDGE, UMTS will start slowly and will have reached no more than 2-3% penetration across Europe by this time.

Content providers can generate revenues from mobile commerce using any or a combination of the following business models:

- Advertising
- Sponsorship
- Revenue sharing with mobile operator (part of airtime or volume generated by the service)
- Subscription-based services
- Transaction fees as for other e-commerce applications (stock trading)

The current situation is that although users are accustomed to paying for mobile services (and an increasing number also for value-added services such as SMS information), web content in general is expected to be free by large parts of the population. Therefore, m-commerce users will also expect a large number of services to be free of charge.

Delivery of free high-value content to the user could be achieved with a business model that is built around advertising and/or sponsoring. We would tend to agree with those in the industry who believe that advertising-only models are not the way forward. New, creative models have to be developed in order to utilise the mobile device in an optimal manner as a one-to-one marketing tool.

We believe that the following aggregated m-commerce revenues per subscriber per month can be achieved on a pan-European basis.

Fugure 18

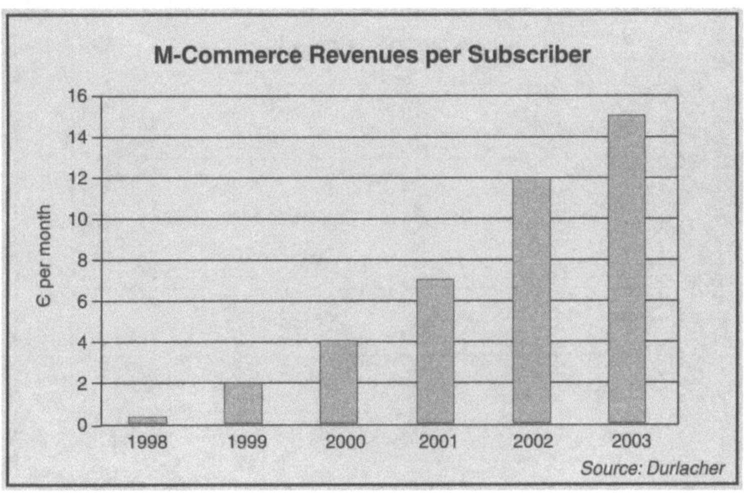

According to Durlacher's projections the European market for m-commerce will increase rom about Euro 323 million in 1998 to Euro 23.6 billion by 2003. The m-commerce market will therefore have a CAGR of 236% until 2003.

The country generating the most revenues from m-commerce will be Italy (Euro 4.8 billion), followed by Germany (Euro 4.1 billion) and the UK (Euro 3.4 billion).

In terms of the leading applications during the forecast period, Durlacher expects advertising to become the key provider of m-commerce revenues (Euro 5.4 billion), once the power of the mobile as marketing tool has been fully discovered. We believe that financial services, such as mobile broking, mobile payment and mobile banking, will be the second largest revenue source with Euro 4.9 billion. Actual mobile shopping, including for example retailing, ticketing, reservations and auctions, will be worth Euro 3.5 billion. It is important to note that those top 3 applications will together make up only 60% of the total European m-commerce revenues in the year 2003. The remaining share is generated by a rather larger selection of applications, such as business applications, telematics, customer care, entertainment, information provisioning and security.

The development of the importance of the various m-commerce applications is indicated below. While in the year 1998 information provisioning was by far the largest contributor to mobile commerce revenues (91%), its relevance is decreasing continually over the years and will be only 5% in 2003. This shows that the market is in a very early stage and the overall market growth will be driven by the arrival of new services.

Fugure 19

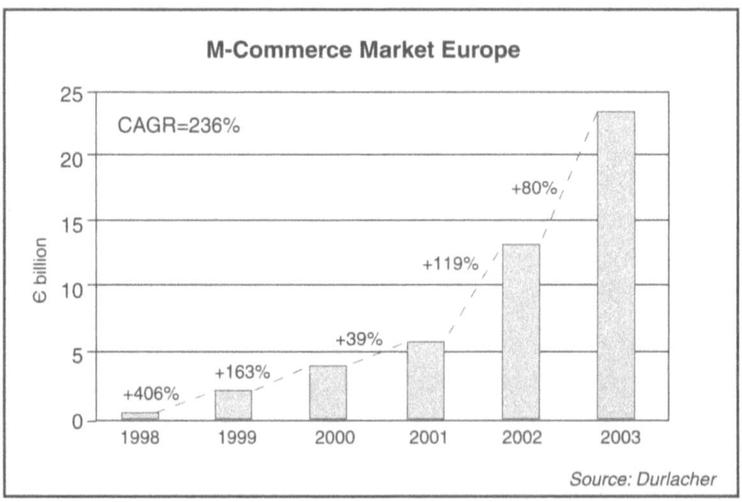

Fugure 20

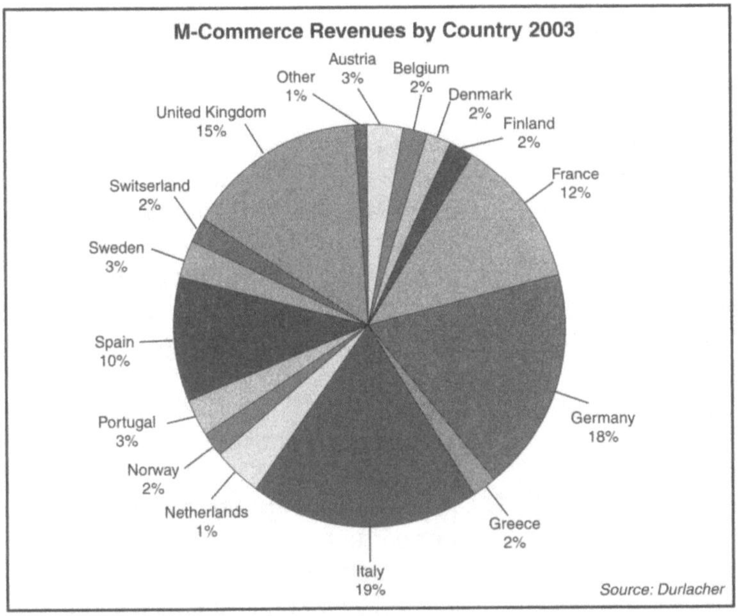

Fugure 21

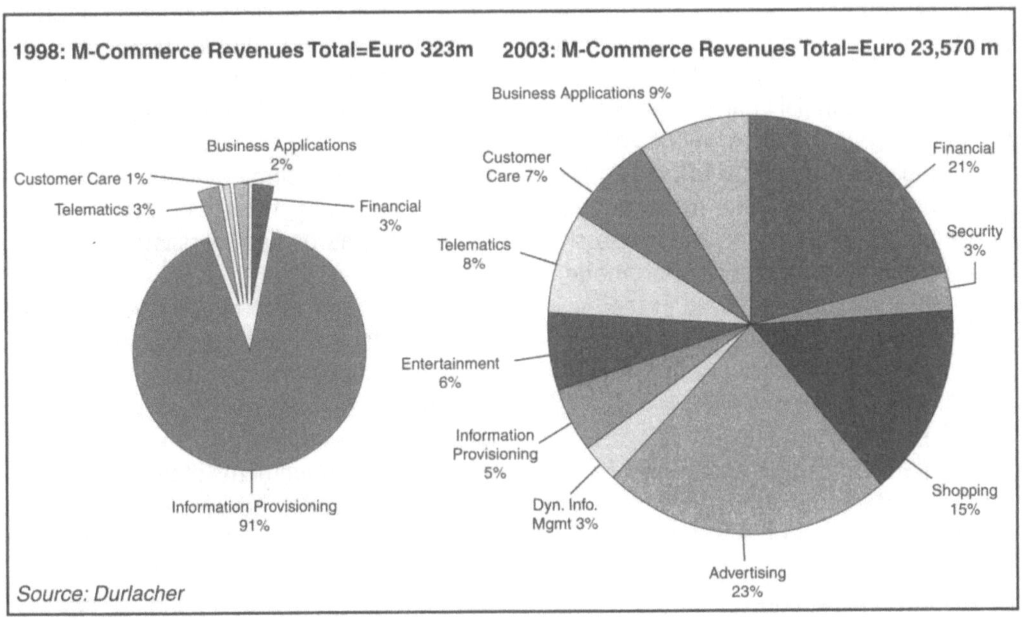

INDUSTRY OUTLOOK

MOBILE PORTALS

The mobile portal today is emerging, in many ways, as a stripped down version of traditional web portals, representing a subset of available content targeted toward the understood preferences of the user base. Relevance becomes critical. Moving forward, we expect many specialised portals to emerge, optimised for and targeted at the mobile world. These will include new mobile-specific services not found on traditional portals.

The key requirements of a mobile portal are to provide the following functionalities:

Personal information management (PIM) includes many of the functions of current PDAs, such as the maintenance of a personal address book, personalised calendar functionality, the management, writing and reading of notes and memos. Synchronisation with desktop applications, e.g. MS Outlookor Lotus Notes, might be required for an initial changeover period. We believe that PIM applications will be increasingly web based (as offered today by Yahoo! and many other major portals) rather than desktop based and wirelessly accessible in an always connected mode, once GPRS is widely deployed in the network and in terminals. Microsoft has announced the availability of its Office software over the net as early as Q1 2000 as a response to Sun's acquisition of German StarOffice. Sun is planning to make this MS Office compatible software package available on the web for free by Spring 2000 in order to boost its thin client network computer concept. Synchronisation with smart phones and communicators will be necessary to ensure anytime/anywhere access to this information.

Personalised content is key using mobile devices, because of the limitation of the user interface. Relevant information must always be only a "click" away, since web access with any existing wireless device is not comparable to a PC screen either by size, resolution nor "surfability". Key personal information includes MyNews, MySports, MyPortfolio, MyWeather, MyHoroscope, MyInterests.

Alert notifications are also an important part of personal content, such as for auctions, betting, classifieds, stock price changes etc. Obviously, the mobile portal ideally communicates with the user in her mother tongue.

Communication facilities within the mobile portal are key applications. Therefore, communicators are being designed that can perform many of the following functions: sending and receiving e-mail, voice mail forwarding, conference calling, faxing, document sharing, instant messaging and unified messaging.

Location-specific information leverages the key advantage a mobile portal has over a traditional web portal by supplying information relevant to the current geographic position of the user. Information requirements may include, for example, restaurant bookings, hotel reservations, nearest petrol station, yellow pages, movie listings. Intelligent inference technology backed into a database works with profile data based on the user's past behaviour, situation, profile and location.

Initially, the market for the mobile portal has to be created, a subscriber base has to be built and airtime must be generated. A good way for doing that is through the use of messaging or e-mail, because customers generally understand these applications and are using them already.

Mobile Operators
Most mobile network operators across Europe are now building portals, mainly with third party data from existing portals or content feeds. The operators are not experienced on the content side, but they have been capturing subscriber data. Mobile operators have the following main advantages over other portal players:

- Existing billing relationship
- Substantial, verified subscriber data
- Location of subscriber

However, location services are not yet widely available as positioning technologies are just starting to get installed by the operators (Sweden's Tele2 is launching the first location services based on Cellpoint's handset-based technology in November 1999).

The European operators that have been among the first to announce or launch a portal service are Sonera (Finland), Omnitel (Italy), TIM (Italy), Telia (Sweden), T-Mobil (Germany), Mannesmann (Germany), Telenor Mobil (Norway), Radio Mobil (Czech Republic), Orange (UK), BT Cellnet (UK) and Vodafone (UK). BT Cellnet has been very early with its Genie portal, gaining more than 200,000 subscribers. Most portals run by the network operator have limited the access to their own subscribers. BT Cellnet decided to give up this approach in order to gain critical mass and opened their Genie in 1999 to customers of other networks as well.

Japan's NTT DoCoMo's iMode is the world's first mobile portal, which has a significant number of subscribers. In October 1999, nine months after the launch, there were more than 2 million people (or 8% of its subscriber base) using the service. It is not based on circuit-switched WAP/GSM technology, but rather uses a proprietary Japanese packet mobile data network, which requires special iMode handsets. More than 200 content providers have developed sites for the handsets, mainly financial services (banks, brokers, credit card issuers) and insurance companies, but also ticket and travel agencies as well as newspapers, CD and book sellers and games and entertainment services. In addition, more than 2000 regular web sites are automatically translated for access with iModeterminals, but their functionality is limited. The network protocol used is comparable to GPRS on GSM networks. Thus, similar services will be available in Europe only from 2001 onwards.

NTT DoCoMo, Telia and Sonera are all interested in transferring their portals into other markets. Sonera is aiming at internationalising its Zed portal in early 2000 primarily into those countries where it already has ventures, such as the US, the UK, Benelux and the rest of Scandinavia. MCI Worldcom subsidiary UUnet is also going to add WAP gateways to its network infrastructure and is offering a portal with e-mail and news services.

Technology Vendors

Mobile network providers Lucent (with its Zingo portal) and Ericsson have both launched a mobile portal during 1999. However, Ericsson's Mobile Internet portalservice is more a demo portal, an accumulation of different applications provided by some application developers, than a managed product. Motorola has launched its Mobile Internet Exchange, where it provides content, such as Worldspan, one of the largest online travel reservation systems.

Through Mobile E-Services Bazaar, which consists of applications and services from more than 30 partners, HP hopes to push its WASP services to the operators. Sharp's Space Town is a portal that will provide relevant content for the entire portfolio of communication products such as PDAs. Spyglass has also announced a demo portal with news for Q1/2000 that should support the marketing of its Prismtechnology, a content delivery platform.

The mobile portal provided by a technology vendor is basically used to gather content that can help in selling WAP gateways, WAP handsets and so forth. We believe that these will represent a temporary effort from the vendor community to help develop the market, which will be terminated as soon as sufficient momentum has been created.

Traditional Portals

Established large portal players have recognised the potential impact that the mobile internet and therefore mobile commerce can bring them. In order to position themselves favourably early on, players such as Yahoo! and MSN have created mobile portals, which are targeted at US subscribers only at this point in time. Both their content and choice of terminal for information access are not relevant for European users. However, we expect that during Q1/2000 many of the major portals will also have established themselves in th largest European markets. Yahoo! is adopting a co-operation strategy with loca partners as they have for instance in Germany (with Mannesmann D2) and in Sweden (with Europolitan). Excite is co-operating in Japan with NTT DoCoMo in the successful iModeportal.

Our view is that regular, "traditional" portals are not providing information that is specific enough for the user of a mobile portal. They are not able to incorporate location-specific information nor do they have the data and knowledge of each customer that the mobile operator has. In many instances, the traditional portal player knows nothing more than an e-mail address, which is basically in the form of, DonaldDuck@hotmail.com or MickeyMouse@Yahoo.com. The true identity of the user might not be known. Moreover, the traditional portal does not usually have a billing relationship with the customer (with the exception of AOL, T-Online, Compuserve and others who also offer internet access).

New Players

New players and start-ups, who are interested in positioning themselves as a mobile portal, start out generally as a technology provider with a view that they might want to offer content to drive the business in the future. Quite commonly they have developed proprietary technology, which can facilitate the process of content providers starting AP enabled services. Examples of this type of company are Breathe and Digital Mobility, who are aiming to establish mobile portals in the UK. Breathe has announced specialised vertical portals for the finance, insurance and accountancy industries using Autonomy's knowledge management technology that can be accessed via a variety of mobile devices. In contrast, Digital Mobility has developed a web hub server, which is a piece of middleware between existing web servers and the WAP world, and they are in the process of designing a portal around it. In Sweden for example, Concurrent Data Dynamics has developed a portal around the company's own Allt.com search engine and its Grail web search software.

The general decision the new players have to make when they are building their portals is whether they want to market the portal themselves under their own brand or sell it as an operator branded product.

Table 4

The Power of Portal Players

Source: Durlacher

Wireless Portal Player	Strenghts	Weaknesses
Mobile Operator	Billing Relationship Location Information Strong National Band	Little content expertise Little partnering experience
Equipment Vendor	Technology understanding Developers contact	Not core business Not core business
Traditional Portal	Strong portal expertise Leading content/apps. Strong partnering experience	No mobile experience No location information No billing usually
New Player	Flexible Niche focus	Technology focused No content expertise

Since mobile operators have learned through the SMS experience that in some cases it can be painful to outsource a core service (Mannesmann provided Dr. Materna with a very lucrative business by outsourcing the entire SMS platform to them), they are less prepared to give the mobile portal away to a third party. Eventually, the new provider might provide the service for a very short time period, until the operator has developed its own offering. Thus, the newer players must also market their own portals independently or with an ISP or another company.

We believe that new portal players can only be successful when they manage to transform themselves into true content players and focus in particular market segments and niches, which are not addressed properly by the existing big guys.

Durlacher believes that the ideal combination for creating a mobile portal would be the mobile operator plus an existing portal player, because they have complementary strengths. Although first a partnering approach might be chosen, acquisition of equity stakes seems to be the likely next step.

KEY SUCCESS FACTORS FOR M-COMMERCE

The successful develoment of the m-commerce market will depend on operators taking advantage of the following capabilities within the mobile environment:

Customer Ownership

Subscriber data, such as billing address, mobile phone number, e-mail address, choice of mobile device and calling patterns, are becoming ever more valuable in the light of mobile commerce. In addition to passive collection of user behaviour and data, companies will be abl to benefit from users actively providing an specifying their own choices and preferences to the portal provider.

The mobile terminal is the perfect vehicle for delivering one-to-one marketing. Mobile firms can link stated individual characteristics with a database, which can extract or infer preferences. Therefore, network operators (especially when they are also service providers) are in pole position to leverage the data warehouses they have built over the years.

The latest takeover wave in the mobile industry is re-characterising the role of the network provider. While Deutsche Telekom acquired Cable and Wireless's One-to-One interest in August 1999 for approximately Euro 4040 per subscriber, Mannesmann has offered Orange as much as Euro 9452 for each subscriber in October of the same year. Established portal players (such as AOL, Yahoo!, and the UK's Freeserve) understand that in the short term it is important to build market share and a large, well-defined subscriber database which can be later leveraged as a platform for m-commerce transactions.

We believe that ultimately the question is not "who owns the customer?", it is "wh does the customer buy into?". In a multi-channel, multi-device, multi-source world, the customer can easily switch access device, internet service and portal. So who does the customer buy into? It will be the mobile portal provider that holds the user's personal data and preferences and uses them to add value to the user's experience. The mobile operators do have a unique position in the mobile portal world, but they must build new competencies quickly.

Personalisation

Personalization is about creating services that customise the end-user experience for the individual subscriber. It is based on one-to-one relationship management and therefore provides the ideal tool for one-to-one marketing. An intelligent personalization platform must be able to learn from both user preferences and past behaviour of the user. The application must be personalised enough to optimise the interaction path, enabling the user to reach the services they want with as few clicks as possible, and presenting information in a compact form optimised for the smartphone or communicator. Companies must also be proactive with respect to service behaviour, i.e. anticipating future requirements of the user and suggesting a likely choice. We believe that personalisation is the difference between a usable application and an unusable application.

Localisation

There are several competing technologies that enable mobile location or positioning services. Location-sensitive information becomes key in mobile commerce. Knowing the location of the user drives the service and application offering to a level that creates significant value to the user. User need local information about their normal local environment. Location specific information is even more valuable in new environments, when travelling.

Ubiquity

The ability to receive information and perform transactions from virtually any location is especially important to time-critical applications, such as stock and options trading as well as betting. Providing mobile users with a similar level of access and information to that available in the fixed line environment is key.

Timeliness

Mobile enables the transmission and use of time-sensitive information whose value is inherent in its immediate delivery. Information transmitted too late can incur significant opportunity costs. It is in such environments that mobile information services come into their own.

Convenience

We are strong believers in technology, but only with the purpose of making life easier for people and taking away the pain of unpleasant tasks and activities. One should always question how a solution could provide added convenience to the user. Technology in itself is exciting, but only its use to increase the quality of life makes it valuable.
Based on the core competencies of the companies that make up the m-commerce value chain, a number of observations can be made as to their evolving roles and implications for the future of this market.

VALUE CHAIN MODELLING

Content Provider Strategies

The content providers in the mobile commerce value chain are mostly concerned with distribution. They want to ensure that their content reaches the public by as many means possible. Reuters, for example, is providing a subset of its information to both Nokia and Ericsson, as well as to mobile operators, who are all offering a mobile portal. Additionally, Reuters is launching a proprietary portal service, which will include the complete portfolio of news, not just a small section, because they believe that the m-commerce market is too important to rely on a single distribution channel. Use of a broad range of distribution channels allows content providers to hedge their bets on the evolution of platforms, ensuring key positioning for their content and building multiple revenue streams.

Content Aggregator Strategies

Content aggregators are trying to stay close to various portal players and ideally sell the same product more than once. The aggregators bring a wealth of useful knowledge to the table because they are used to partnering and negotiating with content owners and other companies in order to develop information applications. Content aggregators might become potential acquisition candidates, as their role is close to the portals while also being close to the original content.

Enabling Technology Provider Strategies

The enabling technology provider benefits from the shift in mobile communications towards m-commerce. They offer a variety of platforms and services that have been earlier run by mobile operators. The software companies and system integrator will increasingly take on projects to build complex applications. IT specialists will provide the middleware, similar to their role in the internet environment.

Mobile Operator Strategies

The mobile operators need to position themselves at the centre of m-commerce development, since they stand to lose

Figure 22

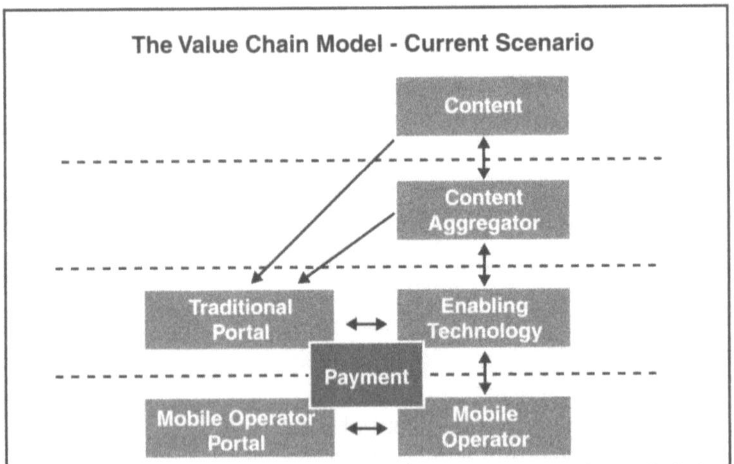

Source: Durlacher

significant revenue streams from the drop in ARPU because of a continuous price decline for their voice services. The mobile operator is fearful of becoming a pure infrastructure supplier, like a vanilla ISP. New mobile commerce enabling applications, such as e-mail and instant messaging, are the first step up the value chain, but ultimately the mobile operator aims to occupy the mobile portal space with a participation in transaction revenue and the handling of the payment side of the transactions. Mobile operators have the strength of owning all their subscriber data, and have access to the location of the customer. However, mobile operators are not experienced in the content business or in partnering with other players, both activities that would be useful for the development of sustainable revenue streams. The operator will use their existing billing relationship with the customer to enhance their current service portfolio; they might use their current valuations to acquire both a banking license and a portal provider, who bring the missing skills to the table. With this amended business model, the mobile operator will also be able to host more applications as a WASP, which can be only done in conjunction with a systems integrator providing the project management experience.

Traditional Portal Strategy

The mobile portal strategy of the traditional portal players often lacks in-depth understanding of national mobile markets and of the specific local dynamics involved in building businesses in a territory. In addition, the differences between a fixed and a more mobile portal model are non-trivial and, as such, lessons learned in the past are not necessarily directly transferable. What they do possess, however, is an existing brand image and an understanding of how to enhance the online experience of the user. We expect that two different types of portal providers, i.e. the traditional portals and the mobile operators, will ultimately join forces and merge. Users do not want to access multiple levels of portals to help them manage their lives; they need only one gateway to the internet world.

Payment Agents

Payment agents, i.e. banks and credit card companies, are trying hard to occupy a

Figure 23

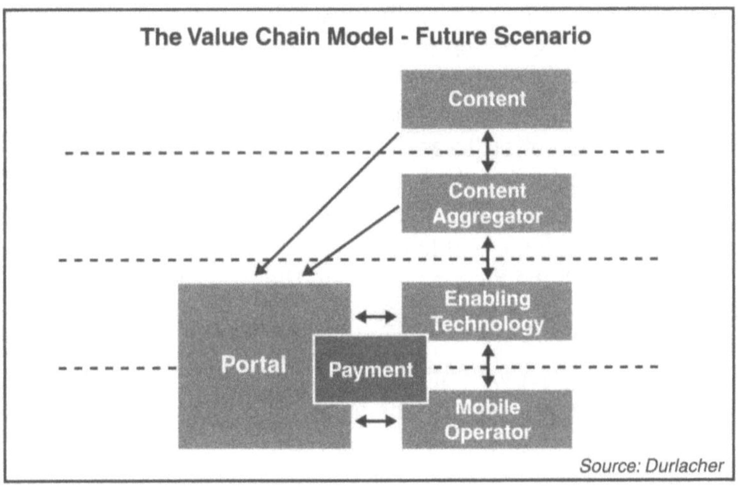

space within m-commerce by piloting many different technology solutions. They are trying to squeeze themselves somewhere into the value-chain in order to defend their right to exist by introducing, for example, payment smartcards. However, in many cases, these physical smartcards are not necessary for any other reason than having the Visa logo and the bank's name on them (thus including these two players in the m-commerce playground). There is the danger that in future mobile operators will include the payment and wallet functionality on the SIM card within the phone. However, mobile operators are not able to offer payment services if they do not own a bank license. Thus, it is likely that mobile operators will move up the value chain and take over the payment agent's role.

INVESTMENT OPPORTUNITIES

Based on the discussion above, we believe that the development of the mobile commerce space will yield some significant investment opportunities over the next four years. However, the opportunities indicated below originate from our subjective view of the market and, although we have undertaken every effort to analyse the upcoming m-commerce market as thoroughly as possible, we clearly cannot guarantee the definite and timely emergence of all the factors necessary. The list is indicative rather than exhaustive.

In the following text we outline a roadmap of opportunities according to the time horizon during which we expect them to take off.

SHORT-TERM OPPORTUNITIES

The market for mobile commerce applications based on WAP (and other protocols) is still relatively undeveloped, because of the continuing existence of a large variety of devices and user interfaces, complex solutions for wireless connection, slow data speeds and untested and in some cases untrusted security architectures. While WAP gateways are now becoming widely available, m-commerce applications are still only available on a limited basis. Applications and services which are localised for the users position, preferences and timeliness requirements will be key in the immediate term.

In the next 12 months we believe that the following investment opportunities will emerge.

First Generation WAP-Only Developers

The first content and applications to be specifically developed for WAP devices have in our view a very significant opportunity to gather market share in the early developmental stages of the market.

Repurposing of E-Commerce Delivery Infrastructure for Mobile Access

Existing applications, which have proven their success in wireline e-commerce, will need to add a mobile distribution component very quickly if they want to ensure participation in portal development. Specialist system integrators will play a major role in the repurposing of applications for mobile use.

Acquistion of Content

The mobile portal provider needs to get closer to content providers and aggregators.

We expect that the current relationships of mobile operator and content provider will become more important and cellular operators may well begin to buy equity stakes in those they deem best of breed content providers in accordance with their overall corporate plan.

Mobile E-mail Platforms

E-mail is the upcoming mobile mainstream application, beyond the existing popularity of mobile voice and SMS. E-mail applications will have to be implemented by all mobile operators, possibily through existing applications being extended to the mobile space by ISPs. Advanced platforms, which allow e-mail access anytime from any device, are required to deliver a superior user experience, no matter whether the e-mail is hosted by an ISP service or resides on a corporate e-mail server.

Personalisation

Developing applications for intelligent, automatic personalisation on mobile devices is a business activity for which there will be plenty of demand. Since e-commerce personalisation technology providers Broadvision and Vignette have just started developing technology solutions for this market, applications are almost non-existent, but our expectation is that the market will be populated very quickly indeed by new mobile specialists and by mutating incumbants.

Mobile Portals

Mobile portals come in a variety of flavours and there will be space for a number of them, next to the mobile operators and the existing general web portals, for particular user groups. Subscriber profile ownership is key in future as it will allow selectively targeted m-commerce and advertising.

Speech Recognition Based Applications

The mobile communications industry urgently needs further developments in speech recognition technology and applications based on it. Benchmarking against Orange's Wildfire product, this will enhance the user experience of the mobile terminal significantly and it will add convenience to accessing information as well as sending and receiving e-mail.

Positioning Technology

Current positioning technologies are just entering commercial availability, but further developments are required to improve the detail and accuracy with which the exact location can be determined.

Compression Technology

Compression technology is increasingly important as bandwidth demand increases with new applications and current technologies are limited to smaller capacity. Compression could provide significant input in levelling out bandwidth shortages.

GPRS Applications

In order to take off by early 2001, GPRS will need applications that utilise the competitive advantage of always on connections at fairly high transmission rates. Specific applications are likely to be built around gambling, stock market and other time critical areas.

Synchronisation
There are only a few players providing services for the synchronisation of smartphones, communicators and PCs, but an increase in demand can be expected in the short-term.
Managing data systematically, so that it has to be entered only once, but remains identical across platforms, requires both sophisticated applications and technologies. Although the underlying applications are already available, there is a market need for applications developed specifically for this space.

Mobile Commerce System Integrators
Mobile system integrators – combining the skills of SI's with mobile experience – are almost non-existent today – this combination of skills is likely to command a premium moving forward. New mobile media consulting is a crucial skill set currently not available to larger vendors and consulting organisations, but one that is necessary to develop company specific applications around existing processes and adapt available technology platforms to an organisation's individual needs. For example, because of the lack of mobile commerce consultants, Nokia is currently developing the mobile broking system for Deutsche Bank. Although this is outside Nokia's core competence, WAP integrators are yet to emerge and Nokia needs to get the market moving.

MEDIUM-TERM OPPORTUNITIES

In about 12 to 24 months m-commerce will provide the following opportunities for investments.

Acquisition of Content
The acquisition model for content is likely to be extended into the mobile space and content providers might be totally acquired in order to secure exclusivity.

Acquisition of a Portal
Mobile operators, who have not been successful in developing their own mobile portal may be forced to acquire a portal player to enable them to extend their service portfolio.

Mobile Advertising
Mobile advertising provides the perfect tool for very targeted one-to-one marketing based on the user profile and location. Appropriate business models have yet to be developed – these may include sponsorship or payment to users for reading advertisements. Applications to facilitate or manage the entire personalised advertising process are currently not available, but again Broadvision's or Vignette's platforms could be used as the basic technology foundation.

Banking Solutions
Acquisitions by mobile operators will most likely include online or mobile banks given their better fit with the mobile business model than a more traditional retail bank. The first step will be to sign those banks up and move them to the WAP environment, before potentially converting

this relationship into an M&A type transaction.

Bluetooth Applications
Once Bluetooth chips are integrated into devices, applications for utilising the possibility to interact from machine to machine must be created. One key area is synchronisation between Bluetooth-enabled terminals, but communication, payment and security applications are also important.

Location-Dependent Information
Currently, several competing technologies have emerged to solve the positioning problem, but the real commercial application of any one technology is still to be proven. Nonetheless this type of information is a key part of the mobile value chain and as such it will be a space which will be populated as the market evolves.

M-Commerce Middleware
There is a place for a variety of middleware providers, who can handle some of the more critical issues, such as security (PKI), mobile payment technology, m-commerce platforms and mobile advertising platform technology.

Business Applications
Integrating the new breed of mobile terminal equipment into corporate business–to-business processes will create opportunities for companies that can provide a link from, say, Siebel's CRM or SAP R/3 software to WAP or GPRS devices and integrate them as a complete solution.

WASPs
Extending the application service provider model into the wireless space opens up a whole range of new opportunities for participants that provide a hosted WAP gateway or the entire wireless front-end to existing operations. There is currently a lack of know-how in many organisations with respect to wireless technology. WASPS providing a managed end-to-end solution can overcome this.

Entertainment Services
We estimate that early technologies and applications that make it possible to create customised content and then repackage these for distribution via the mobile network, will be in demand from 2002 onwards. Similarly, to the market will require application platforms for music downloads into MP3-enabled mobile devices, as well as video streaming and mobile interactive multi-player games.

Infomediaries
New infomediaries for the wireless space will probably emerge who will broker relationships between network owners and content providers. Mobile operators might extend their influence in order to control these content aggregators.

LONG-TERM OPPORTUNITIES

From 24 months onwards, a number of investment opportunities can be expected in the following areas:

Payment solutions
Once PKI infrastructures have taken hold as an integral part of mobile commerce,

mobile operators are likely to position themselves as banks through the acquisition of suitable players.

Mobile Dynamic Information Management

The use of the mobile handset as a device to store extensive personal data will be enabled by the availability of multi-functional SIM cards. This will enable information such as club memberships, frequent flyer points, travel tickets and passes etc. to be stored on the mobile device, alleviating "fat wallet syndrome", from which many persons suffer today.

Mobile Passport

Creating applications that will not only hold a credit card, but also a unique document, such as a passport, on a mobile device, will be enabled through the widespread introduction of security measures, e.g. digital signatures based on PKI.

GLOSSARY

3G	3rd Generation mobile technology according to IMT-2000 standard (e.g. UMTS in Europe)
API	Application Programming Interface
ARPU	Average Revenue Per User
ASP	Application Service Provisioning
ATM	Automatic Teller Machine
Biometric	Identification of a person by a physical or behavioural characteristic (such as the way they sign their name, their fingerprint or the marks on the iris of their eye).
Bluetooth	Chip technology enabling seamless voice and data connections between a wide range of devices through short-range digital two-way radio.
CB	Cell Broadcast
CB	Cartes Bancaires (French credit card issuer)
CDMA	Code Division Multiple Access. Allows reuse of scarce radio resource in adjacent areas. Can give interference
CDMA2000	North American flavour of IMT-2000
CDP	Cellular Digital Packet Data
CAGR	Compound Annual Growth Rate
COO	Cell of Origin
CRM	Customer Relationship Management
DES	Data Encryption Standard (or Data Encryption algorithm): the most widely used method for "symmetric" encryption The main source is ANSI X3.92.
EDGE	Evolved Data for GSM Evolution. Allows networks to meet many of the requirements for UMTS
EMPS	Electronic Mobile Payment System
EMV	Europay-Mastercard-Visa specifications for chip-based payment cards. EMV 1 corresponds with (and generally conforms with) ISO 7816 parts 1-5; the other parts of this specification cover the details of a standard credit/debit application and the requirements for terminals
E-OTD	Enhanced Observed Time Difference
ERP	Enterprise Resource Planning
ETSI	European Telecommunications Standards Institute
FCC	Federal Communications Commission (US regulator)
GAA	GPRS Application Alliance
GPRS	General Packet Radio Services
GPS	Global Positioning System
GSM	Global System for Mobile communications
GSM 1800	GSM operation at 1.8 GHz; formerly DCS 1800

GSM 1900	GSM operating at 1.9 GHz; formerly PCS 1900
HSCSD	High Speed Circuit Switched Data.
IDEN	Integrated Digital Enhanced Network
IM	Instant Messaging
IMT-2000	International Mobile Telecommunications 2000. The IMT-2000 system will provide a seamless, global communications service through small, lightweight terminals
IN	Intelligent Network
IP	Internet Protocol
ISP	Internet Service Provider
ITU	International Telecommunications Union: the international body responsible for telecommunications co-ordination, the successor body to CCITT
IVR	Interactive Voice Response
Java	A high-level object-oriented language, allowing applets (applications) to be written once, run anywhere (whatever the platform is). The aim is to help simplify application development.
KM	Knowledge Management
LAN	Local Area Network
LFS	Location Fixing Scheme
MexE	Mobile Execution Environment
MIM	Mobile Instant Messaging
Mondex	The electronic purse system developed by NatWestBank in the UK, it is now 51% owned by Mastercard International and is licensed to banks in many countries. Mondex is unusual among bank-owned electronic purse schemes in that individual transactions are not reported back to the scheme owner and transactions between purses are allowed, making it closer to a true cash substitute.
MoU	Memorandum of Understanding
MULTOS	Multi-application smart card operating system
NMT	Nordic Mobile Telephone
OTA	Over The Air. The method used to manage applications on a subscriber handset remotely
PDA	Personal Digital Assistant
PGP	Pretty Good Privacy. A popular key algorithm
PHS	Personal Handyphone System (Japan)
PIM	Personal Information Manager
PIN	Personal Identification Number
PLC	Programmable Logic Controller
PKI	Public Key Infrastructure
POS	Point Of Sales
PSTN	Public Switched Telephone Network

OS	Operating System
RAM	Random Access Memory
ROI	Return On Investment
SAT	SIM Application Toolkit
SET	Secure Electronic Transactions: a standard for credit card payment across networks, which does not depend on the security of the network and does not allow the merchant access to the customer's card number. It also links the payment to a specific sale transaction. SET does not require the use of a smart card
SI	Systems Integrator
SIM	Subscriber Identification Module. Smart card holding the user's identity and telephone directory; SMS-Applications may reside on the SIM
SMS	Short Message Service. Facility for sending text messages on GSM handsets
SMSC	Short Message Service Centre
SSL	Secure Socket Layer. A form of data encryption used in computer based transactions
TCP/IP	Transmission Control Protocol / Internet Protocol
TDD	Time Division Duplex
TDMA	Time Division Multiple Access (see also CDMA)
TOA	Time Of Arrival
UI	User Interface
UM	Unified Messaging
UMS	Unified Messaging System
UMTS	Universal Mobile Telecommunications System; the third generation mobile standard
USSD	Unstructured Supplementary Service Data
VAS	Value Adding Service
VXML	Voice Extensible Mark-up Language
WAP	Wireless Application Protocol. Offers internet browsing from wireless handsets
WASP	Wireless Application Service Provider
W-CDMA	Wideband CDMA
WML	Wireless Mark-up Language
WWW:MMM	Mobile Media Mode
XML	Extensible Mark-up Language

Chapter 4: Technical Considerations

Introduction to Wireless Application Protocol and Wireless Markup Language

Title: *Introduction to Wireless Application Protocol and Wireless Markup Language*
Author: *Wireless Developer Network*
Abstract: *This paper exist of three parts; an introduction to Wireless Application Protocol (WAP), an introduction to Wireless Markup Language (WML) and adding client-side logic to WAP using WMLScript. The first part explains what WAP is and what it does. The second part explains what the Wireless Markup Language is and how to understand it. The last part discusses what WMLScript is and how to use it.*

Introduction to the Wireless Application Protocol

Overview

The Wireless Application Protocol is a standard developed by the WAP Forum, a group founded by Nokia, Ericsson, Phone.com (formerly Unwired Planet), and Motorola. The WAP Forum's membership roster now includes computer industry heavyweights such as Microsoft, Oracle, IBM, and Intel along with several hundred other companies. According to the WAP Forum, the goals of WAP are to be:

- Independent of wireless network standard.
- Open to all.
- Proposed to the appropriate standards bodies.
- Scalable across transport options.
- Scalable across device types.
- Extensible over time to new networks and transports.

As part of the Forum's goals, WAP will also be accessible to (but not limited to) the following:

- GSM-900, GSM-1800, GSM-1900
- CDMA IS-95
- TDMA IS-136
- 3G systems - IMT-2000, UMTS, W-CDMA, Wideband IS-95

WAP defines a communications protocol as well as an application environment. In essence, it is a standardized technology for cross-platform, distributed computing. Sound similar to the World Wide Web? If you think so, you're on the right track! WAP is very similar to the combination of HTML and HTTP except that it adds in one very important feature: optimization for low-bandwidth, low-memory, and low-display capability environments. These types of environments include PDAs, wireless phones, pagers, and virtually any other communications device.

The remainder of this overview will concentrate on presenting WAP from a software developer's perspective so that other software developer's can be quickly brought up to speed. Other documents on this site go into much greater detail on development specifics including in-depth reviews and demonstrations using a variety of vendor packages.

WAP and the Web

From a certain viewpoint, the WAP approach to content distribution and the Web approach are virtually identical in concept. Both concentrate on distributing content to remote devices using inexpensive, standardized client software. Both rely on back-end servers to handle user authentication, database queries, and intensive processing. Both use markup languages derived from SGML for delivering content to the client. In fact, as WAP continues to grow in support and popularity, it is highly likely that WAP application developers will make use of their existing Web infrastructure (in the form of application servers) for data storage and retrieval.
WAP (and its parent technology, XML) will serve to highlight the Web's status as the premier n-tier application in existence today. WAP allows a further extension of this concept as existing "server" layers can be reused and extended to reach out to the vast array of wireless devices in business and personal use today. Note that XML, as opposed to HTML, contains no screen formatting instructions; instead, it concentrates on returning structured data that the client can use as it sees fits.

Why Wireless? Why WAP?

Suppose that you work at a large shipyard involved with the construction and repair of commercial and naval ships. Typical projects are discussed in the hundreds or even thousands of man-years. Your organization long ago learned to make us of advances in computing technology by delivering real-time access to information via mainframe terminals on employee desks or on shop floors. As time went on, managers were eventually even able to make the business case for client/server access to mainframe databases from Windows applications. This opened up existing databases to improved reporting, charting, and other user interface features Managers and shop foremen can access parts inventories, repair schedules, shop budgets, and other useful information in order to plan work crew schedules and employee tasking.

It was just another small step from there for management to take advantage of you Web development skills by Web-enabling various mainframe applications (buzzword alert: we now call this Enterprise Application Integration, or EAI). With this information on the Web, information can be shared with parts suppliers and contractors which has greatly reduced ordering times and costs involved. One problem remains, however: out of 10,000 employees and contractors, only about 50(actually interact with the databases. The remainder of the employees continually fill out paperwork, issue reports to their manager, or manually key in data when they return from working on a ship.

Then, you read this article. Imagine if the other 9500 employees actively involved in welding, pipefitting, installing electrical cable, and testing electronics could all wirelessly retrieve and/or edit data when they actually need to! Small, inexpensive devices are given to each employee based on their tasking requirements. Some require handheld devices with built-in barcode scanners, others require keypads, others require simple digital displays. WAP allows a suite of client applications to be built which reuse existing server applications and databases. In addition, these applications can be dynamically downloaded and run on any of these devices. If an electronics tester runs into a bad vacuum tube, he scans the barcode. If a cable installer realizes that 500 more feet of a specific type of cable are required, he selects the "Order Cable" menu option from his wireless phone. If someone installing HVAC ventilation wants to know which pipes or cables run through a specific section of the ship, he enters the query in on his PDA and retrieves either data or imagery information.

In any industry that involves employees stepping out of their office to complete a job, wireless applications will be abundant. In Rifaat A. Dayem's book Mobile Data & Wireless LAN Technologies (Prentice Hall, 1997), he estimates that over 50% of the applications for this type of technology have not even been thought of yet!! WAP helps standardize the applications that will proliferate using wireless communication technologies. Imagine the Web without the combination of HTML and HTTP leaving us instead with "open" specifications from Sun Microsystems, Microsoft, and IBM. I will go out on a limb and say that there is no chance the Web would be where it was today without freely available, vendor-neutral, open standards.

How Does It Work?

WAP uses some new technologies and terminologies which may be foreign to the software developer, however the overall concepts should be very familiar. WAP client applications make requests very similar in concept to the URL concept in use on the Web. As a general example, consider the following explanation (exact details may vary on a vendor-to-vendor basis).

A WAP request is routed through a WAP gateway which acts as an intermediary between the "bearer" used by the client (GSM, CDMA, TDMA, etc.) and the computing network that the WAP gateway resides on (TCP/IP in most cases). The gateway then processes the request, retrieves contents or calls CGI scripts, Java servlets, or some other dynamic mechanism, then formats data for return to the client. This data is formatted as WML (Wireless Markup Language), a markup language based directly on XML. Once the WML has been prepared (known as a deck), the gateway then sends the completed request back (in binary form due to bandwidth restrictions) to the client for display and/or processing. The client retrieves the first card off of the deck and displays it on the monitor.

The deck of cards metaphor is designed specifically to take advantage of small display areas on handheld devices. Instead of continually requesting and retrieving cards (the WAP equivalent of HTML pages), each client request results in the retrieval of a deck of one or more cards. The client device can employ logic via embedded WMLScript (the WAP equivalent of client-side JavaScript) for intelligently processing these cards and the resultant user inputs.

To sum up, the client makes a request. This request is received by a WAP gateway that then processes the request and formulates a reply using WML. When ready, the WML is sent back to the client for display. As mentioned earlier, this is very similar in concept to the standard stateless HTTP transaction involving client Web browsers.

Communications Between Client and Server

The WAP Protocol Stack is implemented via a layered approach (similar to the OSI network model). These layers consist (from top to bottom) of:

- Wireless Application Environment (WAE)
- Wireless Session Protocol (WSP)
- Wireless Transaction Protocol (WTP)
- Wireless Transport Layer Security (WTLS)
- Wireless Datagram Protocol (WDP)
- Bearers (GSM, IS-136, CDMA, GPRS, CDPD, etc.)

According to the WAP specification, WSP offers means to:

- provide HTTP/1.1 functionality:
- extensible request-reply methods,
- composite objects,
- content type negotiation
- exchange client and server session headers
- interrupt transactions in process
- push content from server to client in an unsynchronized manner
- negotiate support for multiple, simultaneous asynchronous transaction

WTP provides the protocol that allows for interactive browsing (request/response) applications. It supports three transaction classes: unreliable with no result message, reliable with no result message, and reliable with one reliable result message. Essentially, WTP defines the transaction environment in which clients and servers will interact and exchange data.

The WDP layer operates above the bearer layer used by your communications provider. Therefore, this additional layer allows applications to operate transparent over varying bearer services. While WDP uses IP as the routing protocol, unlike the Web, it does not use TCP. Instead, it uses UDP (User Datagram Protocol) which does not require messages to be split into multiple packets and sent out only to be reassembled on the client. Due to the nature of wireless communications, the mobile application must be talking directly to a WAP gateway (as opposed to being routed through myriad WAP access points across the wireless Web) which greatly reduces the overhead required by TCP. For secure communications, WTLS is available to provide security. It is based on SSL and TLS.

Introduction to Wireless Application Protocol and Wireless Markup Language

The Wireless Markup Language (WML)

Many references I've come across use terminology such as "WML is derived from HTML" or "WML is loosely based on XML". Warning bells went off in my head when I see statements like this because: (a) it often means that a vendor has added proprietary extensions to some technology and (b) it means that I'm going to have to learn yet another language. Having said that, let me express my relief to find that WML is, in fact, an XML document type defined by a standard XML Document Type Definition, or DTD. The WML 1.1 DTD is defined at: http://www.wapforum.org/DTD/wml_1.1.xml

Much greater detail on XML and WML are given in the WML tutorial located under the WAP training topic, however the following code gives you an example of a simple WML file. If you're familiar at all with XML, you should have no problem understanding its meaning.

```
<?xml version="1.0"? >
<!DOCTYPE wml PUBLIC "-
   //WAPFORUM//DTD WML 1.1//EN"
   "http://www.wapforum.org/DTD/
   wml_1.1.xml">
<wml>
<card id="First_Card" title="First Card">
<p>
Hello World!
</p>
</card>
</wml>
```

The first two lines are required. They give the XML version number and the public document identifier, respectively. From there, all WML decks (one WML file equals one deck) begin and end with the

<wml></wml>

tags. Individuals cards are arranged with the

<card></card>

tags. Also, note that WML, like XML, is case-sensitive!

Included in the WML specification are elements that fall into the following categories: Decks/Cards, Events, Tasks, Variables, User Input, Anchors/Images/Timers, and Text Formatting. See the WML tutorial for specific examples on using these elements to build applications.

Additional Intelligence via WMLScript

The purpose of WMLScript is to provide client-side procedural logic. It is based on ECMAScript (which is based on Netscape's JavaScript language), however it has been modified in places to support low bandwidth communications and thin clients. The inclusion of a scripting language into the base standard was an absolute must. While many Web developers regularly choose not to use client-side JavaScript due to browser incompatibilities (or clients running older browsers), this logic must still be replaced by additional server-side scripts. This involves extra roundtrips between clients and servers which is something all wireless developers want to avoid. WMLScript allows code to be built into files transferred to mobile client so that many of these round-trips can be eliminated. According to the WMLScript specification, some capabilities supported by WMLScript that are not supported by WML are:

- Check the validity of user input
- Access to facilities of the device. For example, on a phone, allow the programmer to make phone calls, send messages, add phone numbers to the address book, access the SIM card etc.
- Generate messages and dialogs locally thus reducing the need for expensive round-trip to show alerts, error messages, confirmations etc.
- Allow extensions to the device software and configuring a device after it has been deployed.

WMLScript is a case-sensitive language that supports standard variable declarations, functions, and other common constructs such as if-then statements, and for/while loops. Among the standard's more interesting features are the ability to use external compilation units (via the use url pragma), access control (via the access pragma), and a set of standard libraries defined by the specification (including the Lang, Float, String, URL, WMLBrowser, and Dialogs libaries). The WMLScript standard also defines a bytecode interpreter since WMLScript code is actually compiled into binary form (by the WAP gateway) before being sent to the client. For more information on WMLScript, see the tutorial on this site.

The Business Case

Pros

WAP's biggest business advantage are the prominent communications vendors who have lined up to support it. The ability to build a single application that can be used across a wide range of clients and bearers makes WAP pretty much the only option for mobile handset developers at the current time. Whether this advantage will carry into the future depends on how well vendors continue to cooperate and also or how well standards are followed.

Cons

It is very, very early on in the ballgame an already vendor toolkits are offering proprietary tags that will only work with their microbrowser. Given the history of th computing industry and competition, in general, this was to be expected. Howeve further differentiation between vendor products and implementations may lead t a fragmented wireless Web.

WAP also could be found lacking if compared to more powerful GUI platform such as Java, for instance. For now, processor speeds, power requirements, an vendor support are all limiting factors to Java deployment but it's not hard to imagine a day in the near future where Java and WAP exist side-by-side just as Jav and HTML do today. In that circumstance, Java would hold a clear advantage over WAP due to the fact that a single technology could be used to build applications for the complete range of operating devices. Of course, on the flip side, the world is not all Java and there wi always be a place for markup languages i lieu of full-blown object-oriented platforms.

Conclusion

Some critics and second-guessers have pondered the need for a technology such as WAP in the marketplace. With the widespread proliferation of HTML, is yet another markup language really required? As we've discussed here, in a word, YES! WAP's use of the deck of cards "pattern" and use of binary file distribution meshes well with the display size and bandwidth constraints of typical wireless devices. Scripting support gives us support for client-side user validation and interaction with the portable device again helping to eliminate round trips to remote servers. WAP is a young technology that is certain to mature as the wireless data industry as a whole matures; however, even as it exists today, it can be used as an extremely powerful tool in every software developer's toolbox.

The Wireless Markup Language (WML)

Introduction

WML is a markup language that is based on XML (eXtensible Markup Language). The official WML specification is developed and maintained by the WAP Forum, an industry-wide consortium founded by Nokia, Phone.com, Motorola, and Ericsson. This specification defines the syntax, variables, and elements used in a valid WML file. The actual WML 1.1 Document Type Definition (DTD) is available for those familiar with XML at: http://www.wapforum.org/DTD/wml_1.1.xml

A valid WML document must correspond to this DTD or it cannot be processed.

In this tutorial, we'll present WML basics and an example. This example will demonstrate events and navigation as well as data retrieval from server CGI scripts. Discussion of client-side scripting and state management will be presented in the WMLScript tutorial.

NOTE: We will only discuss features contained in the WML standard. Information on non-standard WML capabilities added by vendors can be obtained by consulting that vendor's documentation.

Understanding the Wireless Markup Language WML is based on XML, a markup language that has garnered enormous support due its ability to describe data (HTML, meanwhile, is used to describe the display of data...a big difference). While HTML predefines a "canned" set of tags guaranteed to be understood and displayed in a uniform fashion by a Web browser, XML allows the document creator to define any set of tags he or she wishes to. This set of tags is then grouped into a set of grammar "rules" known as the Document Type Definition, or DTD. As mentioned earlier, the DTD used to define WML is located at: http://www.wapforum.org/DTD/wml_1.1.xml

If a phone or other communications device is said to be WAP-capable, this means that it has a piece of software loaded onto it (known as a microbrowser) that fully understands how to handle all entities in the WML 1.1 DTD.

The first statement within an XML document is known as a prolog. While the prolog is optional, it consists of two lines of code: the XML declaration (used to define the XML version) and the document type declaration (a pointer to a file that contains this document's DTD). A sample prolog is as follows:

```
<?xml version="1.0"?>
   <!DOCTYPE wml PUBLIC "-
   //WAPFORUM//DTD WML 1.1//EN"
    "http://www.wapforum.org/
   DTD/wml_1.1.xml">
```

Following the prolog, every XML document contains a single element that contains all other subelements and entities. Like HTML all elements are bracketed by the

<>

and

</>

characters. As an example: <element>datadatadata</element>. There can only be one document element per document. With WML, the document element is <wml>; all other elements are contained within it.

The two most common ways to store data within an XML document are elements and attributes. Elements are structured items within the document that are denoted by opening and closing element tags. Elements can also contain sub-elements as well. Attributes, meanwhile, are generally used to describe an element. As an example, consider the following code snippet:

```
<!-- This is the Login Card -->
<card id="LoginCard" title="Login">
Please select your user name.
</card>
```

In the code above, the card element contains the id and title attributes. (On a side note, a comment in WML must appea between the <!-- --> tags.) We will make use of the WML-defined elements and their attributes later as we build our examples.

Valid WML Elements

WML predefines a set of elements that car be combined together to create a WML document. These elements include can be broken down into two groups: the Deck/Card elements and the Event elements.

Table 1

Deck/Card Elements

wml	card	template	head	access	meta

Event Elements

do	ontimer	onenterforward	onenterbackward	onpick	onevent

Tasks

go	prev	refresh	noop

Variables

setvar

User input

input	select	option	optgroup	fieldset

Anchors, Images, and Timers

a	anchor	img	Timer

Text Formatting

br	p	table	tr	td

Each of these elements is entered into the document using the following syntax:

<element> element value </element>

If an element has no data between it (as is often the case with formatting elements such as <br>), you can save space by entering one tag appended with a \ character (for instance, <br/>).

Building Applications With WML

WML was designed for low-bandwidth, small-display devices. As part of this design, the concept of a deck of cards was utilized. A single WML document (i.e. the elements contained within the <wml> document element) is known as a deck. A single interaction between a user agent and a user is known as a card. The beauty of this design is that multiple screens can be downloaded to the client in a single retrieval. Using WMLScript, user selections or entries can be handled and routed to already loaded cards, thereby eliminating excessive transactions with remote servers. Of course, with limited client capabilities comes another tradeoff. Depending on your client's memory capabilities, it may be necessary to split multiple cards up into multiple decks to prevent a single deck from becoming too large.

Using Variables

Because multiple cards can be contained within one deck, some mechanism needs to be in place to hold data as the user traverses from card to card. This mechanism is provided via WML variables. Variables can be created and set using several different methods. For instance:

- Using the <setvar> element as a result of the user executing some task. The <setvar> element can be used to set a variable's state within the following elements: go, prev, and refresh. The following element would create a variable named x with a value of 123:

  ```
  <setvar name="x" value="123"/>
  ```

- Variables are also set through any input element (input, select, option, etc.). A variable is automatically created that corresponds with the name attribute of an input element. For instance, the following element would create a variable named x:

  ```
  <select name="x" title="X Value:">
  ```

Although we haven't discussed WMLScript yet, it is important to note that WML and WMLScript within a document share the same variables.

Creating A WML Deck

In this example, we'll start by creating a WML deck that allows us to first select a username from a list, enter in a password, then have our selections repeated back to us. This will illustrate the basic handling of user input, events, and variables all within one deck using multiple cards.

Listing 1 - WMLExample.wml

```
<?xml version="1.0"?>
<!DOCTYPE wml PUBLIC "-
  //WAPFORUM//DTD WML 1.1//EN"
  "http://www.wapforum.org
  /DTD/wml_1.1.xml">
<wml>
  <card id="Login" title="Login">
    <do type="accept"
    label="Password">
            <go href="#Password"/>
    </do>
    <p>
    UserName:
    <select name="name"
    title="Name:">
            <option value="John D
            oe">John Doe</option>
            <option value="Paul
            Smith">Paul
            Smith</option>
            <option value="Joe
            Dean">Joe Dean</option>
            <option value="Bill
            Todd">Bill Todd</option>
    </select>
    </p>
  </card>
  <card id="Password"
  title="Password:">
    <do type="accept" label="Results">
            <go href="#Results"/>
    </do>
    <p>
    Password: <input type="text"
    name="password"/>
    </p>
  </card>

  <card id="Results" title="Results:">
  <p>
  You entered:<br/>
  Name: $(name)<br/>
Password: $(password)<br/>
  </p>
  </card>
</wml>
```

As you can see, the prolog of this document contains the XML version number to be used as well as the Document Type Definition to be referenced. Following this comes the wml document element (the deck) that contains three cards: Login, Password, and Results. Each of these cards is defined using the <card> element. Because the Login and Password cards also define events, they use the <do type="accept"> element to define the event to be triggered. Figure 1 shows the initial card loaded in a test browser.

When the "accept" type of the do element is encountered, it is displayed as an option on the WAP device display (see Figures 2, 3, and 4). Selecting this option causes the <go> element to be analyzed.

If you are familiar with the anchor tag (<a>) in HTML, you know that it specifies an href attribute that tells the browser where to link to if this anchor is selected. The WML <go> element's "href" attribute works in the same manner. As with HTML, to link to another card in the document, you simply prepend a # symbol before it. For example, to link to the Results card, we define the following element:

<go href="#Results"/>

This Results card makes use of variables by retrieving and displaying the contents of the name and password variables. Recall

Figure 1

Figure 2

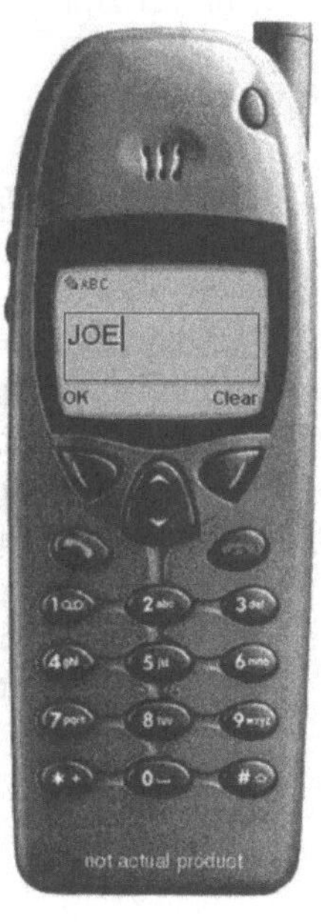

Figure 3

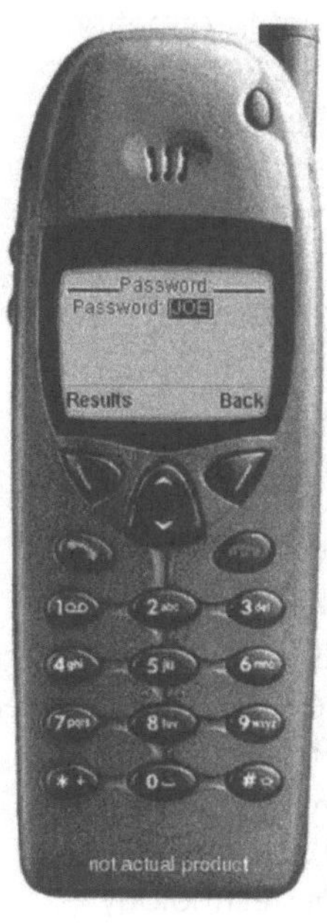

Figure 4

that variables are substituted into a card or deck by using the following syntax:

$(variable_name)

Calling A Server Script

Without the ability to perform server transactions, WML would only serve to provide a standardized way to display text on a client. Adding in the ability to dynamically connect to remote servers opens up every WAP device to the world of Internet messaging, enterprise data, and e-commerce. WAP devices interact with these data sources through a WAP gateway as mentioned in our WAP Overview tutorial. This gateway must interface with a carrier such as CDMA, GSM, or GPRS. However, it is possible to install and test gateway products in conjunction with popular Web servers (such as Microsoft Internet Information Server or Apache) on your LAN. This tutorial won't go into the details of installing and configuring a gateway but to eliminate a very common beginner's error, we'll remind you to be sure to add the following MIME types to your Web server:

WML text/vnd.wap.wml wml
WMLScript text/vnd.wap.wmlscript wmls

Once this has been done, you're ready to go! We'll now create a very simple example which allows the user to select an option and then retrieve data from a server based on that option. For this example, we're using Microsoft Active Server Pages (ASP) technology for the server-side scripting since that is the technology supported by our hosting provider. You could just as easily use other popular serv scripting tools such as Java Servlets, JavaScript, or Perl. Listing 2 gives the WM source code for our new deck. It basically contains a single

<select>

element that gives the user a few options for retrieval. The

<go>

element for this select list calls a server script with the appropriate arguments.

Listing 2 - WMLExample2.wml

```
<?xml version="1.0"?>
<!DOCTYPE wml PUBLIC "-
   //WAPFORUM//DTD WML 1.1//EN"
   "http://www.wapforum.org/DTD/
   wml_1.1.xml">

<wml>

   <card id="Order" title="Query
   Inventory">
      <p>
      <select name="Items" title="Items">
               <option value="Books">
               Books </option>
               <option value="Music">
               Music</option>
               <option value="Video">
               Video</option>
               <option value="Software
               "> Software</option>
      </select>
      </p>
      <do type="accept" label="Query">
```

```
                <go href="http://
                127.0.0.1/WML/
                Inventory.asp"
                method="post">
                        <postfield
                        name="Items"
                        value="$(Items)"/>
                </go>
        </do>
    </card>
</wml>
```

The server script (shown in Listing 3) examines the input and produces WML output to be displayed on the device.

Listing 3 - Inventory.asp

```
<%
Dim Body

If Request.Form("Items") = "Books" Then
   Body = "You selected Books!"
ElseIf Request.Form("Items") = "Video"
Then
   Body = "You selected Video!"
ElseIf Request.Form("Items") = "Software"
Then
   Body = "You selected Software!"
ElseIf Request.Form("Items") = "Music"
Then
   Body = "You selected Music!"
End If

Response.ContentType =
"text/vnd.wap.wml"%>

<?xml version="1.0"?>
<!DOCTYPE wml PUBLIC "-
   //WAPFORUM//DTD WML 1.1//EN"
   "http://www.wapforum.org
   /DTD/wml_1.1.xml">
<wml>
<card>
<p>

<%Response.write(Body)%>

</p>
</card>
</wml>
```

Figures 5 and 6 show the Music option being selected and the resultant screen retrieved from the ASP script.

A few things should be mentioned for those wishing to run this example on their local Web server. You must register the proper MIME types with your Web server so that WML content can be properly sent. The two MIME types that should be

Figure 5

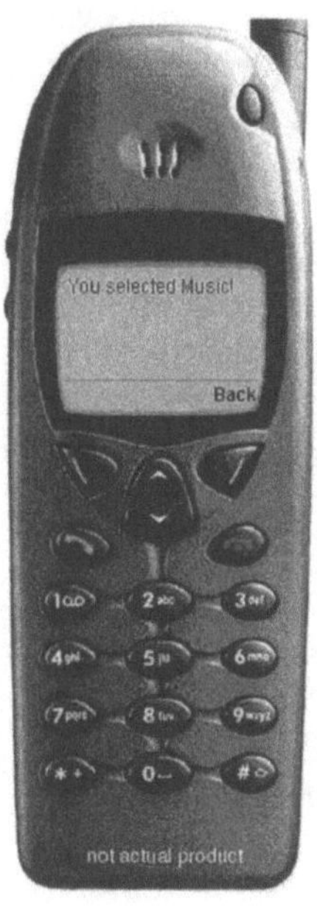

Figure 6

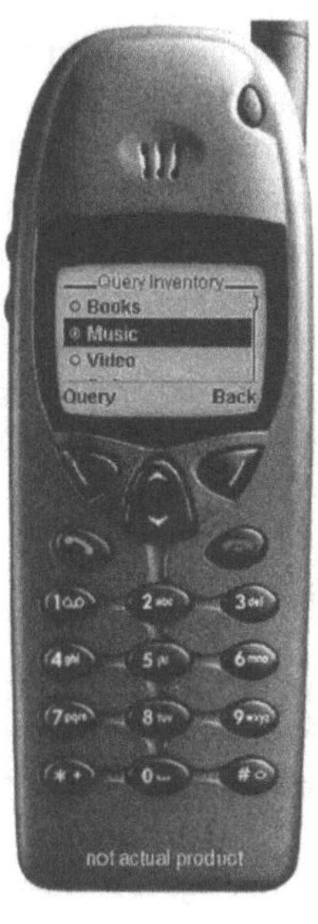

registered are:

.wml text/vnd.wap.wml
.wmls text/vnd.wap.wmlscript

If you'd like to use Wireless Bitmap images (the image format supported by WAP), also add:

.wbmp
image/vnd.wap.wbmp

Finally, I'd like to mention one error I continually received when developing this example using the Nokia WAP Toolkit 1.2. I've seen numerous postings on WAP Development boards concerning this error so I thought I'd explain the problem and solution here. Although I registered the MIME types with IIS 4.0, I still received the message "Mime type not supported." It turns out that even though I was loading the WML source via my local machine's Web server, the Toolkit was switching over to a file://-based URL since the file was local to my machine. When I then attempted to run the script using href="Inventory.asp", I got the error. Switching the href over to http://127.0.0.1/WML/Inventory.asp forced the loading of the script through the Web server which allowed for proper recognition of the WML MIME types.

Conclusion

WML offers software developers an entirely new, exciting platform on which to deploy their applications. With this new platform, however, comes a host of tradeoffs and challenges. A new wrinkle will be added to the design process as things like server round-trips, bandwidth, and display sizes become issues to contend with. While it may take several iterations for developers and vendors to get their product offerings right, there is no doubt that WAP opens the door to a new era in application development and deployment.

Adding Client-Side Logic To WAP Using WMLScript

WMLScript is the WAP corollary to the JavaScript scripting language that was popularized by Netscape Communications. Standardization efforts by Netscape helpe produce the ECMAScript standard, a standard that WMLScript was based on. While JavaScript has since been coopted b server tool vendors (including Netscape an Microsoft), WMLScript is a client-only scripting platform used in combination with WML to provide client side procedur logic. Like WML, WMLScript is compiled vi a WAP gateway into binary form to provid intelligence to mobile clients. In this brief tutorial, we'll discuss what WMLScript is and how to use it. For more information on WMLScript, visit the WAP Forum.

WMLScript Language Syntax

WMLScript syntax is based on the ECMAScript programming language. Unlike ECMAScript, however, the WMLScript specification also defines a bytecode and interpreter reference architecture for optimal utilization of current narrowband communications channels and handheld device memory requirements. The following bullets help summarize some basic syntactical features of the language:

- The smallest unit of execution in WMLScript is a statement and each statement must end with a semicolon (;).
- WMLScript is case-sensitive.
- Comments can either be single-line (beginning with //) or multi-line (bracketed by /* and */). This syntax is identical to both C++ and Java.
- A literal character string is defined as any sequence of zero or more characters enclosed within double (" ") or single (') quotes.
- Boolean literal values correspond to true and false.
- New variables are declared using the var keyword (i.e. var x;)

Data Types

WMLScript is a weakly typed language. This means that no type-checking is done at compile- or run-time and no variable types are explicitly declared. Internally, the following data types are supported:

Boolean
Integer
Floating-point
String
Invalid

The programmer does not need to specify the type of any variable; WMLScript will automatically attempt to convert between the different types as needed. One other point to note is that WMLScript is not object-oriented (such as Java or C++). Therefore, it is impossible to create your own user-defined data types programmatically.

Operators

WMLScript supports a variety of operators that support value assignment operations, arithmetic operations, logical operations, string operations, comparison operations, and array operations. For more information on the wide variety of WMLScript operators, see the WMLScript specification.

Flow Control Statements

The operators and expressions supported by WMLScript are virtually identical to those of the JavaScript programming language so we will not discuss them here. Java does support a number of control statements for handling branching within programs. These include the if-else, for loop, while loop, break, and continue statements.

Functions

Related WMLScript statements can be executed together as a unit known as a function. A function declaration has the following syntax:

extern function
identifier(FormatParameterList) Block ;

The extern keyword is optional and is used to specify a function that can be called from outside the current compilation unit in which the function is defined. A sample WMLScript function declaration looks like this:

```
function RunTime(distance, speed) { var
time = distance / speed; return time; };
```

The above example simply takes two input variables, distance and speed, and uses them to calculate a time variable. The return keyword is then used to return a value.

When calling a function included with one of the WMLScript standard libraries (see below), the library name must be included with the function call. For example, to call the String library's length() function, use the following syntax:

```
var a = String.length("1234567890");
```

The WMLScript Standard Libraries
While WMLScript does not support the creation of new objects via object-oriented programming, it does provide six "pre-built" libraries that aid in the handling of many common tasks. These libraries (with a brief description of each) include:

- Lang - This library contains a set of functions that are closely related to the WMLScript language core. Included in this library are functions for data type manipulation, absolute value calculations, and random number generation.
- Float - The Float library is optional and is only supported on those clients who have floating-point capabilities. Typical functions provided by this library include sqrt(), round(), and pow().
- String - The String library contains a set of functions for performing string operations. Some of the functions included in this library are length(), charAt(), find(), replace(), and trim().
- URL - This library contains a set of functions for handling both absolute URLs and relative URLs. Typical function include getPath(), getReferer(), and getHost().
- WMLBrowser - This library contains functions by which WMLScript can access the associated WML context. These functions must not have any side effects and must return invalid in cases where the system does not support WMLBrowser and where the interpreter is not invoked by the WML Browser. Commonly used functions in this library include go(), prev(), next(), getCurrentCard(), and refresh().
- Dialogs - This library contains a set of typical user interface functions includin prompt(), confirm(), and alert().

Example: Validating User Input Via WMLScript

In the following example, we will build a simple WML card that asks the user to input a social security number (an identification number used by the U.S. Social Security Administration). We will then use WMLScript to verify that the user's input was formatted correctly. Following this verification, we'll alert the user via WMLScript to let them know whether their number was accepted or not. This example, though simple, represents a typical usage of WMLScript on a client.

To build this example, we create a normal WML file containing two cards: an input card and a results card (see Listing 1 below). Accepting the input will result in our validateSSN() function being called. Note that this function is stored in a separate .wmls file (WMLScriptExample.wmls) and is declared within that file using the extern keyword. extern allows a function to be called by other functions or WML events that exist outside of the function's source file. To keep a function "private", simply declare the function without using the extern keyword.

Listing 1 - WMLScriptExample.wml

```
<?xml version="1.0"?>
<!DOCTYPE wml PUBLIC "-
  //WAPFORUM//DTD WML 1.1//EN"
  "http://www.wapforum.org/
  DTD/wml_1.1.xml">

<wml>

  <card id="SSN" title="SSN:">
    <do type="accept" label="Results">
            <go href="WMLScript
            Example.wmls#
            validateSSN($(SSN))"/>
    </do>
    <p>
    Enter SSN: <input type="text"
    name="SSN"/>
    </p>
  </card>

  <card id="Results" title="Results:">
    <p>
    You entered:<br/>
   SSN: $(SSN)<br/>
    </p>
    </card>

</wml>
```

Listing 2 - WMLScriptExample.wmls

```
extern function validateSSN(SSN)
{
   if (String.length(SSN) != 9)
   {
      WMLBrowser.setVar("SSN", "Error:
      String must be 9 digits long.");
   }

WMLBrowser.go("WMLScriptExample.wml
#Results")
;};
```

Figure 7

Figure 8

The WMLScript function shown in Listing 2 makes use of two of the standard WMLScript libraries: WMLBrowser and String. The WMLBrowser.setVar() function sets the value of a WML variable while the WMLBrowser.go() function redirects execution of the script to a card within a WML source file.

W* Effect Considered Harmful

Title: *W* Effect Considered Harmful*

Author: *Rohit Khare, 4K Associates*

Abstract: *The Wireless Application Forum has developed an entire stack of network protocols parallel to, and only marginally compatible with, the existing Internet architecture. They are convinced handheld wireless devices are -- and will remain -- four orders of magnitude less powerful than conventional Internet hosts and thus require optimized transport, applications, and content. At each turn, WAP Forum has chosen to reinterpret existing Internet standards -- often incompatibly. The shift from UDP to WDP, TLS to WTLS, HTTP to WTP, HTML to WML, ECMAScript to WMLScript -- termed 'the W* Effect' -- is disingenuous at best, and at worst, locks in early WAP adopters to today's lowest common denominator. This report presents a summary of WAP, its history and key players, a layer-by-layer tour of its standards, and its market potential for handset providers, network operators, application servers, and content providers.*

1. Executive Summary

The hegemonic spirit of the age inspires industry slogans of "Windows Everywhere!," "Java Anywhere!," and "AOL Everywhere!" While these are mere market visions as yet, it's informative to look back at one technical paradigm in computing that has conquered or co-opted every effective rival to its progress over the last three decades: "IP Over Everything."

The Internet has adapted to radical growth in scale (bandwidth, hosts) and in character (timesharing, PCs, embedded systems). In particular, the death of TCP/IP has been oft foretold, but it has been adapted to cope with greater congestion, greater latency variation, greater bandwidth, link compression, and security requirements.

The latest challenge is integrating handheld wireless devices into today's wired Internet. Such devices have much lower bandwidth, much lower processing power, and much smaller user interfaces.

The "control hypothesis" is that these devices can be accommodated as just another kind of Internet host. That approach is to tweak around the edges by profiling 'smarter' TCP over airlinks, 'compact' HTML, and 'split-proxy' security.

The other approach is to throw everything out and reinvent the protocol stack from scratch. And for good measure, to sprinkle pixie dust over the whole suite by claiming compatibility with those established standards. Unwired Planet, Inc. and its mostly-captive Wireless Application Forum have opted for the latter. In the long run, it cannot succeed.

1.1 Architectural Visions for Handheld Wireless Internet Access

The choice of 'adapt or reinvent' has been mapped onto two different visions of wireless devices. One views them as miniaturized PCs, the other as cellphones-on-steroids. WinCE, Mobile NCs, and Symbian's EPOC operating systems are predicated on the former -- indeed, only a "real computer" would even have an OS onboard. WAP Forum's technology aims for the latter: deploying an absolutely standardized 'microbrowser' environment on today's handsets.

This dichotomy is also entangled with the choice of service model: should the device only access the carrier's value-added services, or any public Internet resource? WAP Forum and its major backers have proclaimed that handheld Internet access is not about 'surfing', and so eschewed generic approaches for custom applications hosted by the carrier. There isn't even an escape hatch -- standard HTML pages do not map directly onto WML.

1.2 Origins & Status of WAP Forum
Unwired Planet was founded in May of 1995 to develop handset-based access to the Internet. By December, it had codified its "insights" as US Patent 5,809,415: "Method and architecture for an interactive two-way data communication network". AT&T Wireless' PocketNet was their most visible early adopter: a commercial disappointment that could barely send e-mail and fetch stock quotes.

From the beginning, their strategy has been to give away its client. As UP's recent S-1 IPO filing states, "we license our UP.Browser software to wireless telephone manufacturers free of per-unit royalties an provide maintenance and support services for an annual flat fee."

WAP's business proposition varies for several membership sectors. Carriers are searching for value-added services to increase airtime sold. Infrastructure manufacturers are trying to jumpstart sales by avoiding fragmentation at any cost. Independent WAP Gateway developers expect niches for multiple vendors, if for no other reason than telecommunications politics. Content providers expect lucrative partnership contracts from the degree of integration and custom development required.

1.3 Key Roles & Players
WAP Forum Chairman Chuck Parrish is als an Executive VP of Unwired Planet -- and tellingly, its second most highly compensated employee. UP is counting on first-mover monopoly profits from their

Table 1

December 1995	Unwired Planet founded
June 1997	WAP public launch
September 1997	WAP 1.0 specs published
December 1997	WAP Forum incorporated
January 1998	Membership opened up
April 1998	WAP 1.0 ratified
March 1999	WAP 1.1 drafts published alone

With that lure, UP convinced Nokia, Ericsson, and Motorola to join in launching WAP Forum, which quickly rubber-stamped a slightly-reworked edition of UP's then-current product. By now, there are 91 member firms (at $27,500 per year). Member manufacturers represent 90% of global handset shipments; commitments to ship 'WAP-enabled products' cover 75%. Member carriers have 100 million subscribers. Dozens of internal working groups have proliferated, including several devoted to WAP marketing.

influence over WAP Forum and three-year head start. Their UP.link server software costs network operators on the order of $1 million for its server software -- plus a per-device licensing fee. Not per-activated-WAP-user -- they expect royalties for every 'WAP-capable device,' regardless. A 2 million subscriber carrier faced a $20 million commitment!

WAP Tools. Server software is the key to deploying WAP, and UP's competitors are still catching up. Nokia and Ericsson have announced servers, as well several independents: APiON (Ireland), CMG (Netherlands; OEM'd from Nokia), and Dr. Materna (Germany). Most of these systems run on Solaris. The largest computer services firms have not committed yet: Lucent, IBM, HP, and Oracle are only rumored to have started work.

Client SDKs for content providers are available from UP, Nokia, Ericsson, and Dynamical Systems Research (UK). Interviewing current vendors of Web translators for handhelds such as AvantGo, Spyglass, Proxinet, and Online.Anywhere uncovered only vague plans.

WAP Devices. It's designed to scale down to two-line, one-way, text pagers, so in one sense there are millions of fielded WAP-compatible devices. In a more conventional vein, there are shipping UP.browser-equipped phones from Motorola (i1000plus), Nokia (7110), and LG (NeoPoint 1000).

WAP support is nonexclusive, though: many of the same firms are partners in Symbian (Psion, Nokia, Ericsson, Motorola, Phillips, IBM, Oracle, NTT, and Sun). Its EPOC32 screenphones can work with mobile IP today and WAP soon. A little further up the complexity scale, WAP support stops: there are no announced plans for WAP-enabled WinCE, PalmOS, or laptop devices.

WAP Services. While a few operators have deployed WAP services (notably Finland) and many more have run trials, the industry will be watching a few high-profile North American launches this year. Nextel has announced Nextel Online by Q499; AT&T Wireless will relaunch son-of-PocketNet; Sprint PCS is uncommitted. Otherwise, carriers may still wait 2-3 years for "third generation" high-bandwidth systems (UMTS) and mobile-IP service to more capable handhelds (with "real" browsers).

Beyond carrier-hosted applications like email and calendaring, content providers have been limited to marquee partners so far: CNNmobile, Reuters, Netcenter, and other consumer brands should be expected. There have not yet been any customer-driven internal applications of the sort envisioned by WirelessKnowledge. As detailed below, WML card 'decks' are sufficiently different from the existing Web to require custom software.

1.4 WAP Component Technologies
With WAP, all the smarts are built into the WAP Gateway at the carrier's premises. To the public Internet, it becomes the termination point: from there to the device, WAP Forum's standards kick in. Claiming that 'wireless is different', WAP

1.1 rewrites almost every Web standard in the book.

Network Layer.
For good measure, there's a low-level control message protocol: WCMP is similar to, but less informative than ICMP. To adapt to the myriad airlink protocols out there, WAP defines its own format for datagram packets. WDP is based on UDP, but with its own port numbering semantics. Security is inserted next. Rather than an an analogy to IPsec, WTLS offers SSL/TLS like transport-layer security, using new elliptic curve cryptosystems for low-powered terminals. What the Internet calls "transport", though is in the next layer: WSP. It's a session protocol roughly equivalent to TCP, but with experimental transaction features.

Application Layer. WSP also folds in what the Internet calls "applications": a creatively reconceived HTTP with binary encoding, push, and stateful access. It's used to fetch 'decks' of 'cards' written in Wireless Markup Language, which is less like HTML pages than a BASIC-like programming language for text screens, complete with scoped functions, variables, and timers. WML is an XML application, to be sure, but it can only be transmitted in a homegrown compressed format called Wireless Binary XML (WBXML). Finally, the 'microbrowser' -- whose user interface has been homogenized to minute detail -- must execute WMLScript, their own, incompatible profile of ECMAScript.

WAE and WTA Servers. Together, these specifications form the Wireless Application Environment (WAE). A WAE server is roughly equivalent to a Web server or proxy. Wireless Telephony Application servers are built on top of WAE, but represent telephone switches instead. To the degree it handles call signaling and user interfaces to conference calls, etc. it diverges from IETF's Session Initiation Protocol and other Voice-over-IP efforts.

1.5 Immediate Market Implications
This NIH attitude requires an incredible amount of engineering effort to follow through. However, the commercial deman has proved sufficient to recreate most of the technology described above. WAP Forum is issuing conformance statements (and, eventually, branding); UP has established a WAP interoperability laboratory; and carriers are signing contracts with integrators. WAP is already success for the near term.

Once deployed, the spotlight shifts to subscribers for 2000. Toeing the industry line that novice users won't expect to surf the Internet, initial product plans focus on built-in applications, hybrid Web-WAP portals, and limited Web access. UP.Sm@rt is already available for carrier-hosted WML based datebook, phonebook, and memo pad applications. Nextel Online with Netcenter is an example of the second offering: enter your appointments and stock portfolio into a Web page, and UP.link will update you on the go. The last category includes reformatted news and adapters for critical Web forms. UP's S-1 filing lists "ABCNews.com, Bloomberg, Reuters, Quote.com, ESPN Sportszone and Travelocity," for example.

How satisfied will customers be in this pasture? These plans do not raise hopes for a grassroots movement of Web sites available in WML format; not even tools easy enough for corporate IS/IT departments to port mission-critical applications to it (see Section 2.6). A phone will essentially be tied to whatever services the carrier hosts when WAP standards fail to take hold on the Web.

How satisfied will carriers be? Carriers won't enjoy running application servers and their tech support. They don't want to miss out on the potential minutes lost because Web content isn't available, either. Look for WAP contracts to include "traditional" wireless data, too: the targeted high-end business customer is precisely the demographic that can also leverage the phone as an IR modem or mobile IP node.

How satisfied will manufacturers be? The entire goal of the microbrowser enterprise is to utterly homogenize handsets -- not just at the user interface, but programmer interfaces as well. Manufacturers don't like selling commodities, which is why so many are hedging their bets with Symbian, WirelessKnowledge, or Mobile NCs. They're planning ahead for HTML microbrowsers, active content, and multimedia -- to break out of the 2-10 line text display WML is predicated upon.

How satisfied will standards committees be? There is an immense investment in each of the existing standards WAP has cribbed from. Conflicts will surface as soon as WAP attempts to promulgate its standards as any but de facto. At IETF, the network layer will remain TCP/IP, tuned for "long, thin pipes". At W3C, the application layer will remain HTTP, XHTML, and DOM. At ECMA/ISO, the scripting language will remain ECMAScript. At ICANN, IANA will not take over registries created by the WAP Interim Naming Authority (WINA).

The bottom line is that WAP is a wrapper around present-day phones and pagers. It will bring a modicum of 'open systems' to the wireless market through interchangeable phones and information services, but it does not appear to have the headroom for truly diverse Internet information services and radically more capable devices. The network effects of treating a phone as just another, albeit dumb, HTML browser, and a just another, albeit slow, TCP/IP device could easily trump all the marginal performance improvements WAP touts.

1.6 Long-Term Implications

This report will argue that WAP should be adopted cautiously. WAP is simply not the long-term answer, and the more it is seen to be immediately inevitable, the longer it will take the wireless-data industry to eventually dig out from that hole.

There is a core myth of 'standards-compliance' at the heart of WAP waiting to be discredited in the press and at several standards bodies. WAP Forum takes pains to appear cooperative, issuing joint white papers with W3C and predicting future iterations of HTML will support WML constructs; XML will adopt a standard binary encoding; HTTP-ng will look like

WSP; WTLS's cryptography will be accepted for TLS; and so on.

Another benefit of WAP is that it's based on existing Internet standards. For example, the Wireless Markup Language specification portion of WAP is based on the HDML (Handheld Device Markup Language) specification and is an XML (Extensible Markup Language)-based language. Microbrowser technology based on WAP allows each handheld device to decide how to best display information received from a server.
-- From WAP Standard to the Rescue, PC Week, October 19, 1998

Instead, the technical public will eventually become aware of the technical issues raised in this report. In particular, these issues will be raised (constructively) at the venues WAP overlaps (Section 4), addressing potential divergence sooner rather than later.

Numerous IETF working groups are addressing long, thin, networks, as well as W3C working groups on XML and HTTP -- a credible reply to WAP's claims of exceptionalism ('wireless is different').
As for Unwired Planet it's worth noting that they know they have a narrow window of opportunity to catapult themselves to the head of the line, and they're not afraid of sticking premium prices to carriers. It's a lucrative market if it's still around in two years.

Figure 1

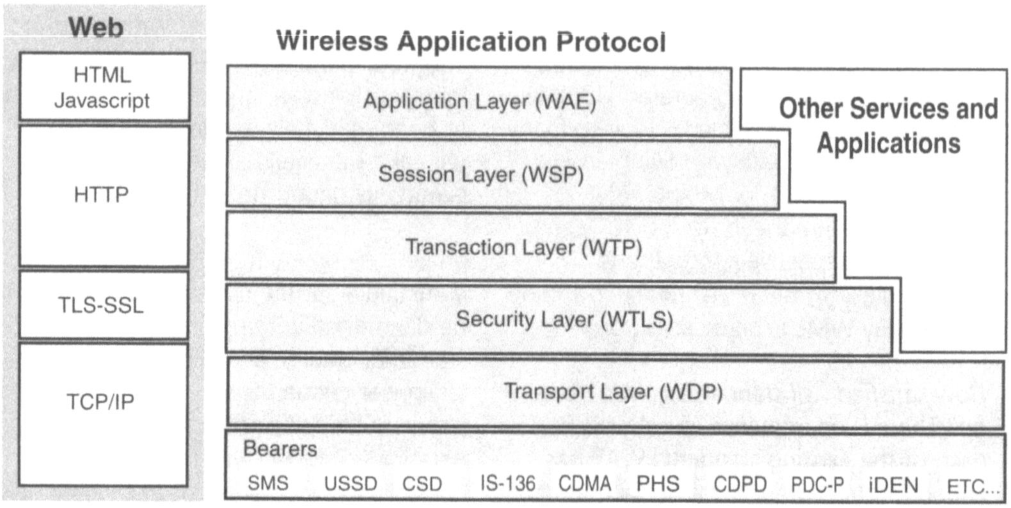

2. WAE, Layer by Layer

At first glance, WAP's architecture appears cleanly layered, inheriting from the OSI and Internet models. The rub is in their claim that it's parallel to the existing Web/TCP/IP stack, as they do in figure 1.

In this section, we present the component specifications of the Wireless Application Environment (WAE) and the Wireless Telephony Application (WTA). We discuss the intent of each WAP proposal, its equivalent 'control' technologies, and relevant market stakeholders.

2.1 WCMP: Control Message Protocol
Internetworking requires interoperable error reporting: a bedrock standard for signaling erroneous addresses, link congestion, and higher-level protocol errors (bad port number, etc). Internet Control Message Protocol (ICMP) thus became the very first step in the transformation of ARPAnet into the Internet.

WAP needs to operate across widely varying airlink "networks", so it also specified an equivalent WCMP. However, no individual carriers are multiprotocol, so there isn't any internetworking going on at all in the WAP market. Because most packet data airlinks include their own error-correction and dropout detection at a lower layer, WCMP is largely irrelevant.

WCMP is not included in the current WAP/1.1 revision. Although WDP/1.1 still references the WAP/1.0 edition as a SHOULD, none of the 20 current airlink adaptations in Section 5.4 of WDP require its use. WCMP is incompatible with ICMP. The official rationale is that "The Wireless Control Message Protocol (WCMP) provides an efficient error handling mechanism for WDP, resulting in improved performance for WAP protocols and applications." WCMP specifies several error codes, which differ from ICMP. WCMP also allows GSM, Flex and ReFlex addresses instead of IP numbers.
"WCMP Type values are different from ICMP Type values. WCMP Type values have been selected by adding 50 to the respective ICMP Type. WCMP Codes are the same [as] in ICMP."

Table 2

WCMP Message	***Type***	***Code***
Destination Unreachable	51	
No route to destination		0
Communication administratively prohibited		1
Address unreachable		3
Port unreachable		4
Parameter Problem	54	
Erroneous header field		0
Message Too Big	60	
Reassembly Failure	61	
Reassembly time exceeded		1
Buffer Overflow		2
Echo Request	178	
Echo Reply	179	

2.2 WDP: Datagram Protocol

WDP is roughly equivalent to UDP. In fact, carriers running mobile IP to the handset (CDPD, iDEN, or circuit-switched PPP) MUST use UDP instead. As with WCMP, the only rationale for reinventing a parallel format was to accommodate airlink addresses ("MSISDN number [handset serial number], IP address, X.25 address or other identifier") and airlink restrictions on packet size and even character sets.

The services offered by WDP include application addressing by port numbers, optional segmentation and reassembly and optional error detection. The services allow for applications to operate transparently over different available bearer services.

Their port numbering strategy is another example of botched reference-by-copy. They correctly placed all of their own service assignments temporarily in the dynamic, private port space, but they haven't even filed for WAP ports in IANA-registered space.

Figure 2: General WDP architecture

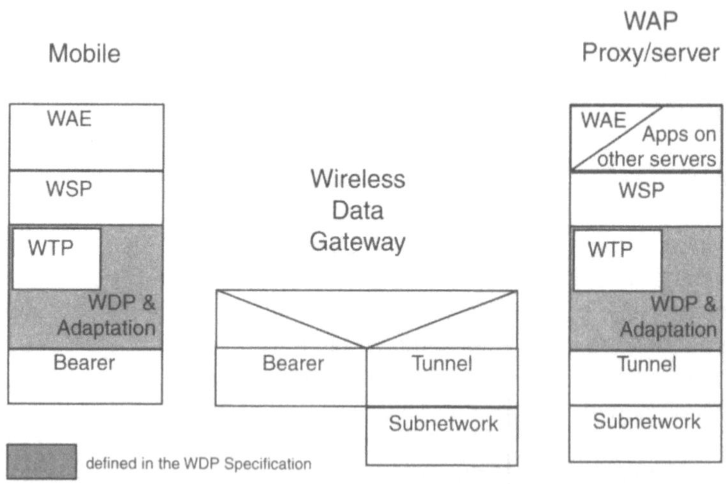

Of course, when they do, IANA will likely bounce half their applications right back, because IESG policy now forbids allocating parallel port numbers for "secure version of X". Basic protocols must be capable of negotiating security upgrades of the same port

WAP is in the process of registering ports for applications in the WAP space. However, at the moment no applications to IANA have yet been made and ports from the Dynamic/Private range are defined. These temporary ports will be changed when ports from the registered range are approved.

Table 3

Application/Protocol	(Temporary) ***Port Number***
Connectionless WAP Browser Proxy Server	49152
Secure Connectionless WAP Browser Proxy Server	49153
WAP Browser Proxy Server	49154
Secure WAP Browser Proxy Server	49155
vCard Receiver	49156
Secure vCard Receiver	49157
vCalendar Receiver	49158
Secure vCalendar Receiver	49159

2.3 WTLS: Transport Layer Security
The first slide in this section claimed that WDP is somehow similar to TCP. This is simply not true: that depiction papered over an inconvenient misalignment of security architectures. The classic Internet stack offers security at the packet- and transport-layers with two separate technologies, IPsec and TLS respectively. WTLS (mis)applies TLS to both individual datagrams and socket connections.

This conflation is only possible because of a deeply embedded assumption that a handset always talks to a single, permanent WAP Gateway. Otherwise, there wouldn't be an advantage to negotiating a single master secret to these different modes -- and it would be impossible to also buffer against lost, duplicated, and reordered packets. Yes, that's right: WAP Forum has TLS doing TCP's job!
Both datagram and connection oriented transport layer protocols must be supported. It must be possible to cope with, for example, lost, duplicated, or out of order datagrams without breaking the connection state.

Since encrypted data cannot be compressed, this layer (like TLS) also provides compression services -- but since it's below the transport, Class 2 and 3 implementations run the risk of double-compression. It's not just the performance loss of recompressing WML/WTP content; it's the possibility of cryptoanalytic weakness since WAP's upper layer compression is not just gzip, but a very predictable tokenization (WBXML).

WTLS defines three levels of security capabilities; only Class 1 is mandatory-to-implement.

Table 4

WTLS Features	Class 1	Class 2	Class 3
Public-key exchange	M	M	M
Server certificates	O	M	M
Client certificates	O	O	M
Shared-secret handshake	O	O	O
Compression	n/a	O	O
Encryption	M	M	M
MAC	M	M	M
Smart card interface	n/a	O	O

WTLS also specifies use of UP partner Certicom's elliptic curve public key encryption. Elliptic curves are more efficient to computer, but not as broadly implemented as stock RSA. This ciphersuite has not been documented on the IETF standards-track either.

To be sure, an RSA key exchange can take as long as 30 seconds on Palm Pilots and other low-power devices. If it's too expensive to have a public-key operation for every transaction, there will have to be a trusted proxy in the middle, effectively splitting the workload. ProxiNet was founded on this insight: devices authenticate themselves to a proxy server once, letting their proxy cluster handle multiple identities and SSL logons to Internet resources. The link itself only requires bulk encryption, a much smaller computational load. Thus, public key generation is done only once, at device provisioning.

It is unclear how effective current TLS ISV's experience will crossover to WAP's own WTLS implementations. Simply preserving the cryptosystems and handshakes enough tô directly reuse TLS libraries to implement WTLS.

Unlike in TLS, the record layer does not fragment information blocks. It is assumed that the transport layer takes care of the necessary fragmentation and reassembly.

For example, there are conflicting values for warning classes:
/* TLS */ enum { warning(1), fatal(2), (255) } AlertLevel;

/* WTLS */ enum { warning(1), critical(2), fatal(3), (255) }
AlertLevel;

Upon careful comparison of the WTLS pdf files and the TLS Internet-Draft, new alerts were also added, for no clear reason:

no_connection(5),
unknown_key_id(52),
disabled_key_id(53),
key_exchange_disabled(54),
session_not_ready(55),
unknown_parameter_index(56),
duplicate_finished_received(57),
user_canceled(90),
no_renegotiation(100),

2.4 WTP: Wireless Transaction Protocol
Extending the confusion between network- and transport- layers in WTLS, WTP also solves a mix of transport- and application-layer problems. It is roughly equivalent to TCP, but without any notions of flow-control or windowing. WTP also adopts several tricks for fast-reestablishment and handshake-minimization to qualify as T/TCP-like (IETF's experimental-track Transactional TCP).

Furthermore, rather than simply expose a streams or socket interface, WTP offers three application message models that seem a lot more like RPC or remote method invocation than the Internet definition of "transport":

Class 0:	Unreliable invoke message with no result message
Class 1:	Reliable invoke message with no result message
Class 2:	Reliable invoke message with one reliable result message

WTP optionally offers segmentation and re-assembly and selective acknowledgements, though much of that ought to remain the purview of lower-layer WDP.

At the same time, WTP radically redefines the meaning of "acknowledgement" to include explict user confirmation. That's like wiring e-mail style read-receipt into TCP.

"If the WTP user does not confirm the indication primitive after a specified time, the transaction is aborted by the provider. Note that this is a much stronger form of a confirmed service than the traditional definition [ISO8509]."

WTP imitated 16-bit TCP port numbering to differentiate the sorts of applications, outstanding concurrent requests, and the direction (using the low-order bit). They did not try more modern approaches such as hashing protocol URIs to map ports to services. WTP added individual transaction-ids, but issues them numerically and

Table 5

WTP User (e.g. WSP)	
WTP	◻ Transaction handling ◻ Re-transmissions, duplicate removal, acknowledgements ◻ Concatenation and separation
[WTLS]	◻ Optionally compression ◻ Optionally encryption ◻ Optionally authentication
Datagram Transport (e.g. WDP)	◻ Port number adressing ◻ Segmentation and re-assembly (if provided) ◻ Error detection (if provided)
Bearer Network (e.g. IP. GSM SMS/USSD. IS-136 GUTS)	◻ Routing ◻ Device addressing (IP address. MSISDN) ◻ Segmentation and re-assembly (if provided) ◻ Error detection (if provided)

requires service in id-order, pessimistically forcing strict ordering in the face of missing or delayed segments.

WTP is the heart of an independent WAP Gateway server project, such as APiON's. The lower levels discussed so far can all be elided by choice of airlink and security level -- but WTP is the lowest layer the microbrowser absolutely requires.

This chapter describes how WTP relates to other WAP protocols. For a complete description of the WAP Architecture refer to [WAP]. The following table illustrates the where the services provided to the WTP user are located.

2.5 WSP: Wireless Session Protocol

At this layer, again, WAP forum scrambles accepted definitions of networking terms. Wireless Session Protocol is actually intended to replace Hypertext Transfer Protocol. The IETF is currently investigating true session-layer support over TCP, with the goal of multiplexing several 'virtual' connections over a single "real" connection. That does not include negotiating upper-layer application protocols, data encoding syntax, or nomadic support for restarting sessions hours later. WSP does all that:

a) establish a reliable session from client to server and release that session in an orderly manner;
b) agree on a common level of protocol functionality using capability negotiation;
c) exchange content between client and server using compact encoding;
d) suspend and resume the session.

On top of that, WSP also yokes together session-oriented and non-session-oriented services. It's the equivalent of a connectionless TCP -- and isn't that known as 'Reliable UDP,' one of the only cardinal sins in Jon Postel's book?

"The currently defined services and protocols (WSP) are most suited for browsing-type applications. WSP defines actually two protocols: one provides connection-mode session services over a transaction service, and another provides non-confirmed, connectionless services over a datagram transport service. The connectionless service is most suitable, when applications do not need reliable delivery of data and do not care about confirmation. It can be used without actually having to establish a session."

Thus, it's no surprise WAP Forum also liberally reinvented HTTP itself (while still using http: as WSP's URL scheme).

- WSP recodes existing RFC-822 style request and response messages in their own binary tokenization format (again, in lieu of general-purpose compression, see WBXML).
- By treating PUT as a POST, WSP cannot support IETF's Proposed Standard for distributed authoring over HTTP (WebDAV).
- WSP does content-negotiation not by Content-Type:, but with explicitly deprecated User-Agent: comment fields.
- WSP has turned a stateless protocol into a stateful one, by implicitly copying headers from previous requests and replies.
- WSP has reversed HTTP's fundamental

semantics by conjuring up a 'push' mode. Their scheme rejects at least three plausible standards-compliant solutions: not by representing push as an HTTP request (to a virtual 'server' on the handset), not by managing group subscriptions to an URL, and not by representing push messages as valid, but unsolicited HTTP response messages.

- And finally, they require support for multiple concurrent downloads -- to a device explicitly designed not for 'desktop' browsing (the #1 scenario for downloading text and graphics concurrently).

"In addition to the general features, WSP offers means to:

a) provide HTTP/1.1 functionality:
   1) extensible request-reply methods,
   2) composite objects,
   3) content type negotiation;
b) exchange client and server session headers
c) interrupt transactions in process
d) push content from server to client in an unsynchronised manner
e) negotiate support for multiple, simultaneous asynchronous transactions."

They have added request headers for the sole purpose of being echoed back in the reply, presumably to let the handset avoid storing even the tiniest amount of per-transaction state (p. 14). They have not justified abandoning TCP RST/FIN (close) in favor of their own suspend/resume semantics -- are there observed usage statistics that motivate such complexity? Push can require human intervention to confirm, tying up connection state for very long intervals, potentially. There are several new code registries they arrogate management of (to their Wireless Interim Naming Authority, WINA):

- a linear (!) list of 40 core MIME types (rather than hashing standard MIME type strings)
- numbers for all current ISO national language tags (saving a mere three bytes!)
- a header-field registry that reinforces the broken X- concept (rather than simply using URIs as header field names, as W3C has long advocated).

In this context, covert-channel hacks like piggybacking time-of-day requests in nonstandard headers with nonsensical values make perfect sense:

"When the gateway receives a WSP method request that includes a header named X-WAP.TOD [with the value set to '0'], it will include that header in the response, with the header value set to the Gateway's current time of day."

This example is not even in the WSP document, but rather tucked away in an ancillary guide to reinterpreting HTTP/1.1 caching [UACaching-11-Feb-1999.pdf]. (Since handsets will barely manage even dozens of web resources in the next year, this seems premature.) Similarly, the rules for copying headers between messages on a stateful WSP connection are in yet a third document [WAEspec-17-Feb-99.pdf].

This clock callback is the kind of over-engineering indicative of telco-mindset overdetermined standardization. In IETF style, local clocks are just that: a platform-specific problem (and with timebases already encoded in virtually every airlink protocol, that would be just fine). Occam's Razor is wielded all too infrequently at WAP Forum.

2.6 WBXML: Binary (Tokenized) XML
At least this layer of WAP's specifications has a one-to-one correspondence to a real standard. However, rather than apply HTTP's zip/deflate compression options in WSP, WAP Forum devised a specific algorithm for XML content. WBXML tokenizes normal XML by preparsing it into a tree structure, extracting common text strings, and transmitting it according to a compression state machine at both ends.

The format is designed to preserve the element structure of XML, allowing a browser to skip unknown elements or attributes. The binary format encodes the parsed physical form of an XML document, ie, the structure and content of the document entities. Meta-information, including the document type definition and conditional sections, is removed when the document is converted to the binary format.

Though WBXML can be applied to merely well-formed XML documents, it does require a public Document Type Definition. The first payload byte encodes the DTD used, locking them into a WINA-controlled 255-slot registry for Formal Public Identifiers (FPIs, see Section 7.2 of the spec). In lieu of something smart, like hashing the FPI, they fixed codes 1, 2, and 3 to WML/1.0, WTA Events, and WML/1.1. Requiring DTD codes forecloses other XML Schema proposals from W3C that are expected to replace DTDs.

On other counts, they did refer to existing practice. So, rather than send the actual charset name, they save four bytes and send an IANA-registered charset numbers (originally only for ASN.1 SNMP MIB definitions, see Section 7.3 of the spec).

The general concept is that there are 255-slot code pages for tag, attribute name, attribute value, and other string tokens. Actual tag references use two high-order bits to encode the content model (empty or open) and presence of attributes, limiting it to 31 tags per code page. Nothing in WBXML speaks to WML's unilateral hijacking of the '$' character. If in fact WBXML was optimized for immediate internal representation, there would have to be hooks for $variable references in the tree.

Dynamical Systems Research surveyed four current WAP software development kits and found notable incompatibility in WBXML encoders for WML:

WML files encoded by the different Encoders appear to be compatible, or least can be made compatible by tweaking the WML source. It only on minor points that incompatibilities arise. However, also these minor points can make a WAP application unusable for an end user.

Claiming that "verbose markup... 1. Wastes network bandwidth and 2. Wastes cache memory", Peter King of Unwired Planet argued for W3C endorsement of WBXML -- though no performance numbers from WAP Forum have ever justified those claims.

W3C's Martin Dürst correctly replied that generic compression should offer comparable or better results without wiring the format to a specific DTD, and that cache representation is an entirely implementation-dependent problem. WBXML can't be pipelined (streamed), to boot: it places a common string table in the preamble -- which wastes precisely the most latency-contributing early bits (streaming delivery allows rendering to begin as soon as the first packet is received).

Peter King later edited a W3C NOTE introducing WML to the XHTML WG that countered by emphasizing random-access (even though an entire deck must be in-memory to use it anyway), the small payloads (even though compression dictionaries could be precompiled), and media-specific compressor proliferation (even though the point is to have only one):

It has been asked why WAP did not opt for generic compression algorithms. Generic algorithms may be more effective at compressing large quantities of data, but WAP content is typically quite small, averaging 200-300 bytes. Also, generic algorithms merely compress, they do not pre-process the content. Thus, without pre-processing, the compiler for each supported content type must reside on the device. Finally, the most effective generic compression algorithms do not allow random access to the compressed content. Thus, to access a compressed resource, the device would be required to decompress the entire resource. This places significant requirements on another resource that's precious in the mobile terminals: RAM. Also, parsing binary data requires less computational energy, compared to parsing text.

Nevertheless, these factors torpedo WBXML as a general-purpose solution, isolating WAP-compliant tools further. Furthermore, W3C has chartered an XML Internal Set WG precisely to investigate common in-memory representations and XML Linking to consider online-access to remote XML trees -- and W3C also owns the trademark to XML...

2.7 WML: Wireless Markup Language

Unwired Planet's initial products introduced "Handheld Device Markup Language," which lives on today in XML-ized form as WAP Forum's 1.1 release of Wireless Markup Language. HDML, however homophonic, has very little relationship to HTML, though. It is ideally suited to today's multiline handset displays -- too perfectly.

One way to think of WML is as programming 3270 screens with a very simple BASIC. An application is delivered to the handset as a 'deck', which allocates a corresponding pool of storage space for variables. Individual 'cards' are executed on entry and exit to render the handset screen: they can fill in (shared) variables, rerender on external event notifications (e.g. call completed), and bind further actions to a small set of buttons. The display calls themselves are essentially character-cell oriented, with tab alignment and bold, italic, underline, big, and small fonts. Tables, frames, colors, and so on are not supported. Semantic HTML markup like <ADDRESS> must be stripped out. Even simple declarative HTML doesn't mean what it used to -- <P> no longer begins a new line. 'Graphics' support is intended mainly for optional icons in place of text menus; no clickable menu maps or such.

WML is really a scripting language. It borrows some DOM-like functions for describing a global navigation history and browser-control operation. It hijacks the '$' character from all other XML and URL escaping rules to indicate variable references (even $$ is no escape: it must be coded as &#24;).

A 'deck' isn't merely a way to bundle independently meaningful cards in a single transmission: it defines a program thread, complete with timers. The WML 'program' shares its stack with other 'decks' on the handset, raising security concerns only partially addressed by lexical scoping. This includes a deck-wide <template> element defining default event bindings and variable values (which can be overridden by similarly-named elements in card-scope).

WML exposes a flat context (ie, a linear non-nested context) to the author. Each WML input control can introduce variables. The state of the variables can be used to modify the contents of a parameterised card without having to communicate with the server. Furthermore, the lifetime of a variable state can last longer than a single deck and can be shared across multiple decks without having to use a server to save intermediate state between deck invocations.

The last part of that quote also refers to WML's ability to 'parameterize' a deck. If, say the only part of a customer-service deck that changes is the name and account number, a WSP server can vend a single resource and let the client substitute those bits of state. Their claims of cache efficiency and reduced traffic are fallacious, since such a deck isn't shared at the handset; it's only shared at the server, which is perfectly capable of computing a custom deck with a CGI script.

Even the basic <A HREF=...> becomes syntactic sugar for <anchor>... <go href="..."/></anchor> (providing <go> hasn't be redefined at deck-scope). To allow a user to traverse a link, one doesn't publish a link source; an URL must be passed as the argument to a <go> task. The equivalent of a <FORM> requires executing internal tasks which escape arguments first in URL, then XML format to be stored in a <var> element for a later submission action. An internal event model for WML browsers executes tasks on first entry, exit, and subsequent entry from a card; and after timer intervals. (This is in part motivated by interstitial advertising and WTA's goals of replacing handset telephony UIs such as "Incoming caller XXX: answer or take message?")
WML uses an enhanced, explicit event binding model. There are three ways to establish an event binding in WML. These mechanisms bind events to explicit tasks (the actions to be taken in response to the event).

- The <a> element creates an anchor in a span of text (similar to HTML). When the anchor is activated, the task bound to the <a> element is executed.
- The <do> element binds a task to a card-level user-interface object and allows the display attributes of that user-interface object to be specified. When that user-interface object is activated, the task bound to the <do> element is executed.
- The <onevent> ("on event") element binds a task to an intrinsic event of its enclosing element.

Within each event binding, the author may specify a task to be performed. In WML 1.0, the defined tasks are:

- <go>--visit an identified resource
- <prev>--return to the previous resource in the history list
- <refresh>--redisplay the current resource
- <noop>--do nothing

To explain how a simple pick list control works now takes five pages. There's inconsistent terminology: an INPUT variable's name is now found in its KEY attribute. The state of the pick list is no longer the list of selected values: it must be computed from an indexed vector by KEY and IKEY. The descriptive text for the control is not bound to it by HTML4's <LABEL> tag, but with a <FIELDSET> instead. The pick list itself can now be internally chunked-- though the OPTGROUP attribute is never demonstrated in the specification.

Needless to say, all these features interact with each other. The result is dramatically different from HTML:

12.5.1 The Go Task

The process of executing a go task comprises the following steps:

1. If the originating task contains setvar elements, the variable name and value in each setvar element is converted into a simple string by substituting all referenced variables. The resulting collection of variable names and values is stored in temporary memory for later processing. See section 10.3 for more information on variable substitution.
2. The target URI is identified and fetched by the user agent. The URI attribute value is converted into a simple string by substituting all referenced variables.
3. The access control parameters for the fetched deck are processed as specified in section 11.3.1.
4. The destination card is located using the fragment name specified in the URI.
   a) If no fragment name was specified as part of the URI, the first card in the deck is the destination card.
   b) If a fragment name was identified and a card has a name attribute that is identical to the fragment name, then that card is the destination card.
   c) If the fragment name can not be associated with a specific card, the first card in the deck is the destination card.
5. The variable assignments resulting from the processing done in step #1 (the setvar element) are applied to the current browser context.
6. If the destination card contains a newcontext attribute, the current browser context is re-initialised as

described in section 10.2.
7. The destination card is pushed onto the history stack.
8. If the destination card specifies an onenterforward intrinsic event binding, the task associated with the event binding is executed and
9. If the destination card contains a timer element, the timer is started as specified in section 11.7.
10. The destination card is displayed using the current variable state and processing stops.

At the same time, the WAP suite does not specify the exact behavior of a WML User Agent. Some guidelines are in Section 5.1.5 of WML and throughout WMLScript. Part of the variability is display size: HDML had to be reengineered so WML could display multiple cards per screen on palmtops. Thus, for the first time a single card could contain multiple input controls.

On the other hand, they explicitly rejected the Web-normative way to handle UA- and content-specific formatting: WML does not support any form of style sheets.

2.8 WBMP: Bitmapped Images
WBMP was quietly slipped into Section 6 and Appendix A of [WAEspec-17-Feb-99.pdf] in WAP/1.1. They have created a new image format inspired by -- but not even interoperable with -- W3C and ISO-standard Portable Network Graphics (PNG). It can support multiple bit planes, palettes, animation, and compression algorithms. For good measure, though, WBMP is only currently defined for Type 0: one-bit, uncompressed icons. All that overengineering in lieu of standard (and dead-simple) X Bitmaps (XBM)...

Lightweight vector graphics ought to be a very useful format for small-scale, variable-geometry displays. There is no mention of W3C's Scalable Vector Graphics WG or existing standard Computer Graphics Metafile (CGM). Current state-of-the-art already embraced FAX viewing on handsets, but G3 TIFF is also absent.

They also botched the usage of image references in WML. The lowsrc approach has been long discredited and never became part of standard HTML. As for offering an alternate local namespace, the handset should simply cache it by URI -- the phone itself will know it has "/images/smiley.gif" onboard by cache etag.

Many handheld devices have a set of small images, or icons, that are used by the native user interface or the browser application. By using the "localsrc" attribute of the WML <img> element the author can re-use those images without having to download new ones. The purpose is similar to that of the HTML "lowsrc" attribute: to save time and bandwidth.

<img localsrc="smiley"
src="/images/smiley.gif" alt="-J " />

2.9 WMLScript
WAP Forum's scripting solution is specified in two parts: WMLScript, "its lexical and syntactic grammar, its transfer format and a reference bytecode interpreter," and the Standard Libraries, " including a language

library, a string library, a dialog library, a floating-point library, a browser library and a URL library."

WMLScript is not fully compliant with ECMAScript. The standard has been used only as the basis for defining WMLScript language. The resulting WMLScript is a weakly typed language. Variables in the language are not formally typed in that a variable's type may change throughout the lifecycle of the variable depending on the data it contains. The following basic data types are supported: boolean, integer, floating-point, string and invalid. WMLScript attempts to automatically convert between the different types as needed. In additions, support for floating-point data types may vary depending on the capabilities of the target device.

WAP Forum has devised yet another tokenization format for 'compiling' WMLScript over the air. URI semantics have been bent to allow special fragment-identifiers to link to WMLScript function calls instead of web resources.

The result is incompatible SDKs, much less devices in the field. DSR' survey again comments:

The Ericsson WMLScript compiler handles variable on the stack in a different way than Nokia's or DSR's WMLScript compilers (and WMLScript VMs). The WAP specifications are actually not very clear on this point. Initial contacts with Ericsson and representatives from WP Forum suggest that both are aware of this problem.

2.10 WTA: Telephony Services

From a network operator's perspective a 'microbrowser' brings a consistent look and feel to value-added information services. WTA brings the same consistency to handset telephony services. A carrier could publish common WML decks for initiating calls, transferring calls, answering or forwarding calls, arranging conference calls, checking voice mail, and two-way paging for its entire subscriber base, reducing customer support costs for all its supported handsets.

WTA provides telephony-specific extensions for call control features, call logging, paging, address book, and phonebook services. It accesses the underlying call signaling of various airlinks. There are specific adaptation layers for features specific to GSM, IS-136 (TDMA), and Pacific Digital Cellular (PDC).

WTA user interfaces are a superset of WAE with push delivery. WTA calling features can be accessed as WMLScript function calls or by resolving URLs representing those function calls.

Unlike a typical WML user agent, the WTA user agent has a very rigid and real-time context management component. For example, the user agent drops outdated (or stale) events, does not place intermediate results on the history stack, and typically terminates after the event is handled.

Within the WTA framework, the client and server co-ordinate the set of rules that govern event handling via an event table. WTA origin servers can adjust the client's rules by pushing (or updating) a client's event table if required .

3. The WAP Marketplace

This section describes the key players and strategic choices WAP raises for several sectors: manufacturers, operating systems, carriers, gateways, and content providers.

3.1 Handset Manufacturers
WAP Forum was co-founded with three manufacturers representing the vast majority of handset sales today. By now, 90% of handset capacity is represented at the Forum, and 75% of that capacity has committed to shipping some form of WAP-compliant phone -- over 20 manufacturers licensing UP.browser alone. Some shipping examples include the NeoPoint 1000 from Lucky Goldstar, the Motorola i1000+, and the Nokia 7110.

But to the degree the microbrowser sets the lowest-common-denominator interface to phones -- even potentially usurping manufacturer-specific UIs for telephony features -- WAP could become a value-destroying proposition. Homogenization is the first step to commoditization.

UP's offer of free UP.browser licenses fills a short-term gap in the product pipeline, but within 18-24 months, the premium providers will return to producing HTML microbrowsers in our opinion. Subsidized client software is not enough to lock-in global handset manufacturers

That explains why so many manufacturers are also hedging their bets. The same three vendors are partners in Symbian (below) and developing other phone concepts such as the Nokia 9000 line. Qualcomm, for example, is taking an opposite tack and pairing with a palmtop vendor to produce the pdQ.

The other strategic decision is the degree to which IP tone makes inroads in the market. Handsets that can double as infrared-linked IP links could be very successful with mobile professionals who also have laptops and palmtops. Microbrowsers become less important to such a device, but without the R&D cost of installing a full OS: let the heavy iron run the dancing hippo applets.

Ericsson-led Bluetooth and Motorola's Piano are less critical diversions. The interest in RF micro-scale networking does not obviate the need for wide-area WAP access. However, to the degree both systems are IP-centric, they reinforce the trend above. (Neither of these standards are particularly open; Bluetooth development kits run to $7,000).

The other side of the hardware market is in base stations and airlink equipment. Since WAP Gateways are designed to operate at a carrier's head-end data center, cellular base stations and telephone switches should not have to be altered. WTA servers will require close integration with switches, but through existing APIs defined by existing airlink signaling. Still, this is why Lucent, Siemens, and the like are joining WAP Forum in 1999.

3.2 Operating Systems
A complete information-management solution on a handheld rapidly outgrows a microbrowser's cache-and-timers platform. A true operating system for low-power, low-resource devices can do a better job of rendering, filing, and synchronizing with PCs.

Psion's Symbian is producing an EPOC32-based operating environment for phones to palmtops to miniature laptops in partnership with Motorola, Nokia, NTT, IBM, Sun, and others. EPOC32 is aimed at ARM processors and graphic displays (rather than pure text devices). It provides true IP Web browsing and email, content translation, and even a Java VM. For this platform, retrofitting WAP support means installing a special-purpose network stack alongside TCP/IP and special-purpose browser. That's exactly what's promised, but we suspect that actual implementation of WAP hooks is not guaranteed, and not a priority.

WirelessKnowledge is an altogether different take on the problem: Microsoft and Qualcomm are focusing on the 'server-side' operating environment. Their goal is an Exchange-derived information routing system for corporate IS/IT. E-mail, faxes, group communications and such would be delivered to an outsource message server, then delivered back to mobile professional by PDA, phone, fax, or, recently rumored, WAP deck.

Microsoft's WinCE is conspicuously absent from this sweepstakes. While telephony might be integrated into handheld computers running WinCE, the reverse does not seem likely. There are no signs WinCE will be adopted for handheld screenphones. Of course, even the Economist noted that "[Bill Gates] is said to regard Symbian as an even bigger threat than the DoJ."

PalmOS currently has several options for wireless Internet access. With a CDPD modem and a native TCP/IP stack, off-the-shelf PalmOS web, mail, and even telnet clients are available. For accessing standards-compliant Web content, though, it helps to offload content extraction, distillation, and security to proxy farms connected to the wired Internet.

ProxiNet got its start by splitting SSL and SSH to locate public key login operations on a trusted proxy and just encrypting the 'last mile' to a small device. They have since developed expertise in provisioning very large-scale, fault-tolerant proxy clusters and a 'building-block' software architecture for reprocessing pages.

AvantGo aims more specifically at push delivery to palmtops, but also offers proxy-based distillation.

Ericsson's WebOnAir proxy is designed for any wireless IP device, but particularly Win9x laptops. It strips out metadata, insignificant whitespace, tidies up HTML, and optionally strips Java applets, JavaScript, downsamples graphics, and compresses the resulting stream.

Online.Anywhere focuses on the third aspect: intelligent content extraction. Customizable site profiles indicate what elements to extract from standard web pages for relayout on different devices (TV, phone, palmtop, etc). This approach brings to layout what webMethods' Web Interface Definition Language (WIDL) brings to data mining.

3.3 Network Operators

Carriers, as ever, need to sell more minutes -- in a soon-to-be-glutted market. With up to six competitors in a metropolitan area, expect serious price erosion in the US market. Churn rates are going to remain a dominant factor in profitability. If 'smartphones' can demonstrably add usage and lock-in users, carriers will flock to WAP.

As it is, though, many carriers are in wait-and-see mode at WAP, and in particular, with Unwired Planet. Amidst all the pilot products, there's little evidence of breakaway success. The larger national carriers are leery of moving first in this area.

Even so, Nextel Online is launching an innovative hybrid portal-phone service with Netscape and UP. The portal could aggregate general news, industry information, and local directories. More to the point, the user's portal is a control panel for subscribing to stock feeds, adding phone numbers, and scheduling meetings. All of these processes could link back to the phone: schedule alerts, news updates, handset access to the portal itself in WML format.

The strategic choices facing carriers are a classic telco ones: capital and reliability. It isn't cheap to acquire WAP Gateway software and proxy farms for operating it -- especially from the one vendor with any kind of operational experience.

UP, as outlined in its S-1 IPO filing, has placed the burden of profitability on the carrier end of its products. They're giving away the handset clients for virtually free, but charging carriers per subscriber. Not per UP.link-using subscriber, but for every device that could run UP.link. Only one carrier has had the negotiating clout to date to force UP to share the risk of subscriber adoption. Others may be paying fees in the millions of dollars for the server -- and several dollars per subscriber as well Given that even two-line alphanumeric pagers can be considered WAP-capable, that can be quite a capitation.

In the longer-term, carriers stand to benefi from the handset homogenization manufacturers fear. UP's white papers for carriers, for example, sketch out the possibility of WML user interfaces for self-service billing and customer care, even provisioning an entire prepaid cellular account from a handset without human intervention.

Thus, carriers may even succeed without "Internet" access per se. While early adopters and heavy users may expect full integration with all possible Web sources, mass consumer markets may be satisfied with a limited menu of carrier-provided information services. Once a solid WAP marketplace exists, deployment may even make economic sense without any information services at all: just for making more abstruse calling features accessible and for phonebook management.

The final strategic lever in carriers' hands is choice of airlink. They are the final arbiters of whether IP-centric wireless data will prevail over WDP/WTP/WSP.

3.4 Application Gateways
The point of establishing a WAP Forum was to catalyze the development of independent, interoperable implementations. UP, Ericsson, Nokia, APiON, and Dynamical Systems have all announced WAP Gateway servers; and IBM, HP, and Oracle are rumored to be interested a year or two further out.

These entrants could have rather diverse business models. UP must make its fortunes on software sales, but handset vendors may subsidize development to spur equipment sales and upgrades. Similarly for computer vendors, whose real motivation may be the Gateway cluster and service contracts. Other independents, like APiON, may make money on software sales simply by virtue of independence: the telecommunications industry is justifiably afraid of competitors buying suppliers.

The strategic question is how Gateway vendors choose to interoperate with existing Internet standards. Development could stop with WSP, leaving it to the carrier to actually produce WML/WMLScript content and update it. A full-service, Internet-oriented approach would yield a proxy farm capable of distilling any Web content -- but then again, perhaps not into WML, which is not well-suited to 'browsing'.

3.5 Content Providers
The ultimate end of the value chain is the origin site itself: what incentives will Web content providers have to publish handheld-savvy content? This is where the network effects of standardization are strongest. Ideally, content ought to be produced in XML and customized with Cascading and Extensible stylesheets to the various browsers out there. Then it's just a matter of adoption to judge when WebTV or WAP phones or voice-response boxes become popular enough to warrant publishing a new stylesheet.

Of course, the vast majority of content won't even be that flexible: at best it will be HTML4-compliant and perhaps follow the W3C's Web Accessibility Initiative Guidelines to ensure it's meaningful in text-only mode.

Carrier partnerships are at the other end of the spectrum. Providing custom, WML versions of news content or WML forms to services like travel agency, stock brokerage, and the like will only make sense at the outset if there is direct compensation in some form or another. Gateways, carriers, and providers will have to work together closely for at least 12-18 months before producing WML decks independently.

WML Developer's Kits are only emerging just now. WirelessDeveloper'99 will be the first conference of its kind this May, focusing on deploying horizontal and vertical applications over several wireless data solutions. 4,500 developers registered to download a copy of UP's SDK through the end of March, and similar numbers ought to be expected for Nokia and Ericsson's WAP IDEs.

There are no promises of standalone, nor open-source implementation of various WAP layers like WBXML, WMLScript, WSP, WTP, WTLS, or WDP that I know of. This is another serious practical impediment to the adoption of WAP Forum standards beyond the Forum.

4. Existing Standards

At almost every juncture, faced by the choice of referencing existing Internet practice by reference or by copy, WAP Forum chose to copy -- and modify. And each standard copied into WAP's specification space is another organizational toe stubbed.

It's a paranoid engineer's game to avoid the uncertainty of moving targets by copying a specification in order to freeze it, or to propose a special purpose solution that takes advantage of tricks legal across layers. This tendency resonates with a separate tendency to overspecify the system, which results in ever-larger, ever-more-complex specifications.

4.1 WAP Forum

This paper has established the degree to which WAP Forum has specified its system -- far beyond merely establishing communication protocols and data formats. For example, it's not enough to stipulate that WSP transports MIME-formatted entities: there are special coders for XML, special format for icons, special rules for explicit-acknowledgment, and special user interfaces to vCards and vCalendar entries.

WAP Forum has consciously adopted this policy, and seems to be making little effort to redress it by promoting its revisions back to their original communities for further review. In other words, they appear to be hoarding change control.

The output of the WAE effort is a collection of technical specifications that are either new or based on existing and proven technologies. Existing technologies leveraged by the WAE effort include:

- Unwired Planet's Hand Held Mark-up Language (HDML),
- World Wide Web's Consortium's (W3C) eXtensible Markup Language (XML) [14],
- World Wide Web's Consortium's (W3C) Hypertext Markup Language (HTML) [5],
- ECMA-262 Standard "ECMAScript Language Specification" [6] that is based on JavaScript™,
- IMC's calendar data exchange format (vCalendar) [7] and phonebook data exchange format (vCard),
- A wide range of WWW technologies such as URLs and HTTP [8], along with
- A wide range of Mobile Network technologies such as GSM call control services and generic IS-136 services suc as send flash.

The resulting WAE technologies are not fully compliant to all of the motivating technologies. Where necessary, modifications were made to better integrate the elements into a cohesive environment and better optimize the network interaction and user interface for small screen limited capability terminals that communicate over wireless networks.

Even WAP's standards-review process has been incompatibly cribbed together. Their maturity levels (Draft, Prototype, Proposed, Approved) correspond neither to IETF's (Internet-Draft, Proposed, Draft, Standard) nor W3C's (Note, Working Draft, Proposed, Recommended). Their documents include IETF definitions of MUST, SHOULD, and MAY, but do not share its commitment to IPR-freedom. Their documents are developed by dues-paying Members behind closed doors like W3C, but without a single public mailing list for discussion and review.

In the subsections that follow, we will examine WAP Forum's exposure to other bodies' institutional interests and its liason efforts, if any.

4.2 W3C

WAP Forum leadership clearly recognizes that the Web Consortium controls the greater part of the standards they rely upon, and have worked since June 1998 to nurture a working relationship. That dialogue culminated in a joint communiqué last September that merely stated areas of mutual interest. There is overlap at almost every level of the protocol stack, but the three most crucial future developments WAP Forum must leverage to remain relevant are 1) the migration to XHTML (nee HTML5), 2) Binary XML, and 3) next generation HTTP.

It is worth noting that the W3C Mobile Interest Group itself does not make that cut. Discussions in that forum to harmonize guidelines for 'compact' HTML4, capability negotiation with CC/PP, and other near-term changes have not made any joint progress.

Regarding XHTML, Unwired Planet helped edit a two-part input document exhorting that group to ensure WML-style state-management, tasks, and variable substitution would remain possible; and to ensure the event model could encompass WTA push events. Neither step is fairly likely, since W3C has enforced an architectural split between declarative markup language issues and operational Document Object Model issues. Almost everything in WML and WMLScript's model of an active client violates that separation.

What XHTML is doing will create a formidable competitor to WML. The Web Consortium is working to recast HTML as a set of XML modules with more clearly structured FORMs and TABLEs; and navigation aids which enable accessibility in a broad sense: for print, audio-only, and other disabilities -- and thus also for small devices. Combined with XSL 'behavior sheets', W3C has already mapped out a migration path to alternative user interfaces and devices. That path makes it easier to envision reliable distillation proxies

Regarding the evolution of HTTP, W3C is pushing the Mandatory extension mechanism to HTTP/1.1, an RDF-based Composite Capability/Preference Profile (CC/PP) content selection framework, and a message multiplexing (MEMUX) session layer. W3C is not in a position to adopt the scrambled WTLS/WTP/WSP substack as a replacement for HTTP. Nor has WAP Forum acted on literally years of advance warning that the W3C HTTP-NG project was aimed precisely at low-bandwidth and wireless devices and its likely architecture. As it stands, there is no possibility of session-layer interoperability, the Mandatory scheme has been rejected by WAP Forum, and the debate between User-Agent and CC/PP (and, for that matter, IETF CONNEG) content negotiation has come to a standstill.

Regarding XML, W3C has separately chartered a working group to recommend a common internal representation (beyond the tree structure already implicit in DOM). It has evinced no interest in custom binary-coding of XML, referring instead to the potential for general-purpose compression as demonstrated in its research papers and code library. Since W3C owns the XML trademark, it may even be in a position to take a more critical view of WAP Forums pseudo-standard. It's not just a product, for which W3C grants use, but a competing definition of XML at stake.

4.3 IETF

The IETF has a massive investment in adapting TCP/IP to 'long, thin networks,' as cellular links are known. There are several research groups focused just on optimizing TCP for radio links, ranging from TCP over Satellites to Mobile Ad-Hoc Networking, which posits handsets as routers.

The most relevant work is a new proposed Performance Implications of Link Characteristics (pilc) working group. It met as a BOF in December 1998 and March 1999. Sun engineer Gabriel Montenegro edited an extensive Internet-Draft cataloging the arguments for TCP/IP all the way to wireless terminals, recommended approaches for typical cellular links, and open research issues [draft-montenegro-pilc-ltn-01.txt]:

"Non-IP alternatives face the burden of proof that IP is so ill-suited to a wireless environment that it is not a viable technology." This is one of the most hotly debated issues in the wireless arena. Here are some arguments against it:

- It is generally recognized that TCP does not perform well in the presence of significant levels of non-congestion loss TCP detractors argue that the wireless medium is one such case, and that it is hard enough to fix TCP. They argue that it is easier to start from scratch.
- TCP has too much header overhead.
- By the time the mechanisms are in place to fix it, TCP is very heavy, and ill-suited for use by lightweight, portable devices

and here are some in support of TCP:

- It is preferrable to continue using the same protocol that the rest of the Internet uses for compatibility reasons. Any extensions specific to the wireless link may be negotiated.
- Legacy mechanisms may be reused (for example congestion control)
- Link-layer FEC and ARQ can reduce the BER such that any losses TCP does see are, in fact, caused by congestion (or a sustained interruption of link connectivity). Modern W-WAN

technologies do this (CDPD, US-TDMA, CDMA, GSM), thus improving TCP throughput.

- Handoffs among different technologies are made possible by Mobile IP [RFC2002], but only if the same protocols, namely TCP/IP, are used throughout.
- Given TCP's wealth of research and experience, alternative protocols are relatively immature, and the full implications of their widespread deployment not clearly understood.

There is active interest in developing session-layer multiplexing over TCP as well. A Requirements for Unicast Transport Session (ruts) BOF was held in December 1998 and a Support for Lost of Unicast Multiplexed Sessions (SLUMS) BOF was held in March 1999. Another half-dozen proposals have emerged in this area, not least of which is W3C's MEMUX. This community is looking for efficient concurrent sessions between hosts as well as fast-setup and teardown -- subsuming parts of WTP and WSP.

And while there is interest in wireless/nomadic application support (such as disconnected IMAP4 email), IETF is not likely to evince much interest in scripting, markup, or graphics formats.

As apathetic as IETFers may seem at the top of the stack, the sense of attendees we spoke to recently was very negative about WAP Forum's low lever redefinitions.

4.4 IANA/ICANN

Section 2 of this paper described a series of registries embedded in WAP specifications. As of October 1998, WAP Forum decided to centralize those registries under a WAP Interim Naming Authority (WINA). Its charter, [WINA-process.doc] states its relationship to IANA (soon to be part of ICANN) thusly:

The Internet Assigned Numbers Authority already manages a number of name and code spaces. These include... Content Types, Port Numbers, [and] Character Sets

WINA is a complement to IANA, and deals with spaces not presently administered by IANA. These are either unique code spaces, or special encoding of already known name spaces.

While it's admirable WAP Forum will ratify an open process to maintain WAP-specific registries, it has not always been clear why they have chose to recode existing namespaces (WCMP Type codes, numbering ISO charsets, numbering FPIs), and encouraging centralized registries where URIs would do (header fields, WMLScript Library IDs). Furthermore, it does not appear WINA has duly attempted to register what should be at IANA with IANA, such as TLS chiphersuite codes, HTTP methods, and WTP port numbers.

While IANA is a purely reactive institution (it runs registries according to definitions in Internet Standards with policies set by the IESG), it does not serve anyone's interests to duplicate -- or approximate -- its work.

4.5 ECMA/ISO/NCITS
These three bodies are involved in the ECMA-262 standardization and evolution process. It does not appear WAP Forum consulted any of them regarding its publications of a derivative language "...not fully compatible with ECMAScript..." The remainder of this section considers the role of the US National Committee for Information Technology Standards (NCITS, formerly X3).

The closest to web standards NCITS has come is PNG and VRML. PNG is relevant insofar as WBMP defines another binary icon format that's very PNG-like in its extensibility. There is no reason to expect NCITS to take action here, though.

Similarly, though they have a role in Java and ECMAScript standardization, the dormancy of the scripting Technical Committee in particular does not indicate it will become a flashpoint for debate.

NCITS represents America's voice in XML/SGML standardization: TC V1 reports to JTC1 / ISO/IEC SC34. There is no conflict between WML and this work per se, since WML is just another well-defined XML application.

WBXML, though, will probably come to a halt here. To the degree their members represent the conventional wisdom of SGML, there is incredible importance placed on preserving the actual stored form of data, and not as much on interactive access. This will come to a head if V1 ever does tackle a tokenized XML -- which in turn would be unlikely unless W3C submits its own work in this area.

There is no Wireless work per se in any Technical Committee at present.

4.6 MNCRS
The Mobile NC reference specification embodies the "control hypothesis" of extending Java and tweaking current TCP/IP on top of existing Web content standards for small wireless devices. As Gabriel Montenegro's IEEE Computer article described in 1997, this team of Japanese manufacturers, Java API developers, and network protocol engineers targeted three classes of devices ranging down to roughly PalmIII levels.

MNCRS mandates TCP/IP over PPP, optionally including NFS, FTP, and Telnet, while looking ahead to IPsec encryption & authentication. The middleware layers are heavily dependent on JavaSoft APIs for directories, user interface, smart cards, an so on. Upper layer application support includes a phonebook API, for example -- but as an arbitrary database it seems too rich for GSM SIM card phone lists and too unstructured for Internet Mail Consortium's vCards (as used by WAP Forum).

The intent, though, is to standardize the lower, not the upper layers. While WAP Forum is seemingly intent on describing a single class of device -- the multiline text handset -- Mobile NCs are a horizontal platform for any sort of Internet client.

4.7 Other Groups
As mentioned above, WAP Forum references the Internet Mail Consortium's documents describing vCard contacts and vCalendar schedules. IMC is committed to

releasing their work as IETF standards, so there should be no problem citing them.

To the degree industry groups like the European Telecommunications Standards Institute (ETSI) and Universal Mobile Telecommunications System (UMTS) Forum touch upon mobile Internet services, it's to reaffirm IP as the fundamental data service of the future. We don't believe WAP can flourish over IP networks in the long-term...

WAP Forum cooperates with several other airlink interface bodies: "the Association of Radio Industries and Businesses (ARIB), Code Division Multiple Access Development Group (CDG), ...Telecommunications Industry Association (TIA), Universal Wireless Consortium (UWC) and Telecommunications Technology Committee (TTC)." These relationships do not touch upon our concerns about WAP's architecture, though.

There have been comments at the last W3C Mobile Workshop regarding an "ETSI MEXI" working group and an "Internet Screen Phone" reference group. No further details can be confirmed at this time.

Coda:

Despite the growth in numbers, one analyst remained skeptical about the long-term viability of WAP. Jane Zweig, senior vice president with Herschel Shosteck Associates Ltd. of Wheaton, Md., said she believed the companies joining the WAP Forum are "hedging their bets."

She said the membership fee was small change for companies that wanted to make sure they participated in wireless access to the Internet. In a recent report on WAP, Zweig said the standard is too limiting because it filters out much of the data available on the Internet so that it be accessed by voice-centric handsets.

Zweig's report said a data-centric approach using handheld computers or personal digital assistants ultimately will provide the kind of full Internet access the broad market wants.

"Although the point is content delivery, WAP will likely fail to provide a meaningful content experience for the mass-market end user," Zweig said.

She added that WAP likely will fill a valuable short-term niche.

-- *From the September 14, 1998 issue of WirelessWeek*

Appendix

BOARD MEMBERS

Unwired Planet
Alcatel
CEGETEL/SFR (Societe Francaise du Communications Network Radio Telephone)
DDI Corporation
Ericsson Mobile Communications AB
IBM
Matsushita Communication Industrial
Motorola
Nokia Mobile Phones
NTT Mobile (NTT DoCoMo)
SBC Communications
Sprint PCS
Telstra Corporation

NETWORK OPERATORS

AT&T Wireless Services
Bell Atlantic Mobile
BellSouth Cellular
Bouygues Telecom
Cellnet Communications
Deutsche Telecom Mobilnet GmbH
France Telecom
Hongkong Telecom Mobile Services
IDO Corporation
Omnitel
One 2 One
Rogers Cantel Mobile Communications
Sonera Corporation
SWISSCOM LTD.
Telefonica Servicios Moviles
Telia Mobile AB
Tokyo Digital Phone
Telecom Italia Mobile
Telenor Mobil Group
TU-KA Cellular Tokyo
Vodafone

DEVICE AND EQUIPMENT MANUFACTURERS

Acer Peripherals
Bosch Telecom Danmark A/S
CMG Telecommunications & Utilities
De La Rue Card Systems
Gemplus
Hewlett-Packard
ICO Global Communications
Intel Corporation
LG Information & Communications
Logica Aldiscon
Lucent Technologies
Mitsubishi Wireless Communications
ORGA Kartensysteme GmbH
Philips Consumer Communications
Qualcomm
RTS Wireless
Samsung Electronics
Schlumberger Industries
Sema Group Telecom
Siemens AG
Sony International (Europe)
Tecnomen Oy
Telital S.p.A.
Toshiba

NEC Technologies (UK)
Nortel
Uniden
Unisys

SOFTWARE COMPANIES

APiON
Bussan Systems Integration Company
Certicom
Comverse Network Systems
CCL
(Computer & Communications Research Laboratories, ITRI)
CTC (Itochu Techno-Science Corporation)
Dr. Materna GmbH
Dolphin Telecommunications
Fujitsu Software Corporation
Geoworks Corporation
Glenayre Technologies
GSM Information Network
M.D. Communications
Oracle Corporation
Puma Technology
RSA Data Security
Sendit AB
Scandinavian Softline
Spyglass
Symbian
Systems Engineering Consultants
Tegic Communications
VTT Information Technology

The DNA of Silicon Valley
The original company of inventors started here. 367 Addison Avenue, the garage out back.
Two young inventors, radicals really, with a few simple tools and a passion to invent.
In this garage more than a company was born.
The spirit that has launched hundreds of companies was born.
Now the original company of inventors is returning here, to this garage, to that spirit.
The original start-up
will act like one again. Watch.
www.hp.com
hp
invent

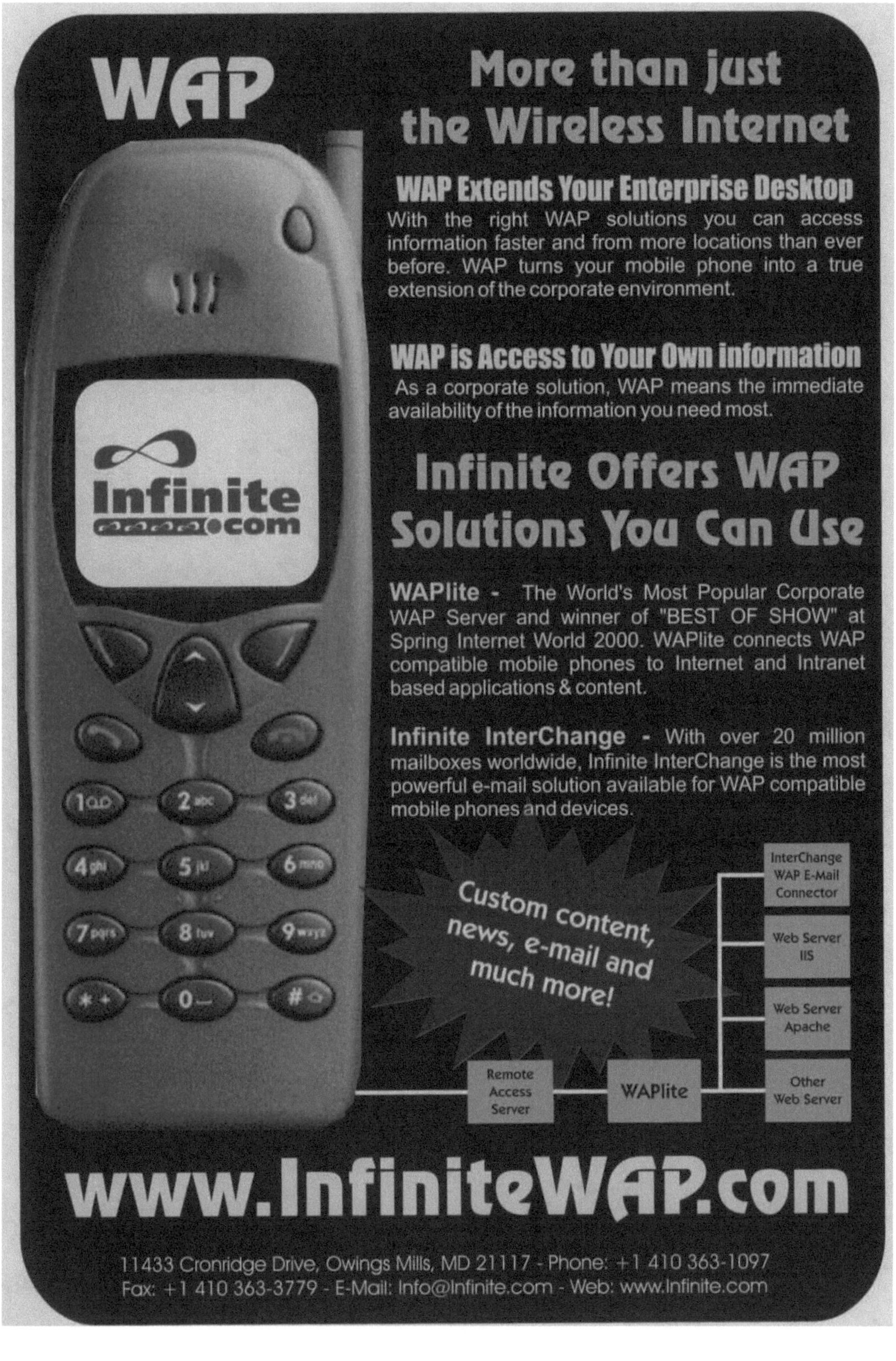
WAP
More than just the Wireless Internet
WAP Extends Your Enterprise Desktop
With the right WAP solutions you can access information faster and from more locations than ever before. WAP turns your mobile phone into a true extension of the corporate environment.
WAP is Access to Your Own information
As a corporate solution, WAP means the immediate availability of the information you need most.
Infinite .com
Infinite Offers WAP Solutions You Can Use
WAPlite - The World's Most Popular Corporate WAP Server and winner of "BEST OF SHOW" at Spring Internet World 2000. WAPlite connects WAP compatible mobile phones to Internet and Intranet based applications & content.
Infinite InterChange - With over 20 million mailboxes worldwide, Infinite InterChange is the most powerful e-mail solution available for WAP compatible mobile phones and devices.
Custom content, news, e-mail and much more!
InterChange WAP E-Mail Connector
Web Server IIS
Web Server Apache
Other Web Server
Remote Access Server
WAPlite
www.InfiniteWAP.com
11433 Cronridge Drive, Owings Mills, MD 21117 - Phone: +1 410 363-1097
Fax: +1 410 363-3779 - E-Mail: Info@Infinite.com - Web: www.Infinite.com

MIX
Papier aus verantwortungsvollen Quellen
Paper from responsible sources
FSC® C105338

If you have any concerns about our products,
you can contact us on
ProductSafety@springernature.com

In case Publisher is established outside the EU,
the EU authorized representative is:
Springer Nature Customer Service Center GmbH
Europaplatz 3, 69115 Heidelberg, Germany

Printed by Libri Plureos GmbH
in Hamburg, Germany